Personalized Medicine Meets Artificial Intelligence

Alfredo Cesario • Marika D'Oria •
Charles Auffray • Giovanni Scambia
Editors

Personalized Medicine Meets Artificial Intelligence

Beyond "Hype", Towards the Metaverse

Springer

Editors
Alfredo Cesario
Fondazione Policlinico Universitario
A. Gemelli IRCCS
Rome, Italy

Marika D'Oria
Fondazione Policlinico Universitario A. Gemelli
IRCCS
Rome, Italy

Charles Auffray
European Institute for Systems Biology
and Medicine
Lyon, France

Giovanni Scambia
Fondazione Policlinico Universitario A. Gemelli
IRCCS
Rome, Italy

ISBN 978-3-031-32613-4 ISBN 978-3-031-32614-1 (eBook)
https://doi.org/10.1007/978-3-031-32614-1

This Springer imprint is published by the registered company Springer Nature Switzerland AG
The registered company address is: Gewerbestrasse 11, 6330 Cham, Switzerland

Introduction

We believe we are not far away from reality if, following the lines of extreme simplification, we say that Personalized/Systems Medicine and, by extension, P4 (Predictive, Preventive, Participatory, and Personalized) Medicine were born when advanced mathematical sciences and bioinformatics entered in full the generation of evidence and the consequent extraction of knowledge from Biomedical Data, mainly from Big Data stemming from omics sciences.

In recent years, mathematical technologies, conjugated with computing capabilities have made (together with more affordable large scale precision analytics) advanced techniques available in a reliable and sustainable way at "the Lab" level paving the way (especially) for Machine Learning powering (and empowering) Computer Systems to become, under the definition of "Artificial Intelligence", an everyday tool to face boldly medical complexity to an extent that was considered unmanageable "just yesterday". These new and enhanced capabilities have made clinical data, in addition to omics data, a target at hand for the synthesis of outcome knowledge from Real-World Experience(s).

Hence, Artificial Intelligence (AI) possibly boosted by the emotional engagement (not necessarily in a "neutral" manner) and intended here with a very pragmatic definition as a "system" working by ingesting large amounts of data through which it trains itself (more or less supervised by a "human touch") and makes analyses oriented at identifying, essentially, correlations and patterns. These are used to make predictions. AI tools are believed to perform, if compared with other systems, more quickly and with relatively few errors when it comes to transform huge volumes of Data into actionable knowledge and information.

Unfortunately, the emotional component inherent with the coupling of the words "Artificial" with "Intelligence" created from one side, possibly, an excessive expectation (possibly as an effect of anthropomorphism and emotional receptivity) that, finally, biomedical complexity was going to be solved at a fingers snap and, from the other, a hype that created (some) confusion with regard to the pragmatic and accountable benefits AI could bring in understanding the mechanisms at the basis of the transition from a status of health to one of disease. The reliability of models that mimic diseases in real world scenarios is currently tested in AI-driven and simulated environments.

This hype has been fuelled, as well, by "consumer" level "weak AI" (yes, one of the ways to categorize AI is "weak" versus "strong", whereas weak AI or "narrow" AI refers to systems trained to execute one specific task and "strong" AI, or "Artificial General Intelligence" (AGI), put simply, can mimic cognitive capabilities of the human brain and thus be capable, beyond the task for which it is initially trained, to solve, autonomously, unfamiliar ones) because weak AI is at the base of the personalized virtual assistants we use every day to find a route on a navigator, to connect our position to a cab, and in "n" other situations with which we are very familiar.

A different story is when we deep dive into the exercise of finding pragmatic, substantial, and structured situation in which AI has become "handy" in Personalized Medicine (PM). This is the focus of this book that is structured into three sections. We have pivoted on the editors' personal and professional trajectories to focus on the use and usefulness of AI in the field of PM from the extraction of actionable biomarkers' profiles from omics Data to make computer vision reliable and actionable into Surgery, passing through the extraction of criteria potentially useful for clinical decision support from Lifestyle (behavioural) Data to solve complex phenotypes in cancer, chronic diseases, and beyond.

Finally, we elaborate on the concept of Metaverse, from a cross-industry perspective analysing where Health Care and Research stand in terms of uptake and experimentation to the forecast of building entirely new models for R&D not only in the deep science content but, as well, in everyday design and discovery operations with initial real-life deployment of "metaverse like" (still in the augmented virtual reality endeavour) solutions into clinical trials.

In Part I, several perspectives provide a comprehensive state of the art that includes scientific, technological, and regulatory challenges of using AI in PM. Chapter 1 by Cascini and colleagues focuses on how PM is developing by integrating AI solutions through the lenses of Public Health, which interacts with the main audience of healthcare: all citizens. Chapter 2 by Laino and Savevski dives into the practical evolution to Digital Health, providing a broad scenario in which AI is used in the biomedical field and the main challenges encountered worldwide. Chapter 3 by Giorgianni offers a lucid vision of the current status of data flow in Europe, introduces the professional figure of the Data Protection Officer, and envisions a way forward for data protection. Chapter 4 by Recchia and Gussoni draws a distinction between Digital Health, Digital Medicine, and Digital Therapeutics (DTx) while introducing why the latter can potentially transform the treatment of chronic diseases, without neglecting practical issues such as clinical evidence, data safety, and patients' literacy. The section closes with Chap. 5 by Tagliaferri and colleagues that considers current methodological constraints of using AI in PM from an academic point of view, while proposing the avenue of Deep Humanism as a way through which patients can be considered more holistically with digital and technological instruments.

Part II is entirely focused on consolidated evidence from a stratified landscape of AI-based solution applications in PM practice. In Chap. 6, Kyriazakos and colleagues show a detailed application of how big data and AI algorithms can improve clinical research with a specific focus on DTx. In Chap. 7, Villoslada and colleagues explain how network analysis of signalling pathways can improve the development of combination therapy in multiple sclerosis, as a beautiful example of how AI can boost the progress of Systems Pharmacology for immunotherapy. In Chap. 8 by Surendran and colleagues, we enter in the field of virtual trials: a meticulous methodology leads the reader to understand how virtual patient cohorts can be generated with an application in the field of oncology. Chapter 9 by Giacò and colleagues explores the issue of Deep Phenotyping and how it can be achieved nowadays by taking COVID-19 variants as prominent example. On the other hand, Chap. 10 by González-Colom and colleagues provides the example of the Synergy-COPD project to demonstrate multiscale modelling to predict chronic obstructive pulmonary disease (COPD) towards Systems Medicine and AI. In Chap. 11 by Tagliaferri and colleagues, AI is used to ameliorate radiotherapy workflows, especially in high-complexity organizations such as research hospitals. This section closes with Chap. 12 by Mascagni et al., in which several applications of AI in surgery are provided, highlighting the novel scenario of Surgical Data Science.

Part III provides a broader perspective on upcoming technologies with resourceful insight about the near future. In Chap. 13, Bellina and Jungmann explore the dialogue between start-ups and established organizations as a virtuous collaboration through which healthcare and PM could find a sustainable innovation process. In Chap. 14, Neumann and colleagues elucidate how AI augments the field of Medtech, especially with Software in a Medical Device (SiMD) and Software as a Medical Device (SaMD) solutions, explained with a strong regulatory framework. In Chap. 15, Patarnello explores the fascinating field of Quantum Computing and how it will help enhancing precision medicine, while indicating current challenges that still need to be addressed. In Chap. 16, Pagliai depicts an interesting representation of how the Metaverse Continuum is evolving health care to the next level and leads to re-imagine this field. On the other hand, Chap. 17 addresses the very interesting topic of human-like interactions while using anthropomorphic technologies in health care, such as chatbots, robots, and virtual avatars.

With Chap. 18, Manto and D'Oria wrap up the main aspects of using AI in PM by focusing on some ethical and educational issues that regard individual and algorithmic "identity". Finally, Chap. 19 by D'Oria and colleagues prepares the seeds for thinking about a broader perspective on the relationship between AI possibilities and human limits, to critically reflect on how this relationship affects AI-guided PM.

With the hope that this book will be helpful to envision the current and future scenarios of medical research and care, we warmly thank Dr. Ina Karen Stoeck at Springer Nature, whose precious care has guided us in this beautiful journey we are really proud about, and Mr. Bibhuti Bhusan Sharma at Springer Nature, for his constant dedication in managing this international project. We are also glad to all the staff at Springer Nature involved in making this dream possible.

Fondazione Policlinico Universitario
A. Gemelli IRCCS
Rome, Italy
Alfredo Cesario

Fondazione Policlinico Universitario
A. Gemelli IRCCS
Rome, Italy
Marika D'Oria

European Institute for Systems
Biology and Medicine
Lyon, France
Charles Auffray

Fondazione Policlinico Universitario
A. Gemelli IRCCS
Rome, Italy
Giovanni Scambia

Contents

Part I

State of the Art

Personalized Medicine Through Artificial Intelligence: A Public Health Perspective

1

Fidelia Cascini, Stefan Buttigieg, Roberta Pastorino, Walter Ricciardi, and Stefania Boccia

Abstract

Digital technologies are leading the transformation of the healthcare sector and are bringing new models of health service delivery that are mostly focused on prevention and based on personalization and precision. From a Public Health perspective, personalized medicine implies a transformation of health systems that facilitates improved targeting of healthcare services based on specific population sub-group needs, to maximize their effectiveness and relevance. The digitalization of healthcare systems and the use of health data-driven approaches allow the development of novel, targeted Public Health solutions, which are necessary in the complexity of the healthcare ecosystem. The growing amount

F. Cascini
Section of Hygiene, Department of Life Sciences and Public Health, Università Cattolica del Sacro Cuore, Rome, Italy

Italian Ministry of Health, Digital Health Expert, Rome, Italy

S. Buttigieg
European Public Health Association, Utrecht, The Netherlands

R. Pastorino (✉)
Public Health Area, Department of Woman and Child Health and Public Health, Fondazione Policlinico Universitario A. Gemelli IRCCS, Rome, Italy
e-mail: roberta.pastorino@unicatt.it

W. Ricciardi
Section of Hygiene, Department of Life Sciences and Public Health, Università Cattolica del Sacro Cuore, Rome, Italy

S. Boccia
Section of Hygiene, Department of Life Sciences and Public Health, Università Cattolica del Sacro Cuore, Rome, Italy

Public Health Area, Department of Woman and Child Health and Public Health, Fondazione Policlinico Universitario A. Gemelli IRCCS, Rome, Italy

A. Cesario et al. (eds.), *Personalized Medicine Meets Artificial Intelligence*,
https://doi.org/10.1007/978-3-031-32614-1_1

of health data generated every year makes big data platforms an essential tool for data management and analytics. Although various different artificial intelligence systems are being developed with the aid of such platforms, they generally share similar objectives: to assist health systems' response to specific emerging health demands; to design healthcare services that are able to scale their provision according to the growth of populations; to improve public health resilience and responsiveness, to promptly control epidemic-related emergencies; to better differentiate patient communities through risk stratification and to inform individual decision-making, both essential for the personalized medicine movement. In this chapter, we present the development of artificial intelligence and its promising applications allowing targeted Public Health interventions, and current limitations to address as well.

Keywords

Public Health · Personalized medicine · Artificial intelligence · Healthcare

1.1 The International Landscape of Artificial Intelligence for Healthcare Purposes

Artificial Intelligence (AI), defined as a system's ability to correctly interpret external data, to learn from such data, and to use those learnings to achieve specific goals and tasks through flexible adaptation, can be considered one of the major drivers of the digital transformation challenge we are faced with.

Digital technologies, machine learning algorithms, and AI are transforming medicine, medical research, and Public Health. The United Nations Secretary-General has stated that safe deployment of new technologies, including AI, can help the world to achieve the United Nations Sustainable Development Goals, which would include the health-related objectives under Sustainable Development Goal 3 (United Nations 2022). AI could also be the key to improve global commitments on achieving universal health coverage.

The EU's coordinated approach to make the most out of the opportunities offered by AI and to address the challenges that it brings is based on its Digital Single Market, where rules and regulations on various related topics (data protection, business development, etc.) create an environment for growth without leaving single countries lagging (European Commission 2022). In its seminal *White Paper on Artificial Intelligence*, the European Commission (2020a) describes the characteristics of the policy framework necessary to develop trustworthy and secure AI applications for all areas, including healthcare. The EU must be:

- An "ecosystem of excellence", starting in research and innovation, where the right incentives accelerate the adoption of solutions based on AI, including by small- and medium-sized enterprises.

- An "ecosystem of trust" in which compliance with EU rules and regulations is enforced, including the rules protecting fundamental rights and consumers' rights, especially for AI systems operated in the EU which pose a high risk.

The use of AI technologies to improve healthcare systems holds a promising future, with progress already being made in various health-related fields, such as drug discovery, medical imaging, screening/prevention. AI could be fundamental in assisting healthcare providers, helping to avoid errors, and allowing clinicians to focus on providing care and solving complex cases. However, from a Public Health perspective, to maximize the benefits for society, the legal, ethical, regulatory, economical, and social constraints of AI must be addressed rapidly.

AI technologies are usually designed by companies or through public–private partnerships (PPPs). To strengthen European competitiveness in the PPPs on AI and to engage various stakeholders and investors in the technological development of the EU, the European Commission spent between 20 and 50 million € per year to fund partnerships on AI Data and Robotics from 2014 to 2020, with an overall 1.1 billion € spent under the Horizon 2020 research and innovation programme, which included big data (Pastorino et al. 2019) and healthcare (OECD 2019).

Some of the world's largest technology companies are developing new applications and services, which they either own or invest in. The potential benefits of these technologies and the economic and commercial potential of AI for health care are warranting an ever-greater use of AI worldwide (WHO 2021a).

The European Commission categorizes AI applications as "generally high-risk" because they are both employed in sectors where significant risks are expected to occur and in such a manner that increases the amount of predictable risk to be taken (these include healthcare, transportation, the energy sector, etc.). This is particularly true for healthcare systems, where applications can have unpredictable and far-reaching consequences, which might affect patient safety, access to care, quality of care, and certain fundamental human rights. The European Commission has continued to build on this work and has adopted the risk stratification process, varying from "Unacceptable Risk" to "Minimal Risk" in the ongoing discussion on the harmonized rules on Artificial Intelligence or what is better known as the "AI Act" (European Commission 2021). In parallel, this legislation is furthermore supported by the work being done on the "Data Governance Act", "The Open Data Directive", and the initiatives under the "European strategy of data". In the European strategy of data, there is a focus on the relevant deployment of data infrastructures, tools, and computing for the European Health Data Space which will determine the availability of high-quality data essential for training and further developing Artificial Intelligence systems (European Commission 2020b). This specific space will facilitate non-discriminatory access to health data and the training of artificial intelligence algorithms on those datasets, in a privacy-preserving, secure, timely, transparent, and trustworthy manner supported by institutional governance.

In fact, through this regulatory framework, the European Commission is seeking to ensure common normative standards for all high-risk AI systems. The health sector's involvement in high-stakes situations which require the use of sophisticated diagnostic systems and systems supporting human decisions need to be reliable and accurate. Building on this aspect is the important element of "Human Oversight" outlined in Article 14 of the "AI Act", where human oversight will play an important role in preventing or minimizing the risks to health that may emerge because of the implementation of a high-risk AI System.

Different organizations throughout the world are building on this important discussion. The Pan American Health Organization (PAHO) offers its own guiding principles for all AI applications in healthcare or Public Health (PAHO 2021):

- People-centred: AI technologies must respect the rights of the individual.
- Ethically grounded: progress and discussion must be made in compliance with the principles of human dignity, non-maleficence, and justice.
- Transparent: when developing AI algorithms, having clear objectives and goals is mandatory.
- Data protection: data privacy and security are paramount in AI development.
- Demonstrates scientific integrity: AI applications must be reliable, reproducible, fair, honest, and accountable, according to the best practices.
- Open and shareable: openness and shareability must be the founding principles of every AI development process.
- Non-discriminatory: AI for Public Health must be based on fairness, equality, and inclusiveness.
- Human-controlled: all automated decisions must be reviewed by human beings.

These guiding principles will be fundamental in developing global cooperation initiatives centred on AI in Public Health and navigating the complex legislative and policy environment that is being developed for the safety of our increasingly interconnected society.

1.1.1 Artificial Intelligence Systems Suitable for Public Health Applications

AI encompasses many fields of scientific enquiry, and its objective of mimicking human cognitive functions has many facets. To deploy and implement AI systems in healthcare settings they need to undergo a process of training, in which various types of input data, depending on the use case, will be "fed" to the algorithm and return associations or predictions. In this way, clinical data, diagnosis data, or screening results can be used to predict individual or population trends, help diagnose disease or create associations between various features of care (FDA 2013).

Machine Learning (ML) is based on algorithms that improve automatically through experience, feedback, and use of data. The trained algorithm then generates rules that can be used to classify new data or predict future data; this has many applications in a Public Health perspective, as it can be used to understand the complex connections between genetics, environment, and diseases or to predict illness. ML algorithms can be divided into two major categories: unsupervised learning and supervised learning, with the latter being based on labelled data (the machine has some positive and negative examples of what it should be able to identify). Unsupervised learning is best applied to extract patterns and features from data, whilst supervised learning is more suitable for predictive modelling, given its ability, for example, to build relationships between the patient traits (age, gender, etc.) and the outcome of interest (i.e., cardiovascular disease) (Jiang et al. 2017).

Deep Learning (DL) leverages algorithms as networks of decisions to learn from data. These networks are called neural networks or deep neural networks, depending on the number of layers in the network. DL can identify diseases thanks to imaging and can predict health status relying on health records (e.g. diabetic retinopathy in retinal fundus photographs). Its main advantage over other learning algorithms is its returned performance over bigger databases, being able to draw patterns in an abundance of unlabelled data, making it highly scalable.

Different learning methods can then be used to create disparate types of AI that have applications in Public Health. Natural language-related AI is a subfield of AI that aims to bridge the divide between the languages that humans and computers use to operate. Specifically, Natural Language Processing (NLP) automates the ability to read and understand human language, and by doing so, it ensures behaviour and sentiment analysis through social media and consumer-generated data. Natural Language Understanding (NLU) understands human writing using a coded comprehension of grammar, syntax, and semantics; this might be employed, for example, in the identification of loneliness or depression in older adults based on the content and patterns of their text messages. Finally, Natural Language Generation (NLG) transforms structured data into plain language or text, which can be useful to automatically remove identifiers and sensible information from electronic medical records or to produce automated medical reports given certain exam results as input.

Automated scheduling and automated planning are a branch of AI focused on organizing and prioritizing the activities required to achieve the desired goal, and expert systems (also known as knowledge-based systems) are AI programs that have expert-level competence in solving specific problems. Possible functions of expert systems and management systems include identifying and eliminating fraud or waste, scheduling patients, predicting which patients are unlikely to attend a scheduled appointment, assisting in the identification of staffing requirements, optimizing the allocation of health-system resources by geographical location according to current health challenges, and using administrative data to predict the length of stay of health workers in underserved communities (NHS UK 2019). Digital Decisioning Platforms represent the evolution of automated scheduling and planning expert systems. Generative AI, powered by advanced models like GPTs and stable diffusion, is rapidly developing. Public health has been involved in this process mainly regarding tasks such as analyzing patient data, creating informative presentations, and writing public health messages. However, more research is needed to fully understand generative AI potential and limitations, also in the Public Health field.

Other domains of AI can benefit Public Health indirectly. One such example is Cognitive Search, which employs AI systems to merge and understand digital contents from different sources by deriving contextual insights from conceptual data, improving the relevance of the results generated from a user search, for example in a search engine. This could help evaluate the quality and intent of information distributed during health emergencies, and sift through emerging information based on source and credibility. During the COVID-19 pandemic, this has been applied to the most utilized search engines to give citizens the most up-to-date information on prevention and medication use, filter obsolete information and reduce confusion (Microsoft New Zealand News Centre 2021).

1.1.2 Improving Healthcare Services Using Artificial Intelligence

One of the main focuses of AI development in healthcare is creating a support system to improve the early diagnosis of various diseases. This is being particularly explored in the field of oncology, where AI is being evaluated for use in radiological diagnoses, such as in whole-body imaging, colonoscopies, and mammograms. AI can also aid in optimizing radiological treatment dosing, recognizing malignant disease in dermatology or clinical pathology, and guiding RNA and DNA sequencing for immunotherapy (Bi et al. 2019).

In general, AI developments in early diagnosis are being studied in most health-related fields, such as in the early diagnosis of diabetic retinopathy, cardiovascular disease, liver disease, and neurological disorders (Kamdar et al. 2020). Currently there are only a handful of prospective clinical trials on the effectiveness of AI in early diagnosis, with some showing promise of equivalent detection ability to human professionals in specific tasks, with even fewer focusing on the potential benefits of human–machine partnerships. One of the risks in relying excessively on AI and machine learning algorithms is the development of an automation bias, where medical practitioners might not consider other important aspects in patient care and overlook errors that should have been spotted by human-guided decision-making (The Swedish National Council on Medical Ethics 2020).

AI can also be used to digitalize and store traditional paper medical records and process large amounts of data from images and other types of inputs or signals (such as motion data or sound data). Steps in image and signal processing algorithms typically include signal feature analysis and data classification using tools such as artificial neural networks, which work via complex layers of decision nodes (Wahl et al. 2018). Medical imaging is one of the most rapidly developing areas of AI application in healthcare. Whilst improving automated image interpretation and analysis is a priority, other important aspects of AI application to medical imaging are being explored, such as data security and user privacy solutions for medical image analysis, deep learning algorithms for restoration/reconstruction and segmentation of complex imaging and creation of fuzzy sets or rough sets in medical image analysis (Xia 2021).

Furthermore, with health systems, in general, growing more complex every year, administration and management of care are becoming increasingly laborious. AI can be used to assist personnel in complex logistical tasks, such as optimization of the medical supply chain, to assume mundane, repetitive tasks or to support complex decision-making (Schwalbe and Wahl 2020). This is made possible by a combination of AI advancements in the fields of natural language processing, automated scheduling and planning, and expert systems (PAHO 2021).

Many AI tools can also be used in specific public health programmes or in wide public health approaches to improve wellbeing. AI can be used for health promotion or to identify target populations or locations with "high-risk" behaviour concerning communicable and non-communicable diseases. AI can improve the effectiveness of communication and messaging specifically directed to certain sub-populations, both in terms of its ability to recognize priority groups and in its adaptiveness in creating tailor suited messages to benefit population health (micro-targeting) (Privacy International 2021). One example of such application is micro-targeting individuals or communities with technological, linguistic, or cultural barriers to better communicate the importance and safety of vaccinations, such as the COVID-19 vaccination (NBC News 2021). AI tools could therefore be adapted to improve access and equity of care, furthering the development of truly personalized medicine.

AI can also have a leading role in performing analyses of patterns of data for health surveillance and disease detection (Russell and Norvig 2010; Alcantara et al. 2017; Morgenstern et al. 2021; CDC Foundation 2022): AI tools can be used to identify bacterial contamination in water treatment plants, identify foodborne illnesses in restaurants or hospitals, simplify detection and lower the costs. Sensors can also be used to improve environmental health, such as by analysing air pollution patterns or using machine learning to make inferences between the physical environment and healthy behaviour (Roski et al. 2019).

Another application of AI in public health surveillance is evidence collection and its use to create mathematical models to make decisions. Although many public health institutions are not yet making full use of all possible sources of data, some fields, such as real-time health surveillance, are steadily improving. This has improved the public health outlook on pandemic preparedness and response, though the long-term ramification of such important changes will only be evident in the future (Whitelaw et al. 2020).

The development of public health policy also proves to be fertile ground for artificial intelligence where attempts at analysing argumentation on food quality in a public health policy was attempted. Models with new recommendations based on stakeholders' arguments by target specific audiences have been consequently generated (Bourguet et al. 2013). Healthcare has always depended in part on predictions, prognoses, and the use of predictive analytics. AI is just one of the more recent tools for this purpose, and many possible benefits of prediction-based health care rely on the use of this technology. For example, AI can be used to assess an individual's risk of disease, which could be used for the prevention of diseases and major health events (OECD 2019).

Various studies suggest that artificial intelligence may improve several pathologies, such as heart failure, utilizing predictive models and telemonitoring systems for clinical support and patient empowerment. For example, given the expected increase in the number of heart failure patients in the future due to the ageing of the population, predicting the risk of a patient having heart failure could prevent hospitalizations and readmissions, improving both patient care and hospital management, which would have a high impact on costs and time (Larburu et al. 2018).

Machine learning is also increasingly being applied to make predictions related to population health: using novel big data resources, ripe with different data types, may allow for improvements in prediction algorithms necessary to navigate the complex health data ecosystem successfully (Alcantara et al. 2017). A good example of this, is the integration of data types to better understand complex associations between genetics, environment, and disease. The Harvard group has been using large administrative datasets to untangle the relationship between genetics and environment in all diseases recorded in health insurance claims data (Lakhani et al. 2019). Using biobanks and their massive datasets allows scientists around the world to discover new genetic variants (e.g. through genome-wide association studies) and novel risk factors associated with disease more efficiently and with higher sensitivity and specificity compared to traditional "one-at-a-time" methods (CDC 2019).

Using electronic medical record data, machine- and deep-learning algorithms have been able to predict many important clinical parameters, including suicide, Alzheimer's disease, dementia, severe sepsis, septic shock, hospital readmission, all-cause mortality, in-hospital mortality, unplanned readmission, prolonged length of stay, and final discharge diagnosis (Topol 2019).

All in all, predictive models have been used much more widely by clinicians than by public health professionals. However, on closer inspection, any application improving patient care at any level can be considered relevant to the field of public health. The ability of clinicians and healthcare providers to make better informed decisions on patient health will be improved by context-specific algorithms, that use massive quantities of clinical, physiological, epidemiological, and genetic data. Precision Medicine will further benefit from these advanced algorithms, as their accuracy, timeliness, and appropriateness in clinical care improve over time, decompressing our reliance on human resources. This advancement, however, still necessitates computer-literate physicians, who are up to date with new generation data-driven approaches. The key to a complete incorporation of AI into clinical care will therefore be the integration of human clinical judgement with advanced clinical machine learning algorithms (Khemasuwan et al. 2020).

Table 1.1 shows the most common AI applications in healthcare:

1.2 Current Limitations of AI Applications in Public Health

AI poses major technological, ethical, and social challenges, which need competent professionals to address. In fact, beyond the many opportunities, artificial intelligence presents some critical issues that could slow down the adoption of these applications. With Public Health interventions targeting entire populations, the introduction of AI might either improve or worsen health inequities on a large scale (Weiss et al. 2018); as assessed by many ethics and policy guidance

Table 1.1 Most advanced AI applications in the field of healthcare

Diagnosis and clinical care uses	Administration and management of healthcare services	Predictive models for public health purposes
Medical Imaging: – Interpretation – Analysis – Data security – Restoration – Fuzzy/Rough Sets. – Image quality Oncology: – Radiological Diagnosis – Treatment dosing – Clinical pathology classification – Immune therapy gene sequencing	– Guiding complex logistical tasks – Reduce the burden of mundane/repetitive tasks – Support complex decision-making – Optimization of medical supply chain – Time and cost management – Communicating information in a timely and cost-effective manner	– Micro-targeting – High-risk behaviours profiling – Predict water contamination – Identify patterns in foodborne illnesses – Predict air pollution patterns – Real-time health surveillance – Population Health predictions – Genome-wide associations

documents, the promotion of AI deals with personalized recommendations and individual action, but this should not threaten the importance of collective action to take care of social and structural determinants of health (Panch et al. 2019).

Moreover, like all public health interventions, AI has the potential to create enduring benefits but will require not just a broad coalition of support and partnership between the public and private sector but also the trust and enduring support of patients.

Currently, despite their wide testing, AI-based prediction algorithms that affect patient care have not reached a sufficient level of accuracy needed for precise long-term predictions. This poses a serious challenge for healthcare workers, as long-term predictions of limited reliability could impact an individual's life for years in the future: for example, both false-positive and false-negative predictions on an essential diagnosis could affect the level of risk clinicians are willing to undertake in order to treat a health condition, thus heavily impacting health outcomes. Furthermore, these prediction algorithms could be biased towards or against certain population sub-groups (e.g., ethnic groups, religious groups), both in terms of potentially discriminatory health practices suggested and issues about individual autonomy on personal data use and informed consent. These potential pitfalls of AI-based algorithms and their long-term health inferences raise essential ethical concerns that have to be addressed by all the stakeholders involved (WHO 2021b).

One of the main and most obvious implications of AI introduction in Public Health is the risk of inequalities in accessing technologies, in the opportunity to benefit from them and in the burdens generated by them (IEEE Standards Association 2019). An example is represented by developing countries that depend on AI-based platforms developed by other richer countries, which also leads to a significant financial burden. Most AI developments in healthcare respond to the needs of high-income countries (HICs), where most research is conducted, however low- and middle-income countries, where workforce shortages and limited resources constrain the access to quality of care, could also benefit from the implementation of AI in Public Health.

Another element that exacerbates this digital divide is the lack of availability and accessibility of Internet services: Mobile Health apps, which make heavy use of Artificial Intelligence, are of no use for people living in areas that do not have Internet access. Not only could this gap manifest between population groups, but also between researchers, public and private sectors and even health systems (Smith et al. 2020). A similar difference could present between those who choose to actively use AI technologies and those who do not. AI systems might be programmed according to certain values and judgements that could create or worsen health inequities, for instance, those related to gender, race or belonging to minority groups (Norris 2001). Moreover, there could be disagreement about the system of values that inform AI systems (Caliskan et al. 2017).

Through harmonized standards and requirements, both at the research stage and in the evaluation phase, it is essential to ensure effectiveness for the patient as well as safety in use. The current scientific landscape will see an increase in the number of clinical trials that verify the effectiveness and efficacy of AI in Public Health; proper development of a precautionary supervised system of trials will ensure that they are ethical, legal, and inclusive in scope. Nevertheless, the next decade will shed light on how the broad political, economic, and cultural global framework in which these technologies are developed will transform public health through the use of AI (Larburu et al. 2018).

The biggest challenge for AI in these healthcare domains will be ensuring their adoption in everyday clinical practice. Widespread adoption of AI systems must be approved by regulators, integrated with digital health platforms through health data pipelines, standardized at a level such that similar products perform similarly, taught to clinicians, accepted by the patients, paid for by public or private payer organizations, and updated over time. The change is expected to be based on multilevel—from local to supranational—collaborations, and supported by regulatory bodies acting for Public Health interests (Davenport and Kalakota 2019).

References

Alcantara MF, Cao Y, Chang L et al (2017) Improving tuberculosis diagnostics using deep learning and mobile health technologies among resource-poor communities in Perú. Smart Health 1–2: 66–76. https://doi.org/10.1016/j.smhl.2017.04.003

Bi WL, Hosny A, Schabath MB et al (2019) Artificial intelligence in cancer imaging: clinical challenges and applications. CA Cancer J Clin 69(2):127–157

Bourguet J-R, Thomopoulos R, Mugnier M-L, Abécassis J (2013) An artificial intelligence-based approach to deal with argumentation applied to food quality in a public health policy. Expert Syst Appl 40(11):4539–4546

Caliskan A, Bryson JJ, Narayanan A (2017) Semantics derived automatically from language corpora contain human-like biases. Science 80(356):183–186

CDC (2019) Data science and machine learning in public health: promises and challenges. Source: https://blogs.cdc.gov/genomics/2019/09/20/data-science/ Last consultation: 22/08/2022

CDC Foundation (2022) What is public health? Source: https://www.cdcfoundation.org/what-public-health Last consultation: 22/08/2022

Davenport T, Kalakota R (2019) The potential for artificial intelligence in healthcare. Future Healthc J 6(2):94–98. https://doi.org/10.7861/futurehosp.6-2-94

European Commission (2020a) White paper on artificial intelligence – a European approach to excellence and trust. Source: https://ec.europa.eu/info/publications/white-paper-artificial-intelligence-european-approach-excellence-and-trust_en Last consultation 22/08/2022

European Commission (2020b) A European strategy for data. Source: https://eur-lex.europa.eu/legal-content/EN/TXT/PDF/?uri=CELEX:52020DC0066&from=EN Last consultation: 05/05/2022

European Commission (2021) Regulatory framework on AI | Shaping Europe's digital future. Source: https://digital-strategy.ec.europa.eu/en/policies/regulatory-framework-ai Last consultation 05/05/2022,

European Commission (2022) Communication from the commission to the European parliament, the European Council the European economic and social committee and the committee of the regions, Artificial Intelligence for Europe. Source: https://op.europa.eu/en/publication-detail/-/publication/fc930f14-d7ae-11ec-a95f-01aa75ed71a1/language-en Last consultation 22/08/2022

FDA (2013) Guidance for industry electronic source data in clinical investigations. Source: https://www.federalregister.gov/documents/2013/09/18/2013-22645/guidance-for-industry-on-electronic-source-data-in-clinical-investigations-availability Last consultation: 22/08/2022

IEEE Standards Association (2019) IEEE releases ethically aligned design, version 2 to show "Ethics in Action" for the Development of Autonomous and Intelligent Systems (A/IS). Source: https://standards.ieee.org/news/2017/ead_v2.html Last consultation: 22/08/2022

Jiang F, Jiang Y, Zhi H et al (2017) Artificial intelligence in healthcare: past, present and future. Stroke Vasc Neurol 2. https://doi.org/10.1136/svn-2017-000101

Kamdar JH, Jeba Praba J, Georrge JJ (2020) Artificial intelligence in medical diagnosis: methods, algorithms and applications. In: Jain V, Chatterjee J (eds) Machine learning with health care perspective. learning and analytics in intelligent systems, vol 13. Springer, Cham

Khemasuwan D, Sorensen JS, Henri G, Colt HG (2020) Artificial intelligence in pulmonary medicine: computer vision, predictive model and COVID-19. Eur Respir Rev 29(157): 200181. https://doi.org/10.1183/16000617.0181-2020

Lakhani CM, Tierney BT, Manrai AK et al (2019) Repurposing large health insurance claims data to estimate genetic and environmental contributions in 560 phenotypes. Nat Genet 51(2): 327–334. https://doi.org/10.1038/s41588-018-0313-7

Larburu N, Artetxe A, Escolar V et al (2018) Artificial intelligence to prevent mobile heart failure patients decompensation in real time: monitoring-based predictive model. Mobile Information Systems 1546210. https://doi.org/10.1155/2018/1546210

Microsoft New Zealand News Centre (2021) The search to save lives: how machine learning is fighting covid-19 with data. Source: https://news.microsoft.com/en-nz/2021/05/06/the-search-to-save-lives-how-machine-learning-is-fighting-covid-19-with-data/ Last consultation: 22/08/2022

Morgenstern JD, Rosella LC, Daley MJ et al (2021) AI's gonna have an impact on everything in society, so it has to have an impact on public health: a fundamental qualitative descriptive study of the implications of artificial intelligence for public health. BMC Public Health 21

NBC News (2021) As vaccine demand slows, doctors shift to 'microtargeting' vulnerable communities. Source: https://www.nbcnews.com/news/us-news/vaccine-demand-slows-doctors-shift-microtargeting-vulnerable-communities-n1266031 Last consultation: 22/08/2022

NHS UK (2019) Artificial intelligence: How to get it right. Source: https://www.nhsx.nhs.uk/media/documents/NHSX_AI_report.pdf Last consultation: 02/08/2022

Norris P (2001) Digital divide: civic engagement, information poverty, and the internet worldwide. Cambridge University Press, Cambridge (MA)

OECD (2019) Public-private partnership on AI, data and robotics. Source: https://stip.oecd.org/stip/policy-initiatives/2019%2Fdata%2FpolicyInitiatives%2F26437 Last consultation: 05/05/2022

PAHO (2021) Artificial intelligence in public health. Source: https://iris.paho.org/handle/10665.2/53732 Last consultation: 22/08/2022

Panch T, Pearson-Stuttard J, Greaves F, Atun R (2019) Artificial intelligence: opportunities and risks for public health. Lancet Digit Heal 1:e13–e14

Pastorino R, De Vito C, Migliara G et al (2019) Benefits and challenges of Big Data in healthcare: an overview of the European initiatives. Eur J Pub Health 29(Suppl._3):23–27. https://doi.org/10.1093/eurpub/ckz168

Privacy International (2021) Micro-targeting. Source: https://privacyinternational.org/learn/microtargeting Last consultation: 05/05/2022

Roski J, Chapman W, Heffner J et al (2019) How artificial intelligence is changing health and health care. In: Matheny M, Thadaney Israni S, Ahmed M, Whicher D (eds) Artificial intelligence in health care: the hope, the hype, the promise, the peril. National Academy of Medicine, Washington, DC

Russell S, Norvig P (eds) (2010) Artificial intelligence: a modern approach, 4th edn. Pearson, London

Schwalbe N, Wahl B (2020) Artificial intelligence and the future of global health. Lancet 395:1579–1586

Smith MJ, Axler R, Bean S et al (2020) Four equity considerations for the use of artificial intelligence in public health. Bull World Health Organ 98:290–292

The Swedish National Council on Medical Ethics (2020) In brief: artificial intelligence in health care. Source: https://smer.se/en/2020/05/28/in-brief-artificial-intelligence-in-healthcare/ Last consultation: 22/08/2022

Topol E (2019) High-performance medicine: the convergence of human and artificial intelligence. Nat Med 25:44–56. https://doi.org/10.1038/s41591-018-0300-7

United Nations, Dept. of Economic and Social Affairs Sustainable Development (2022) Developmental Goal 3: Ensure healthy lives and promote well-being for all at all ages. Source: https://sdgs.un.org/goals/goal3 Last consultation 22/08/2022

Wahl B, Cossy-Gantner A, Germann S et al (2018) Artificial intelligence (AI) and global health: how can AI contribute to health in resource-poor settings? BMJ Glob Health 3:e000798

Weiss D, Rydland HT, Øversveen E et al (2018) Innovative technologies and social inequalities in health: a scoping review of the literature. PLoS One 13:e019544

Whitelaw S, Mamas MA, Topol E, Van Spall GC (2020) Applications of digital technology in COVID-19 planning and response. Lancet Digital Health 2:e435–e440

WHO (2021a) Ethics and governance of artificial intelligence for health – WHO Guidance. Source: https://www.who.int/publications/i/item/9789240029200 Last consultation: 22/08/2022

WHO (2021b) WHO Consultation towards the development of guidance on ethics and governance of artificial intelligence for health – Meeting report. Source: https://www.who.int/publications/i/item/who-consultation-towards-the-development-of-guidance-on-ethics-and-governance-of-artificial-intelligence-for-health Last consultation: 22/08/2022

Xia Y (ed) (2021) Artificial intelligence for medical image analysis – special issue. BioMed Res Int. Source: https://www.hindawi.com/journals/bmri/si/617603/ Last consultation: 22/08/2022

2 Artificial Intelligence and Digital Health: An International Biomedical Perspective

Maria Elena Laino and Victor Savevski

Abstract

With the scientific and technological progress achieved through -omic sciences (e.g., genomics, proteomics) and the development of sophisticated Artificial Intelligence (AI)-based solutions, Personalized Medicine has reached new opportunities for patient prevention and care in a new clinical avenue called "Digital Health" (DH). Investments in this field are rapidly increased worldwide. The chapter shows how AI is used in DH by elucidating four principal applications reported by the literature. AI-based solutions can support in retrieving big data, in analyzing Real-World Data (RWD) to produce Real-World Evidence (RWE), in predicting prognostic outcomes and risks, in personalizing clinical diagnostics, while customizing therapy development and monitoring patients' adherence. The chapter finally summarizes some challenges that still need to be addressed by the stakeholders involved in the field of DH.

Keywords

Artificial Intelligence · Digital Health · Patient management · Predictive model · Virtual ward · Patient support program

2.1 Introduction

The scientific progress achieved with the -omics (e.g., genomics, proteomics, metabolomics, radiomics, etc.), and the technological advances brought by Information Technology (IT), bioinformatics, and data sciences have rapidly increased the prognostic and diagnostic opportunities for Personalized Medicine (PM), aiming at

M. E. Laino · V. Savevski (✉)
AI Center, IRCCS Humanitas Research Hospital, Rozzano (MI), Italy
e-mail: victor.savevski@humanitas.it

A. Cesario et al. (eds.), *Personalized Medicine Meets Artificial Intelligence*,
https://doi.org/10.1007/978-3-031-32614-1_2

delivering "*the right treatment to the right patient at the right time*" (Cesario et al. 2021). The boost of innovation in biomedical research has rapidly increased life expectancy worldwide, leading healthcare systems to face a higher demand for their services and workforce, to meet patients' and citizens' specific needs. Several challenges affect this demand such as changes in patients' expectations, a shift in lifestyle choices, or the continuous innovation of services, technologies, and clinical possibilities (Spatharou et al. 2020). For example, it has been estimated that by 2050, one in four people in Europe and North America will be over the age of 65, meaning that the health systems will deal with more chronic and complex needs.

The management of such needs is therefore expensive and requires systems to transform their organizations in a more proactive and long-term way (Spatharou et al. 2020). To this aim, several investments have been made in the field of biomedicine regarding Artificial Intelligence (AI)-based solutions to address several challenges such as remote patient management and disease prediction. Consequently, biomedical infrastructures are going to integrate novel evidence-based AI solutions to promote a new clinical avenue called "Digital Health" (DH) that can potentially revolutionize this field by addressing some of the mentioned challenges (Cesario et al. 2022).

This chapter will introduce the role of AI in DH by giving an overview of the main global trends in using AI solutions for biomedical purposes, without neglecting current limitations and challenges that still need to be overcame.

2.2 The Role of Artificial Intelligence in Digital Health

Artificial Intelligence is a discipline that outlines how computers can simulate, reproduce, and eventually enhance human intelligence mechanisms. According to the Medical Subject Heading (MeSH) Browser, AI is defined as:

> Theory and development of Computer Systems which perform tasks that normally require human intelligence. Such tasks may include speech recognition, learning; visual perception; mathematical computing; reasoning, problem solving, decision-making, and translation of language (MeSH Browser 2022).

On the other hand, Digital Health is a discipline that includes digital and technological care programs to enhance the efficiency of healthcare delivery and make medicine more personalized and precise; it uses remote monitoring tools and AI-driven solutions to facilitate understanding of health problems and challenges faced by individual citizens and subpopulations of patients (WHO 2021).

To take a glance at the international framework, a search performed within the PubMed database on June 16, 2022, shows the trend represented in Fig. 2.1, for a total of 357 results. This trend is likely to increase exponentially.

Even if the AI's impact and efficacy for DH is still at its roots, we could expect interesting evolutions by looking at the pipeline of industrial and academic ideas on

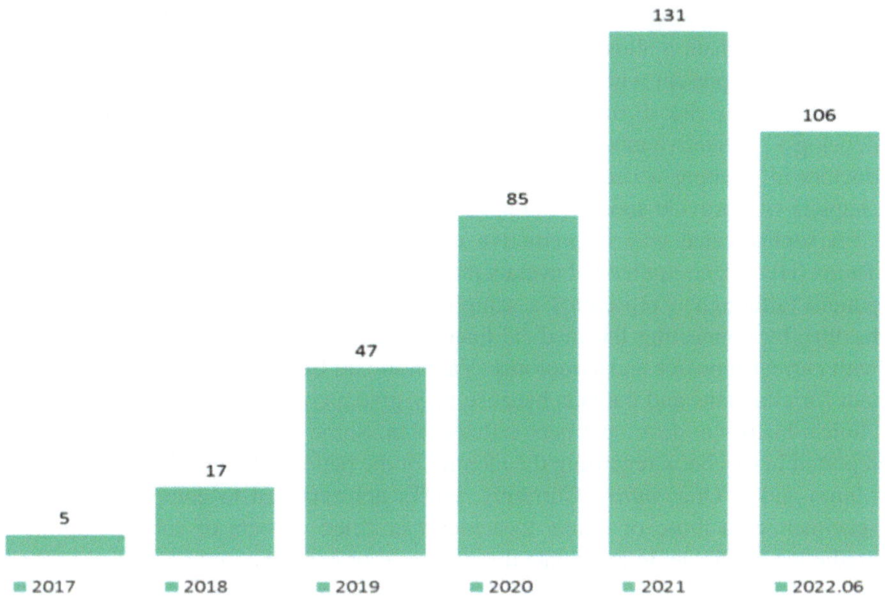

Fig. 2.1 Trend of scientific publications on PubMed for the query "artificial intelligence" [Title/Abstract] AND "digital health" [Title/Abstract], without setting any filter for year, language, or publication type

the global market. Currently, we envision four main directions of using AI solutions within the international medical field:

a. Support for prognostic outcomes and risk prediction
b. Support for clinical diagnostic personalization
c. Support for therapy development or improvement
d. Support for data collection and analysis [synthetic data, Real-World Data (RWD), Real-World Evidence (RWE)]

These directions will be elucidated in the following paragraphs.

2.2.1 Prognostic Outcomes and Risk Prediction

Prognostic outcomes and risk prediction is a process that involves the classification of individuals with certain characteristics or conditions and their classification according to stage, severity, and other clinical variables (European Parliament 2022).

One of the first fields in which AI-aided detection and prediction is applicable (but not limited to) is medical imaging. For example, starting from a database containing a considerable number of images of a neoplasm that are then compared with images of healthy cells and tissues, AI algorithms can be trained to early detect cancer by comparing images among them (Bi et al. 2019). The comparison of massive information allows the algorithm to identify if key features of that specific neoplasm are present within the patient's diagnostic images. The recognition helps anticipating the risk of disease manifestation/recurrence. Algorithms can therefore develop a "predictive model" (e.g., probabilistic and/or statistical analysis) that help doctors in creating a dedicated preventive path. In Part II of this book, several chapters will provide specific examples.

A sophisticated way in predictive modeling is the creation of Human Digital Twins (HDTs), computerized avatars that simulate the information ecosystem of the patient (although it represents a sampling of data and not the totality of his/her health), by connecting the medical history (including family history, if available) with current illnesses and symptoms (Valentini and Cesario 2021). HDT are important for clinicians and patients because they give a comprehensive overview of the clinical history to develop a personalized plan. Some interesting reflections will be reported in this book regarding the identity of the patient and the ethical aspects (see Manto and D'Oria, *infra*). Currently, HDTs are expected to compare the patient specimen with those of others to achieve stratified clusters of subpopulations in predictive medicine, to anticipate disease onset or exacerbation.

2.2.2 Clinical Diagnostics Personalization

AI-based algorithms show their greatest utility in multifactorial analyses of big data and RWD to create accurate predictive models for diagnostics and the reduction of clinical errors (Miller and Brown 2018). Predictive modeling starts from large datasets (based on the information retrieved from a patient, subpopulations with related characteristics, populations with the same disease, or seemingly distinct populations), whose data are classified according to similarities, rules, connections, neural networks, statistics, or probabilities.

The main tasks of these algorithms for diagnostics are as follows:

- Risk prediction and diagnosis of several diseases in their types, features, and levels of complexity (Kourou et al. 2014).
- Integration of omics and multi-omics data for complex diseases personalized diagnostics (Fornecker et al. 2019).
- Discovery of novel associations between diseases, including comorbidities and multimorbidities (Deparis et al. 2018).
- Observation of different pathology-specific therapeutic outcomes to identify key instances (e.g., mutations or genetic alterations) that originate new cancers (Zhang et al. 2017).

AI-based algorithms are prevalently used for disease prevention and therapeutic outcome prediction in several fields (e.g., radiology, radiation therapy, ophthalmology, dermatology (Naylor 2018), gastroenterology, gynecologic oncology, senology (Kourou et al. 2014), hematology (Radakovich et al. 2020), and infectious diseases

(Peiffer-Smadja et al. 2020; Ozsahin et al. 2020). Specifically, their use helps personalizing preventive diagnostics and therapeutic choices by gathering interpersonal features within the same cluster of patients, and therefore identify discordance from intrapersonal variability.

2.2.3 Therapy Development or Improvement

Another example that demonstrates how AI is utilized in the biomedical field regards the development/improvement of drugs and therapies to predict their outcomes (including the assessment of possible toxicity or the onset of adverse events) and personalize them according to each patient's characteristics (Bhinder et al. 2021). Such models, like those based on quantitative structure–activity relationship (QSAR) approaches (Golbraikh et al. 2012; European Parliament 2022), can help predicting large numbers of new compounds for various biological end points. These models can further facilitate greater understanding of molecules' behavior (e.g., potential antimicrobial activity) by screening a large volume of molecules and virtually test them to identify antibacterial compounds structurally distant from known antibiotics (Stokes et al. 2020; European Parliament 2022).

Besides, the current way of manufacturing pharmaceutical products and related existing logistic solutions is not mature for this revolution, since one of the current challenges is cybersecurity. A potential solution could be the envisioned in the concept of "cryptopharmaceuticals" elaborated by Nørfeldt et al. (2019), where pharmaceutical products are connected in a patient-specific blockchain of individual dosage units for personalized medication and, potentially, avoid counterfeit products.

AI-based models are also used for randomized controlled trials (RCTs) design, to increase the success rates or personalize patients' enrollment (Harrer et al. 2019), or to assess the risks and benefits of medical interventions. Sometimes, undertaking a RCT is not always possible under certain clinical conditions; therefore, Machine Learning (ML)/Deep Learning (DL) algorithms can be used in designing *in silico* clinical trials (ISCTs), computerized simulations often customized on HDTs that study the development or regulatory evaluation of a virtual drug, a device or a therapeutic intervention (the chapter of Surendran and colleagues—*infra*—elucidates very well the approaches to generating virtual patient cohorts in oncology). To this aim, studies with real patients may be reduced in favor of sophisticated simulations that predict, for example, the safety and efficacy of a treatment on a specific patient [the so called "N-of-1 clinical trials" (Lillie et al. 2011)], or a subset of patients with similar clinical pathophenotype.

2.2.4 Deep Data Collection and Analysis

AI solutions in clinical practice must be based on RWE from clinical trials, especially to implement clinical decision-support tools. Important preconditions for AI to deliver its full potential in global medicine and healthcare are the integration of broader data sets across organizations, a strong governance to improve data quality, and a greater confidence from practitioners, and patients in AI solutions and the ability to manage the related risks. Recently, many studies have been published in the field of synthetic data generation for healthcare, which can be applied and applied across a variety of domains (Goncalves et al. 2020). Indeed, generative neural networks—AI models that produce realistic data from a training dataset—are listed between the most innovative ideas proposed in the last decades (Laino et al. 2022). They represent promising tools in protecting patient privacy, diversifying datasets, for training and educational purposes, and in accelerating clinical research (Arora and Arora 2022). Synthetic data could help in any field of healthcare in coping with the data lack, anonymity, and quality (Elazab et al. 2020). Furthermore, when applied to medical imaging, generative neural networks have proven to be useful in many fields, such as:

- The generation of different images within the same modality [i.e., generating T1 images from T2 images at Magnetic Resonance (MR)] or from different modalities [i.e., producing MR from Computed Tomography (CT) scans], which could lead to a reduction in radiation exposure and an improvement in image interpretation (Kossen et al. 2021).
- The reconstruction of images, with reduction of artifacts and improvement of image quality, which could help in reducing CT radiation exposure and MR acquisition time.
- The data augmentation and the improvement of data availability and quality at a low cost, which is especially helpful for rare diseases (Laino et al. 2022).

However, since the type and quality of generated data are strictly dependent on the training dataset, and since it carries a lot of problems for their implementation in the clinical practice—as hallucination, deepfake, misdiagnosis of medical conditions (Laino et al. 2022)—the generation of synthetic data finds at the moment few clinical applications, while remaining a very active topic in biomedical research (Sorin et al. 2020).

2.3 Digital Health for Personalized Patient Compliance and Flow Management

Two main applications of DH can improve and personalize the clinical experience:

a. Monitoring patients' symptoms and adherence to therapy
b. Digitalization of clinical pathways

Both applications will be described below and require the integration of AI solutions with the hospital and home settings, specifically when DH focuses on several specialties (e.g., oncology, cardiology, neurology) to provide a holistic care.

2.3.1 Monitoring Patients' Symptoms and Adherence to Therapy

Patients can actively provide healthcare professionals with some information regarding their daily biometrics (e.g., hours of sleep, heart rate, and steps taken in a day), as well as their psychological state through dedicated questionnaires. This kind of information, obtainable through apps, sensors, and wearable devices (e.g., electronic bracelets), is known as "Internet-of-Medical-Things" (IoMT)—a subset of information deriving from the broader Internet-of-Things (IoT)—and represents the patients' data ecosystem to facilitate remote and real-time monitoring of certain parameters which, otherwise, could not be retrievable except with a physical visit.

According to Al-kathani and colleagues (Al-Kahtani et al. 2022), many technologies are rapidly evolving in healthcare to assist patients, and to collect/interpret data such as: "ambient intelligence communications technologies" with embedded human–computer interaction (HCI); remote monitoring with wearable devices (including smartphones, necklaces, etc.); Augmented Reality (AR) technologies for immersive medical education and training; smart robots to automatize some routinary activities of the patient or to transport the patient from a location to the another in the hospital. The urgency of remote monitoring has been accelerated during the COVID-19 pandemic (Khan et al. 2021), since social distancing required healthcare professionals to find alternative and smart solutions to communicate with patients, to transmit clinical data in a secure way, while improving the accuracy of care. A comprehensive example of patients remote monitoring is represented in the chapter of Kyriazakos and colleagues in this book [see *infra*].

2.3.2 Digitalization of Clinical Pathways

DH solutions are likely to address repetitive and largely administrative tasks that absorb significant time of doctors and nurses, optimizing healthcare operations and increasing their adoption. For example, they can facilitate the transition from hospital-based to home-based care toward virtual assistants, remote monitoring, and personalized alerts since patients are progressively increasing ownership about their care. Embedding AI in clinical workflows through a proactive engagement of several stakeholders (e.g., regulatory bodies, healthcare providers) requires combining well-designed DH solutions with the existing and the newest technologies, as well as managing cultural change and capability building within the organization.

a. *Virtual wards.* Virtual wards are a remote care experience to follow patients who cannot go to the hospital remotely (Hutchings et al. 2021). If a cancer patient is unable to travel from home (for example, due to an infection such as COVID-19), he/she can still be followed by doctors by checking his/her own biometric parameters (e.g., oxygen levels) after being adequately informed and trained for the correct use of the measuring instruments. Information is delivered (via video call, call, or messaging) to healthcare professionals who are in daily contact with the patient, offering telemedicine services and recalling him/her to the hospital for observation or treatment if necessary. Algorithms can support in data processing to predict the future course of the patient's condition and help the physician to make more personalized data-driven decisions.

b. *Patient Support Programs* (PSP). A broader way through which provide care continuity to patients beyond the outpatient setting are PSP. By making use of multiple remote monitoring tools (apps, wearable devices, questionnaires, etc.) provided to the patient, healthcare professionals can check his/her health status in real time, be in constant contact with him/her, and intervene when appropriate. PSP are based on AI algorithms that can predict the risk of relapse (either physical or psychological), and they are specifically designed to monitor patients in the time distances between two clinical encounters (i.e., when he/she is not physically present in the hospital and may live outside the region). Patients are trained in using properly the monitoring tools and have access to apps/portals for educational materials on disease management. Additionally, they can request virtual coaching sessions with dedicated health professionals (psychologist, nutritionist, etc.).

Both solutions can strengthen patients' adherence to treatments (Su et al. 2022), and could also improve the therapeutic alliance with the whole team of experts that follows the patient.

2.4 Investments in the AI and DH Environment

The race to build and adopt AI tools as well as investing in AI start-ups is fast growing, with commercial uses of AI becoming an important reality worldwide (Mou 2019). Companies in every sector—from retail to agriculture—are more and more including AI software into their products or administrative management. This explains why AI is actually leading technology investments. The landscape of healthcare AI startup companies is progressively expanding with solutions aimed to solve some of the industry, hospital, or healthcare providers' most pressing problems (Young 2022).

This growing significance of AI and DH solutions to deliver precise and effective healthcare solutions has promoted the number of venture capital investors along with start-ups to invest in healthcare. Enormous potential and wide range of application in this sector could enhance patient management while minimizing costs, thus expediting market growth. For example, in the first half of 2022, Digital health ventures headquartered in Europe captured US 2.43 billion across 124 deals, with an average deal size of US 19.6 million.

In the future, the global market size of AI for healthcare is expected to reach US $45 billion by 2026 and 67.4 billion by 2027 [Artificial Intelligence in Healthcare Market Worth $45.2 Billion by 2026 (Markets and Markets 2020)]. On the other hand, AI integration in daily life will have a significant impact on major economic sectors, leading to the need to confront the possibility of job losses and rearranging our life habits (Mou 2019).

2.5 Further Considerations

According to the Organization for Economic Co-operation and Development (OECD) "*High expectations must be managed, but real opportunities should be pursued*" when considering the potential of AI in medicine and healthcare (OECD 2020). Indeed, some dimensions must be carefully examined to achieve sustainable goals for all patients and communities.

For example, although DH and AI can enable rapid and inexpensive information sharing, some concerns about privacy and personal health data still exist. Regulatory challenges go from data to privacy protection (see Giorgianni, *infra*), touching the framework on Digital Therapeutics (see Recchia and Gussoni, *infra*), and the approval of Regulatory Bodies about the introduction of medtech for personalized patient treatment (see Neumann et al., *infra*).

Another concern regards the possibility of augmenting the digital divide among minority groups in using and trusting DH systems, and therefore delivery jeopardized access to PM opportunities. The gap between low- and high-income populations should be managed effectively by fostering digital health literacy skills. In facts, it is not guaranteed that the healthcare professional and the patient would be able to master and understand DH tools, hence education about the technologies and the wording of the world of AI is a pivotal asset to consider. Specific programs on DH could be implemented to empower communities to embrace digital transformation with AI and IoMT in a sustainable manner (WHO 2021; Lin and Wu 2022).

Third, risk prediction should be considered as a "probability" and not a "prophecy": HDTs do not represent "the patient" but a sampling of the information available about the person (Valentini and Cesario 2021). Besides, every result coming from the calculation of an algorithm—even when precise, evident, and fully accurate—must always be contextualized and carefully explained by the healthcare professional to patients to increase their understanding and awareness. This perspective cannot be separated from the promotion of a transparent communication between all the stakeholders involved in this field that elucidate the real benefits that these technologies can bring to patients.

References

Al-Kahtani MS, Khan F, Taekeun W (2022) Application of internet of things and sensors in healthcare. Sensors 22(15):5738. https://doi.org/10.3390/s22155738

Arora A, Arora A (2022) Synthetic patient data in health care: a widening legal loophole. Lancet 399(10335):1601–1602. https://doi.org/10.1016/S0140-6736(22)00232-X

Bhinder B, Gilvary C, Madhukar NS, Elemento O (2021) Artificial intelligence in cancer research and precision medicine. Cancer Discov 11(4):900–915. https://doi.org/10.1158/2159-8290

Bi WL, Hosny A, Schabath MB et al (2019) Artificial intelligence in cancer imaging: clinical challenges and applications. CA Cancer J Clin 69(2):127–157. https://doi.org/10.3322/caac.21552

Cesario A, D'Oria M, Bove F et al (2021) Personalized clinical phenotyping through systems medicine and artificial intelligence. J Pers Med 11(4):265. https://doi.org/10.3390/jpm11040265

Cesario A, D'Oria M, Simone I, Patarnello S, Valentini V, Scambia G (2022) Open innovation as the catalyst in the personalized medicine to personalized digital medicine transition. J Pers Med 12:1500

Deparis S, Pascale A, Tommasi P, Kotoulas S (2018) An analytical method for multimorbidity management using Bayesian networks. Stud Health Technol Inform 247:820–824

Elazab A, Wang C, Gardezi SJS et al (2020) GP-GAN: Brain tumor growth prediction using stacked 3D generative adversarial networks from longitudinal MR Images. Neural Netw 132:321–332. https://doi.org/10.1016/j.neunet.2020.09.004

European Parliament (2022) Artificial Intelligence in healthcare. Applications, risks, ant ethical and societal impacts. Source: https://www.europarl.europa.eu/RegData/etudes/STUD/2022/729512/EPRS_STU(2022)729512_EN.pdf Last consultation: 26/08/2022

Fornecker LM, Muller L, Bertrand F et al (2019) Multi-omics dataset to decipher the complexity of drug resistance in diffuse large B-cell lymphoma. Sci Rep 9(1):895. https://doi.org/10.1038/s41598-018-37273-4

Golbraikh A, Wang X, Zhu H, Tropsha A (2012) Predictive QSAR modelling: methods and applications in drug discovery and chemical risk assessment. In: Leszczynsk J, Kaczmarek-Kedziera A, Puzyn T, Papadopoulos MG, Reis H (eds) Handbook of computational chemistry. Springer, Cham

Goncalves A, Ray P, Soper B et al (2020) Generation and evaluation of synthetic patient data. BMC Med Res Methodol 20(1):108. https://doi.org/10.1186/s12874-020-00977-1

Harrer S, Shah P, Antony B, Hu J (2019) Artificial intelligence for clinical trial design. Trends Pharmacol Sci 40(8):577–591. https://doi.org/10.1016/j.tips.2019.05.005

Hutchings OR, Dearing C, Jagers D et al (2021) Virtual health care for community management of patients with COVID-19 in Australia: observational Cohort study. J Med Internet Res 23(3): e21064. https://doi.org/10.2196/21064

Khan MM, Mehnaz S, Shaha A et al (2021) IoT-based smart health monitoring system for COVID-19 patients. Comput Math Methods Med 2021:8591036. https://doi.org/10.1155/2021/8591036

Kossen T, Subramaniam P, Madai V et al (2021) Synthesizing anonymized and labeled TOF-MRA patches for brain vessel segmentation using generative adversarial networks. Comput Biol Med 131:104254. https://doi.org/10.1016/j.compbiomed.2021.104254

Kourou K, Exarchos TP, Exarchos KP et al (2014) Machine learning applications in cancer prognosis and prediction. Comput Struct Biotechnol J 13:8–17. https://doi.org/10.1016/j.csbj.2014.11.005

Laino ME, Cancian P, Politi LS et al (2022) Generative adversarial networks in brain imaging: a narrative review. J Imaging 8(4):83. https://doi.org/10.3390/jimaging8040083

Lillie EO, Patay B, Diamant J et al (2011) The n-of-1 clinical trial: the ultimate strategy for individualizing medicine? Per Med 8(2):161–173. https://doi.org/10.2217/pme.11.7

Lin B, Wu S (2022) Digital transformation in personalized medicine with artificial intelligence and the internet of medical things. OMICS 26(2):77–81. https://doi.org/10.1089/omi.2021.0037

Markets and Markets (2020) Artificial intelligence in healthcare market. Source: https://www.marketsandmarkets.com/Market-Reports/artificial-intelligence-healthcare-market-54679303.html?gclid=CjwKCAjwu5yYBhAjEiwAKXk_eO8oAKGpGFX4EkP2-_i4aIZDw8_qElriAtv7rCMJxkZ-bn586guaahoCfGcQAvD_BwE Last consultation: 26/08/2022

MeSH Browser (2022) Artificial Intelligence. Source: https://meshb.nlm.nih.gov/record/ui?ui=D001185 Last consultation: 26/08/2022

Miller DD, Brown EW (2018) Artificial intelligence in medical practice: the question to the answer? Am J Med 131(2):129–133. https://doi.org/10.1016/j.amjmed.2017.10.035

Mou X (2019) Artificial intelligence: investment trends and selected industry uses. International Finance Corporation. Source: https://documents1.worldbank.org/curated/en/617511573040599056/pdf/Artificial-Intelligence-Investment-Trends-and-Selected-Industry-Uses.pdf Last consultation: 26/08/2022

Naylor CD (2018) On the prospects for a (deep) learning health care system. JAMA 320:1099–1100. https://doi.org/10.1001/jama.2018.11103

Nørfeldt L, Bøtker J, Edinger M et al (2019) Cryptopharmaceuticals: increasing the safety of medication by a blockchain of pharmaceutical products. J Pharm Sci 108(9):2838–2841. https://doi.org/10.1016/j.xphs.2019.04.025

OECD (2020) Trustworthy artificial intelligence in health. Source: https://www.oecd.org/health/trustworthy-artificial-intelligence-in-health.pdf Last consultation: 26/08/2022

Ozsahin I, Sekeroglu B, Musa S et al (2020) Review on diagnosis of COVID-19 from chest CT images using artificial intelligence. Comput Math Methods Med:9756518. https://doi.org/10.1155/2020/9756518

Peiffer-Smadja N, Rawson TM, Ahmad R et al (2020) Machine learning for clinical decision support in infectious diseases: a narrative review of current applications. Clin Microbiol Infect 26(5):584–595. https://doi.org/10.1016/j.cmi.2020.05.020

Radakovich N, Nagy M, Nazha A (2020) Artificial intelligence in hematology: current challenges and opportunities. Curr Hematol Malig Rep 15(3):203–210. https://doi.org/10.1007/s11899-020-00575-4

Sorin V, Barash Y, Konen E, Klang E (2020) Creating artificial images for radiology applications using Generative Adversarial Networks (GANs) – a systematic review. Acad Radiol 27:1175–1185. https://doi.org/10.1016/j.acra.2019.12.024

Spatharou A, Hieronimus S, Jenkins J (2020) Transforming healthcare with AI: The impact on the workforce and organizations. Source: https://www.mckinsey.com/industries/healthcare-systems-and-services/our-insights/transforming-healthcare-with-ai Last consultation: 26/08/2022

Stokes JM, Yang K, Swanson K et al (2020) A deep learning approach to antibiotic discovery. Cell 180(4):688702.e13

Su PH, Malik S, Jheeta A et al (2022) Investigating the impact of the TUITEK® patient support programme, designed to support caregivers of children prescribed recombinant human growth hormone treatment in Taiwan. Front Endocrinol 13:897956. https://doi.org/10.3389/fendo.2022.897956

Valentini V, Cesario A (2021) Oltre la persona: cos'è lo 'Human Digital Twin' nella medicina personalizzata. In: Anelli F, Cesario A, D'Oria M, Giuliodori C, Scambia G (eds) Persona e Medicina: Sinergie sistemiche per la medicina personalizzata. FrancoAngeli, Milano, pp 70–76

WHO (2021) Global strategy on digital health 2020-2025. World Health Organization, Geneva (CH)

Young AS (2022) AI in healthcare startups and special challenges. Intelligence-Based Med 6. https://doi.org/10.1016/j.ibmed.2022.100050

Zhang C, Peng L, Zhang Y et al (2017) The identification of key genes and pathways in hepatocellular carcinoma by bioinformatics analysis of high-throughput data. Med Oncol 34(6):101. https://doi.org/10.1007/s12032-017-0963-9

GDPR as a First Step Towards Free Flow of Data in Europe

3

Francesco Giorgianni

Abstract

The purpose of this short essay is to describe how the European Union is willing to promote a true European Market for Data. Data is more than the new oil: oil is going to end one day; data multiplies as much as humans "produce" them. To properly monetize data, it is paramount to protect them. The international success of the General Data Protection Regulation (GDPR) stems from this vision. We analyse the key pillars of GDPR: Data Protection Officer (DPO), security measures, and sanctions. GDPR stays at the core of data protection regulation. We touch base with Digital Service Act and Digital Market Act acknowledging that they are aimed at eliminating obstacles to data free flow in Europe and allow GDPR architecture to truly succeed in a competitive scenario, by eliminating barriers to entry of new digital business, especially in the health digital medicine market where cross selling and interlocking of customers by large gatekeepers is more than a risk.

Keywords

Europe · Data protection · GDPR · Digital market · Regulation

3.1 The GDPR Revolution

The General Data Protection Regulation (GDPR) is a European Regulation adopted in 2016, and entered into force in May 2018, which has profoundly innovated the compliance systems of companies. The "Mother" Directive 95/46/EC on privacy had been adopted by the various countries of the European Union (in Italy with the

F. Giorgianni (✉)
Of Counsel LCA-amm Law Firm, Rome, Italy
e-mail: francesco.giorgianni@giorgianni.eu

A. Cesario et al. (eds.), *Personalized Medicine Meets Artificial Intelligence*,
https://doi.org/10.1007/978-3-031-32614-1_3

Privacy Code) in a very uneven way, creating interpretative gaps, the same that forced the Court of Justice, on several occasions, to intervene to restore a "level playing field" adequate to the circulation of data in Europe.[1]

In fact, the European regulatory breakthrough inaugurated with the GDPR was then recently strengthened and clarified through the approval of the Digital Service Act and the Digital Market Act. The Union, on the one hand, demands a strict protection of the data of its citizens, but on the other hand starts a season aimed at breaking the oligopoly of the GAFAM (Google, Apple, Facebook, Amazon, Microsoft, or the "gatekeepers") and opening the platforms for non-discriminatory access by the large European players of goods and services, which would find an obstacle precisely in the impossibility of monetizing the circulation of data to their advantage.

The GDPR is grafted on a solid European privacy regulation tradition, which must be revised to the needs imposed by the digital transformation of the global economy. In fact, many contents of the Mother Directive are found in EU Regulation 2016/679 GDPR, but it is precisely the setting of the GDPR that overturns the philosophy of personal data protection (Table 3.1).

In summary: companies that offer digital medicine services must be able to equip themselves with their own privacy risk assessment system, without counting, as in the past, upon legal review by the Supervisory Data Protection Authority (SDPA). With the GDPR, in the event of a breach of personal data (data breach), or inspection following complaints from the interested parties, the SDPA will verify whether the company has adopted "adequate" security measures to mitigate the privacy risk. The burden of proof of these security measures is proportional to the quality of the "compliance tools" adopted, and therefore to the ability to generate reliable KPIs to quantitatively and qualitatively measure the excellence of the preventive activity carried out by the data controller.

The first pivot on which a GDPR compliance programme leverages is, in fact, the principle of accountability ("active responsibility") of the data controller. The second pivot is the figure of the Data Protection Officer (DPO).

---

[1] Please, see the milestone judgement of the European Court of Justice of May 13th, 2014 (*Personal data—Protection of individuals with regard to the processing of such data—Directive 95/46/EC—Articles 2, 4, 12 and 14—Material and territorial scope—Internet search engines—Processing of data contained on websites—Searching for, indexing and storage of such data—Responsibility of the operator of the search engine—Establishment on the territory of a Member State—Extent of that operator's obligations and of the data subject's rights—Charter of Fundamental Rights of the European Union—Articles 7 and* 8). In Case C-131/12, REQUEST for a preliminary ruling under Article 267 TFEU from the Audiencia Nacional (Spain), made by decision of 27 February 2012, received at the Court on 9 March 2012, in the proceedings Google Spain SL, Google Inc. v Agencia Española de Protección de Datos (AEPD), Mario Costeja González.

Table 3.1 Principal innovations between the Directive 95/46/EC and the GDPR REG UE 2016/679

DIRECTIVE 95/46/EC	GDPR REG UE 2016/679
– Data processing must be "regulated" and sometimes "authorized" by the Supervisory Data Protection Authority (command & control regulation). – Sanctions are National (Eu Member States opted for negligible or minimal sanctions). – Security measures must be minimal. – The international transfer of data by cloud providers to countries deemed "inadequate" must be guaranteed at least by Standard European Contractual Clauses.	– Data processing is justified according to the principle of "accountability" (self-regulation) of the data controller and the data processor. – The sanctions are European and reach up to 4% of the group's global turnover. – Security measures must be adapted to the risk inherent in the processing of data. – The international transfer of data by cloud providers to countries deemed "inadequate" is allowed, provided that in the country to which the data are transferred the rights of the interested parties are guaranteed. – Mandatory registration of the processing of personal data and data breach. – Mandatory (for some treatments) of the data protection impact assessment. – Mandatory appointment of the Data Protection Officer (e.g. for large corporate groups or even small entities that process very sensitive data).

3.2 The DPO According to the GDPR and the Guidelines of the European Data Protection Board

The DPO is appointed according to professional qualities, specialized knowledge of privacy legislation, and the ability to fulfil the following duties:

- Providing advice on the mapping of treatments (and applications) and identify the main areas of the company exposed to privacy risk
- Defining a processing log model, data breach register, and a risk analysis methodology aimed at building the company's burden of proof
- Providing advice to the competent business areas on technical and organizational security measures and verify the effectiveness of the minimum ones adopted
- Supporting the General Counsel in the negotiation of contracts related to international data transfers (e.g. cloud contracts)
- Supporting the General Counsel in contracts and in the drafting of privacy policies, as well as in the management of the rights of the interested parties (verifying the effectiveness of business processes in compliance with the times and methods provided for by the GDPR)
- Training employees and management on the obligations arising from the GDPR (accountability)
- Monitoring the application of the GDPR, define company policies and appointments on the subject

- Carrying out data protection audits on infrastructures, applications, and treatments used by the company and third parties
- Supporting the company in the "privacy by design & by default" processes for the construction of apps, customer services, HR process management, and in all activities aimed at the digital transformation of business processes
- Providing opinions on the Data Protection Impact Assessment (DPIA) and collaborate in the definition of the methodology as well as in the drafting of the DPIA themselves
- Relating to the SDPA, not only in inspections and events of personal data breaches, but also maintain with the authority a correct relationship aimed at representing regulatory and provisional solutions on privacy issues of corporate and sector interest

With reference to large industrial groups, or companies managing sensitive data like in the digital medicine business, the European guidelines suggest the adoption of a team of DPOs that can be inspired by the company's organizational model, without being a slavish photocopy. Because the DPO must be placed—from an organizational point of view—"close" to the processing activities of personal data and must have visibility on the life cycle of the data, precisely to ensure respect for the rights of the interested parties to access it, request its correction, inhibit or limit its processing, and finally request its oblivion or cancellation.

Another very lively discussion was recorded in the imminence of the entry into force of the GDPR and concerned the issue of the independence of the DPO and its authority, or to be able to report directly to the top management. Over time, the issue of independence has been measured by the SDPAs in proportion to the budget and resources assigned to the DPO to meet its tasks, but also according to the company assignment. It is therefore desirable that the DPO can report not only to the Chief Executive Officer, but also to the Board of Directors, in compliance with the governance rules chosen by the group.

The solutions that provided for the placement of the DPO in the Audit, ICT, or Corporate Protection functions were immediately discarded: these functions perform tasks that are in conflict of interest with the delicate mission entrusted to the DPO. The only corporate function that appears the most suitable to "host" the DPO is the Legal area, provided that the hierarchical carryover does not go to the detriment of the autonomous judgment capacity that the DPO must have, as expressly required by the GDPR. It is also permissible that the Lawyer can count on a privacy structure at the service of the company, but it is always suggested that the DPO can integrate all the privacy skills into his team, precisely to avoid professional competition and duplication of roles that are difficult to explain in complex contexts such as industrial groups. Not speaking about small medium enterprises that cannot find efficient to manage privacy issues through the diarchy DPO—Privacy Corporate Lawyer.

The DPO can also be "external" to the company organization, and this choice can be useful and functional to a transitional path aimed at reorganizing the DPO function of a large industrial group. Whereas for small medium companies the external DPO can be a solution provided that the Board of Directors is made aware and that there is solid data protection governance within the company.

3.3 Security Measures and "Privacy" Crimes

The GDPR provides that security measures must be "technical" and "organizational". The latter consist of operational indications for management and employees contained in policies and operating instructions, assignment of privacy roles accompanied by training activities on data protection issues. Technical security measures, on the other hand, involve both physical and logical security measures. The DPO has a key role in monitoring, alongside the data controller, in maintaining the measures, negotiating with cloud providers data protection agreements, and verifying that the standard controls prepared by the provider of both cloud and hybrid infrastructure are adequate from the point of view of privacy legislation.

The digital platforms containing the register of processing, register of incidents and data breach and data protection impact assessment of applications and data processing, basically constitute not only the burden of proof of the data controller but also an additional technical and organizational security measure.

The GDPR provides (art. 84) that it is the task of the individual Member States to regulate criminal matters, with sanctions that are "effective, proportionate and dissuasive". The Italian legislator has revised the criminal cases provided for by the Privacy Code (Legislative Decree 196/2003 and subsequent amendments), introducing the provision of damage as a characterizing element as an alternative to the purpose of profit. Therefore, it will not only be held against the economic profit of the offender but also against the damage caused to the interested parties, including the damage to the image and reputation of the victim.

The new privacy offences are as follows:

- Unlawful processing of data (art. 167)
- Unlawful communication and dissemination of personal data subject to large-scale processing (art. 167 bis)
- Fraudulent acquisition of personal data subject to large-scale processing (art. 167 ter)
- Falsehood in the declarations to the SDPA and interruption of the execution of the tasks or the exercise of the powers of the SDPA (art. 168)
- Non-compliance with provisions of the SDPA (art. 170)
  - Infringements of remote controls on workers (Art. 171)

Unfortunately, these crimes do not fall into the category of ground crimes provided for by Law 231/2001 and, therefore, by the related corporate crime prevention model. It therefore becomes fundamental to evaluate the ability to prevent these forms of crime by the corporate Privacy Framework and the related appointments letters to Privacy Referent (broadly speaking the first- and second-lines management).

3.4 "Sensitive" Personal Data

Art. 4 of the GDPR includes data relating to health among the sensitive data: "personal data relating to the physical or mental health of a natural person, including the provision of health care services, which reveal information relating to his state of health". Considering this, the following must be reflected as health-related data:

- Information about the person collected during patient registration before receiving health services
- The pseudonyms (numbers or other pseudo-identification tools) attributed to patients for the management of their healthcare path within the structure
- The results of examinations, diagnostic checks, medical visits, anamnesis, diagnoses, etc.
- Any information collected on the patient who will compose his health file, including financial data, payment systems, etc., which the Italian privacy code attracts to the health data regime, considering them sensitive also in this sense

As part of the activity of providing health services, it will also be necessary to collect and process other sensitive data such as data relating to sex life, or identity—sexual preferences; genetic data (RNA/DNA resulting from analysis of biological samples; unique information on the physiology of the individual and hereditary characteristics).

Biometric data (physical and physiological characteristics, facial features, photographs, or fingerprints that allow the unique identification of the individual), or data on racial or ethnic origin, data relating to religious beliefs and consequences in the health field regarding limitations of health treatments, etc.

3.5 "Data Monetization"

Technological advances and big data analytics capabilities, through artificial intelligence and machine learning, have made it easier to profile and automat automated decision-making, with potentially significant repercussions on the rights and freedoms of individuals. The widespread availability of personal data on the Internet and those that can be obtained from Internet-of-Things (IoT) devices, associated with the ability to find correlations and create links, can allow the determination, analysis, and prediction of aspects of a person's personality, behaviour, interests, and habits. Profiling and automated decision-making can be useful for individuals and organizations, offering them benefits such as: efficiency improvements and resource savings. They also have many commercial applications: for example, they can be used to better segment markets and tailor services and products to individual customer needs.

Integrated mobility and intermodality services offer indisputable advantages and opportunities to customers and to the community in general, as they allow an optimization of routes, a reduction in the resources used, and a reduction in significant emissions. However, the guidelines of the European privacy SDPAs consider that "profiling and automated decision-making can entail significant risks to the rights and freedoms of natural persons, which require adequate safeguards. These processes can be opaque. Natural persons may not know that they are being profiled or do not understand the consequences".

In fact, data controllers in the automotive or transport sectors could be able to exploit big data relating to the movements of millions of customers on their network and carriers, but all this must be done in a manner consistent with the privacy regulations in force. In this sense, the role of the DPO is fundamental to drive data monetization in a legitimate and sustainable way. This can be accomplished using rigorous and consistent privacy by design and default procedures.

3.6 The Way Forward

In the Strategic Plan of the European Commission of 2022, there is a key assumption that is worth to highlight:

> We tackled the social and economic impact of the pandemic, together with Member States, through a series of ambitious, far-reaching programmes and instruments. The combined firepower of the Union's long-term budget and NextGenerationEU will deliver EUR 2.018 trillion to boost our economy and rebuild a post COVID-19 Europe that is greener, fairer, more digital and more resilient.

The European Union is aiming at a green, resilient, and digital Europe. The resources dedicated to these objectives have an unprecedented financial dimension and this will be igniting a process of revision of the Stability Pact and the Treaties, in order to adapt the European decision-making process to the new challenges ahead. This is coherent with a previous strategic statement by Ursula Von Der Leyen: "balance the flow and use of data while preserving high privacy, security, safety and ethical standards" (von der Leyen 2019).

Coherently with her statement of 2019, on April 23rd of this year, once the Trialogue (European Commission, European Parliament, and European Council of Ministers) found the agreement on the Digital Service Act, she stated:

> Today's agreement on the Digital Services Act is historic, both in terms of speed and of substance. The DSA will upgrade the ground-rules for all online services in the EU. It will ensure that the online environment remains a safe space, safeguarding freedom of expression and opportunities for digital businesses. It gives practical effect to the principle that what is illegal offline, should be illegal online. The greater the size, the greater the responsibilities of online platforms. Today's agreement—complementing the political agreement on the Digital Markets Act last month—sends a strong signal: to all Europeans, to all EU businesses, and to our international counterparts. (European Commission 2022a).

Moreover:

> The digital services impacted by the new discipline are:

- Intermediary services offering network infrastructure: Internet access providers, domain name registrars
- Hosting services such as cloud computing and webhosting services
- Very large online search engines with more than 10% of the 450 million consumers in the EU, and therefore, more responsibility in curbing illegal content online
- Online platforms bringing together sellers and consumers such as online marketplaces, app stores, collaborative economy platforms, and social media platforms
- Very large online platforms, with a reach of more than 10% of the 450 million consumers in the EU, which could pose risks in the dissemination of illegal content and societal (European Commission 2022a)

All these players are the turbines that generate data upon data, analysed by powerful machine learning, supporting multi profiling of our preferences in the market for goods and services. Not to speak about the influence of gatekeepers into the democratic processes and the polarization of public opinion on sensitive controverted topics.

Concretely, the DSA contains:

1. Measures to counter illegal goods, services, or content online, such as:
   - a mechanism for users to easily flag such content and for platforms to cooperate with the so-called trusted flaggers
   - new obligations on traceability of business users in online marketplaces
2. New measures to empower users and civil society, including:
   - the possibility to challenge platforms' content moderation decisions and seek redress, either via an out-of-court dispute mechanism or judicial redress
   - provision of access to vetted researchers to the key data of the largest platforms and provision of access to NGOs as regards access to public data, to provide more insight into how online risks evolve
   - transparency measures for online platforms on a variety of issues, including on the algorithms used for recommending content or products to users
3. Measures to assess and mitigate risks, such as:
   - obligations for very large platforms and very large online search engines to take risk-based action to prevent the misuse of their systems and undergo independent audits of their risk management systems
   - Mechanisms to adapt swiftly and efficiently in reaction to crises affecting public security or public health
   - New safeguards for the protection of minors and limits on the use of sensitive personal data for targeted advertising
4. Enhanced supervision and enforcement by the Commission when it comes to very large online platforms. The supervisory and enforcement framework also confirms important role for the independent Digital Services Coordinators and Board for Digital Services (European Commission 2022a)

GAFAM online platforms act as "gatekeepers" in digital markets. That is why together with the Digital Service Act, the Digital Markets Act approved by the European Commission on December 15th, 2020, aims to ensure that these platforms behave in a fair way online. Together with the Digital Services Act, the Digital Markets Act is one of the centrepieces of the European digital strategy (European Commission 2022b).

The sanctions for the gatekeepers? Fines of up to 10% of the company's total worldwide annual turnover, or up to 20% in the event of repeated infringements. It also provides periodic penalty payments of up to 5% of the average daily turnover and, in case of systematic infringements of the DMA obligations by gatekeepers, additional remedies may be imposed on the gatekeepers after a market investigation. Such remedies will need to be proportionate to the offence committed. If necessary and as a last resort option, non-financial remedies can be imposed. These can include behavioural and structural remedies, e.g., the divestiture of (parts of) a business (European Commission 2020a).

GDPR stays at the core of data protection regulation. Digital Service Act and Digital Market Act are aimed at eliminating obstacles to data free flow in Europe and allow GDPR architecture to truly succeed in a competitive scenario, by eliminating barriers to entry of new digital business, especially in the health digital medicine market where cross selling interlocking of customers by large gatekeepers is more than a risk.

We are at the beginning of an interesting European regulatory season which has seen Antitrust and Data Protection Portfolio converging to the same Commissioner (Vestager)—but mostly, the season of accountability opened by GDPR continues its pathway. Self-regulation—coupled with high sanctions—is the best way to address digital transformation in a global market for data where Europe wants to count not only as a rich market that produces personal data to be exploited by others (European Commission 2020b).

References

European Commission (2020a) The Digital Markets Act: ensuring fair and open digital markets. Source: https://ec.europa.eu/info/strategy/priorities-2019-2024/europe-fit-digital-age/digital-markets-act-ensuring-fair-and-open-digital-markets_en Last consultation: 22/08/2022

European Commission (2020b) Commission Staff working document executive summary of the impact assessment report accompanying the document proposal for a regulation of the European parliament and of the council on contestable and fair markets in the digital sector (Digital Markets Act). Source: https://competition-policy.ec.europa.eu/system/files/2021-09/DMA_summary_of_impact_assessment.pdf Last consultation: 22/08/2022

European Commission (2022a) Digital Services Act: commission welcomes political agreement on rules ensuring a safe and accountable online environment. Source: https://ec.europa.eu/commission/presscorner/detail/en/ip_22_2545 Last consultation: 22/08/2022

European Commission (2022b) Digital Markets Act (DMA). Source: https://competition-policy.ec.europa.eu/sectors/ict/dma_en Last consultation: 22/08/2022

von der Leyen U (2019) A Union that strives for more – My agenda for Europe, Political guidelines for the next European Commission 2019–2024. Source: https://ec.europa.eu/info/sites/default/files/political-guidelines-next-commission_en_0.pdf Last consultation: 22/08/2022

Digital Therapeutics: Scientific, Technological, and Regulatory Challenges

4

Giuseppe Recchia and Gualberto Gussoni

Abstract

Digital Health Technologies have the potential to improve human health and well-being, optimize the quality and safety of care, increase access to treatment, making health services more efficient and reducing overall health care costs. Within the framework of Digital Health and Medicine, Digital Therapeutics are an emerging class of therapeutics that offer interventions driven by high-quality software programs, based on scientific evidence obtained through methodologically rigorous confirmatory clinical investigation, to prevent, manage, and treat a broad spectrum of physical, mental, and behavioural conditions. Similar to drugs, Digital Therapeutics consist of active ingredients and excipients. While the "digital active ingredient" is primarily responsible for the clinical outcome, "digital excipients" (virtual assistant, reminders, reward systems, etc.) are necessary to ensure the best user experience to the patient and to allow the prolonged use of the therapy. The research and development process of Digital Therapeutics may be divided into different phases, like research, discovery, pilot, and full clinical development. The confirmatory randomized controlled clinical trials are critical to generate evidence of benefit for regulatory approval, reimbursement, and prescription. Digital Therapeutics have the potential to transform the management of chronic diseases and to represent the first therapeutic option offered by doctors to their patients with chronic disease and dependence. In order to fulfil this transformative potential of Digital Therapeutics, it is necessary to address several barriers that prevent their uptake in medical practice, such as clinical

G. Recchia (✉)
daVi DigitalMedicine srl, Verona, Italy

Fondazione Smith Kline, Verona, Italy
e-mail: giuseppe.recchia@davidigitalmedicine.com

G. Gussoni
Fondazione FADOI, Milan, Italy

A. Cesario et al. (eds.), *Personalized Medicine Meets Artificial Intelligence*,
https://doi.org/10.1007/978-3-031-32614-1_4

evidence, reimbursement, data safety and privacy issues, information and education of patients and healthcare professionals.

Keywords

Digital therapeutics · Digital health · Digital medicine · Research & development · Reimbursement

4.1 Introduction

The research for new therapies depends on the availability of new science and new technologies. The potential to improve patients' health outcomes, in particular for chronic diseases, has led in recent years to a growing interest in digital technologies and in particular in Digital Therapeutics (DTx).

However, the lack of agreed definitions has created considerable confusion about their role and partly hindered their development (Recchia et al. 2020). The taxonomy released by the Digital Medicine Society, Digital Therapeutics Alliance, and others has recently proposed a classification of products (applications, systems, platforms, and others) based on digital technologies that can represent a useful basis for in-depth analysis and discussion (Table 4.1).

4.2 Digital Health

Digital Health includes technologies, platforms, and systems that engage consumers for lifestyle, wellness, and health-related purposes; capture, store, or transmit health data; and/or support life science and clinical operations (Goldsack 2019). Digital Health applications (*apps*) are primarily of interest to consumers who aim at improving their well-being, for example by enhancing/optimizing certain physiological functions. Among the products represented in this category there are most of the more than 350,000 digital applications that can be downloaded from virtual stores (IQVIA 2021) and for which—in most cases—no clinical evidence is available.

As it is increasingly common for doctors to recommend such apps for managing specific aspects of well-being, there is a need to guide doctors, patients, and citizens on the quality of the proposed health apps. Among the most interesting experiences, the American Psychiatric Association (APA) has developed AppAdvisor (APA 2022), a model for the assessment of digital mental health apps, based on the analysis and evaluation of several app characteristics, namely access and basic information, data privacy and cybersecurity, clinical evaluations, usability and data integration with other health systems.

Table 4.1 Definitions and characteristics of Digital Health, Digital Medicine, and Digital Therapeutics (Goldsack 2019)

	Digital health		
		Digital medicine	
			Digital therapeutics
Definition	Digital health includes technologies, platforms, and systems that engage consumers for lifestyle, wellness, and health-related purposes; capture, store, or transmit health data; and/or support life science and clinical operations.	Digital medicine includes evidence-based software and/or hardware products that measure and/or intervene in the service of human health.	Digital therapeutic (DTx) products deliver evidence-based therapeutic interventions to prevent, manage, or treat a medical disorder or disease.
Clinical evidence	Typically do not require clinical evidence.	Clinical evidence is required for all digital medicine products.	Clinical evidence and real-world outcomes are required for all DTx products.
Regulatory oversight	These products do not meet the regulatory definition of a medical device and do not require regulatory oversight	Requirements for regulatory oversight vary. Digital medicine products that are classified as medical devices require clearance or approval. Digital medicine products used as a tool to develop other drugs, devices, or medical products require regulatory acceptance by the appropriate review division.	DTx products must be reviewed and cleared or certified by regulatory bodies as required to support product claims of risk, efficacy, and intended use.

4.3 Digital Medicine

Digital Medicine is a subset of Digital Health and includes evidence-based software and/or hardware products that measure and/or treat in the service of human health (Goldsack 2019). These products are about the patient, the doctor and therefore the disease dimension. In most cases, they are applications for mobile devices, with both measurement and treatment functions.

Digital Medicine products for measurement can be used to support the process of diagnosis, to monitor the progression of a disease or therapy, to guide the treatment of the disease with a drug or a medical device. Data can be generated by the patient either in passive mode through the use of wearable, ingestible or implanted sensors, or in active mode by filling in questionnaires or performing online tests.

Digital devices developed to prevent or treat a disease can deliver therapy, rehabilitation, or patient support in different modes, such as:

- *Digital Self-Management, Education & Support*: applications that provide patient education, instructions, and proven guidance on how to manage different aspects of diseases such as diabetes mellitus (Nkhoma et al. 2021), high blood pressure (Alessa et al. 2019), asthma (Khusial et al. 2020). Education and support may be a 'digital active ingredient' of a DTx.
- *Digital Drug Supports*: applications providing the optimal conditions for the use of a drug with which they are associated or combined. This is addressed through the reminder of the assumption, the drug dosage calculation, the support to the management of adverse events, the connection with the doctor or other patients with previous experience of the same therapy, etc. (Brittain et al. 2022).
- *Digital Rehabilitation*: digital motor, cognitive, pulmonary, cardiologic, or other rehabilitation systems, aimed at recovering compromised functions and capacities through measurements (e.g. with t-shirts equipped with inertial sensors) and treatment (e.g. with serious games engaging the patient in different motor exercises) (Seron et al. 2021). They can be considered a type of DTx.
- *Digital Connected Devices*: devices such as subcutaneous pumps that deliver the right dose of drug (e.g. insulin) at the right time in response to an algorithm that processes information received from sensors (e.g. continuous glucose measurement sensors) (Nimri et al. 2020).
- *Digital Therapeutics* (see below).

Since the data generated by these products can have an impact on the patient's pathway and on the physician's decisions, these products are generally classified as medical devices from a regulatory perspective, subject to the various regulations (European Union 2017). Requirements for regulatory oversight vary. Digital medicine products that are classified as medical devices require clearance or approval.

4.4 Digital Therapeutics

As a subset of Digital Medicine, DTx are medical devices aimed at therapeutic intervention, in most cases designed to modify the dysfunctional behaviours of the patient, developed through randomized controlled clinical trials, approved for use in clinical practice by regulatory bodies, ideally reimbursed by insurance companies or health services, prescribed by the physician (although some therapies that meet these criteria could be offered in a similar way to over-the-counter drugs).

What differentiates DTx from drugs is the mechanism of action and the active ingredient, i.e. the element responsible for the clinical effect, which is a chemical or biological molecule for drugs, while being an algorithm for DTx (Recchia et al. 2020). DTx are mainly indicated for the treatment of chronic diseases, in particular mental and metabolic diseases. Although only a limited number of therapeutics have been approved to date, many of them are under active development, and this category of products is evolving rapidly (Patel and Butte 2020).

4.4.1 Definition

According to the above-mentioned taxonomy (Goldsack 2019), these products deliver evidence-based therapeutic interventions to prevent, manage, or treat a medical disorder or disease. However, in recent years, some doubts emerged both on their meaning (often confused with Digital Drug Supports) and on the quality of clinical evidence needed to support their use. Updates to this definition have therefore been proposed, to emphasize the quality of the confirmatory evidence and the independence of their benefit from the drug, such as:

> Technologies that offer therapeutic interventions driven by high-quality software programs, based on scientific evidence obtained through methodologically rigorous confirmatory clinical investigation, to prevent, manage, and treat a broad spectrum of physical, mental, and behavioural conditions. (Gussoni 2021).

This definition underlines that DTx deliver an independent therapeutic intervention through confirmatory RCTs.

4.4.2 Digital Forms

Like the active ingredient of the drug, which has to take a pharmaceutical form (tablet, vial, cream, etc.) to deliver the clinical benefit, the DTx algorithm has to take a form that can interact with the patient. This can be an app for the smartphone, a serious game for a console, or a program for virtual reality visors. The choice of the most appropriate digital form depends both on the characteristics of the patient (adult, child, elderly) and on the therapeutic indication.

4.4.3 Mode of Action

In most cases, the modification of the patient's dysfunctional behaviour by DTx is achieved through the delivery of education and support, cognitive-behavioural therapy (CBT) specific to the indication of interest (e.g. chronic insomnia (Baglioni et al. 2020), addiction/abuse, eating disorders, etc.) or other psychotherapeutic interventions, such as motivational interviewing, psychoeducation, and others. In these latter cases, DTx can be considered as an alternative mode of delivering a proven therapeutic intervention (NICE 2018).

In other cases [such as Attention Deficit Hyperactivity Disorder (ADHD) or autism], digital therapy is based on serious games to promote the learning of specific functions or the development of certain activities, probably by inducing remodelling of synaptic connections in the brain or multiple features of attentional control in children with ADHD (Gallen et al. 2021).

4.4.4 Composition

A Digital Therapeutic may consist of several elements, functionally integrated with each other. The *Digital Active Ingredient* is the element responsible for the clinical effect, whether positive (clinical benefit) or negative (adverse effect). It represents, for example, the flow of activities in a CBT that from the first meeting with the patient are performed, in order to achieve the improvement of health outcomes such as preliminary request of information about the health status; analysis of the patient's answers; provision of information to the patient about the disease and the therapy; collection of information from the patient about his health status; reports on the therapeutic progress, and other.

In addition to the active ingredient, digital excipients may be the other building block of the patient-facing application (Ambrose et al. 2020). In general, the purpose of excipients is to enhance the uptake of the active ingredient, making it as bioavailable as possible. In the case of DTx, the aim of the digital excipient is to involve efficiently the patient in the use of the application for the expected duration of the therapy, making the active ingredient "digitally bioavailable". The various digital excipients include modules for rewarding the patient, reminders to take the digital therapy and complementary therapies, modules to connect the patient with his/her doctor and with other patients with the same disease, social support, etc.

It is therefore conceivable that the same active ingredient may have different therapeutic effects depending on the digital excipients, which may make it more or less digitally bioavailable to the patient. DTx also include a platform for downloads and a dashboard for the prescribing physician, who can supervise and control the patient's therapy and aggregate the information of different patients.

4.4.5 Regulatory Classification

While, in the USA, the FDA Digital Health Software Precertification Program is a pilot scheme that takes the novel approach of regulating the company rather than the product, in the European Union DTx must comply with the Medical Device Regulation 2017/745 (MDR) standard (European Union 2017). Under this regulation, the majority of apps that can self-classify as class I (and therefore do not required authorization by a Notified Body) under the previously active Directive 2007/47/CEE (MDD) will be "up-classed" to class IIa under MDR, meaning they will need to meet more stringent criteria to receive a CE mark.

4.4.6 Research & Development (R&D)

The R&D programme for new DTx is closely dependent on the product objectives, which are to achieve certification as a medical device (a necessary but not sufficient condition to qualify the product as DTx) and to achieve reimbursement and competitive advantage. Although there are many similarities with the R&D of drugs, the process with DTx is characterized by some specific aspects that must be considered both in the design phase and in its execution.

4.4.6.1 Research and Discovery

The first step, performed in a "*pre-clinical laboratory environment*", is to develop the therapeutic intervention that will be used in the software design. This intervention can be either selected from the scientific literature (as for many CBT) or developed *ex novo*, using elements of therapies described in the literature that are combined with each other on the basis of the personal experience of patients, specialists, and/or practitioners involved in the team of experts.

The combination of digital active ingredient and digital excipients, in a digital form suitable for the best use by a patient, is the output of this pre-clinical phase of development which is aimed at providing a minimum detail for DTx necessary to activate the clinical development phase (in terms of identification of the intended use, usability, minimization of risks, adherence to the best standards for privacy and cybersecurity, and possibly simulated clinical verification).

4.4.6.2 Pilot Development

Once the software prototype is available, the pilot development aims to produce the data and information that will allow an appropriate assessment of the opportunity to complete or not the development through confirmatory RCT. Since full development is by far the most expensive phase in terms of resources and time, the decision to continue the development is crucial and must be based on sufficiently supportive data, in order to minimize the risk of wasting money and time in a therapy that does not have the opportunity to offer a real benefit to the patient.

The first test to perform in this phase concerns the evaluation of the usability and acceptability of the application by the patient. This test may be performed on healthy volunteers or on patients with the disease to be treated, in order to have a better representativeness of the results, and should involve subjects with different levels of technology literacy. Then a pilot clinical study should be performed on a limited number of patients, with a trial design that may involve both the presence and absence of a control and the possible use of patient self-administered measurements, e.g. in the case of depression.

4.4.6.3 Full Development

The results of the pilot phase may lead to a review of the software, either by fixing the digital active ingredient to improve the expected therapeutic outcomes or to upgrading the excipients with new value-added utilities to improve the user experience. Based on these results, the decision to commit to full development can be taken, considering both technical aspects (quality of the software, interface, utility, etc.) and strategic and commercial aspects (competitors, costs, reimbursement issues, etc.).

The full development phase consists of one or more confirmatory RCTs that must generate the full evidence of benefit of the candidate DTx for its approval, reimbursement, prescription, and use. The design of the RCTs depends on the expected use of the Digital Therapeutic. If used in combination with a specific drug, the trial should demonstrate the therapeutic superiority of the combination over the drug alone and non-inferiority with respect to tolerability. In the case of standalone DTx that may be prescribed in addition (add-on) to the patient's usual therapy, the trial should demonstrate the therapeutic superiority of the candidate DTx over therapy as usual.

In these trials, the digital placebo is represented by an app (or the appropriate digital form) comparable with the DTx being tested, in terms of content (the placebo carries all the information provided by the DTx, but using static rather than interactive interfaces) or in terms of graphic presentation (it presents the same interfaces and introduces the same user routine, but without the elements responsible for the clinical effect) (Ritterband et al. 2017).

In our opinion, the digital active ingredient tested in the confirmatory clinical trial cannot be changed or modified during the trial, except within "windows of possible modification" (Torous et al. 2019), identified among the standards and ethical principles of the development of these therapeutics. This is however a critical point, due to the frequent advisability/need to update the technology. While the confirmatory randomized controlled design is the best condition for the clinical trial, the operational and logistic execution of the study could benefit from the opportunities offered by the Decentralized Clinical Trial approach. This modality could reduce the time and costs of DTx development, and at the same time improve patient involvement and data quality (Khozin and Coravos 2019).

4.4.6.4 Post-Marketing Surveillance

As occurs with any effective therapy, potential benefits may be accompanied by unintended and/or adverse effects in the short or long term. In the STARS-Adjunct clinical trial of AKL-T01, for example, 18% of participants experienced a device-related adverse event during the 12-week trial (Kollins et al. 2021). Therefore, after product approval and launch, the DTx post-marketing surveillance is necessary to identify new potential adverse effects and/or evaluate the benefit/risk profile of the product. More generally, post-marketing surveillance studies provide an excellent opportunity to collect important information in terms of treatment compliance and user experience, while potentially providing new information with a view to enhancing and updating software.

4.4.7 Digital Therapeutics and Drug Therapy

With respect to drug therapy, a standalone DTx with the same indication could either replace or be added to the patient's existing or newly prescribed drug therapy. Another option could be the joint development of a DTx–drug combination. In the latter case, the confirmatory clinical development of Digital Therapy must be performed in combination with a specific drug, and the results of such development can only be referred to this combination and are not transferable to other compounds, even if similar. Due to a convergence of opportunities, several agreements have been signed between drug companies and DTx startups, with the final aim to develop a new bio-digital pipeline of products (Table 4.2).

4.4.8 Challenges for Adoption in Medical Practice

DTx have the potential to transform the management—both clinical and economic—of chronic diseases, which represent a major health emergency in most countries in the world. However, despite the claimed benefits in terms of efficacy and tolerability, DTx use is still very limited. Although some DTx are already available and some are reimbursed in countries like the USA, Germany, France, and others, that does not imply that they have already entered into medical practice.

A number of barriers exist to the introduction and adoption of DTx in medical practice, and can be related—as for other innovations—to healthcare professionals, institutions, and patients (Table 4.3) (Frederix et al. 2019). To overcome these barriers, it is necessary to address and remove the main critical issues and to create

Table 4.2 Some agreements between drug companies and DTx startups

DTx Company	Pharma Company	Year	Therapeutic Indication
GAIA AG	Servier	2015	Depression
Voluntis	Sanofi	2017	Diabetes
Click Therapeutics	Sanofi	2018	Various indications
Pear Therapeutics	Novartis	2018	Schizophrenia
Voluntis	AstraZeneca	2018	Oncology
Voluntis	Abbvie	2018	Immunology diseases
Akili Laboratories	Shionogi	2019	ADHD—Autism
Click Therapeutics	Otsuka	2019	Depression
Voluntis	Novartis	2019	Oncology
Noom	Novo Nordisk	2019	Obesity
Wellthy Therapeutics	Bayer	2019	Various indications
Welldoc	Astellas	2019	Diabetes
Voluntis	BMS	2020	Oncology
Click Therapeutics	Boehringer Ingelheim	2020	Schizophrenia
KAIA AG	Chiesi Farmaceutici	2020	COPD
daVi DigitalMedicine	Polifarma	2022	Insomnia—Hypertension

Table 4.3 Some barriers to adoption in healthcare and medical practice (The Lancet 2018)

Main barriers to large-scale deployment	Key measures on how to address the barriers
Stakeholder resistance to adopt digital health based care: – Lack of patient motivation and digital health literacy skills. – Lack of healthcare provider belief in digital health care.	*Stakeholder resistance to adopt digital health based care:* – Establish patient digital health education programmes. – Redesign contemporary workflow models.
Legal, ethical, & technical barriers: – Mobile data privacy, security & liability concerns. – Lack of interoperability.	*Legal, ethical, & technical barriers:* – Establish European-wide digital health certification programmes. – Assure compliance to applicable digital health directives. – Assure interoperability of digital health services.
Other barriers: – Lack of health economical evaluations. – Lack of reimbursement.	*Other barriers:* – Encourage economical evaluations of digital health based care. – Inform health insurance industry & policy makers. – Stimulate digital health-related knowledge and experience sharing.

the necessary conditions for the introduction of DTx and their benefits in medical practice.

4.4.8.1 Clinical Evidence

The use of these technologies requires clinical evidence, which can have different levels of intensity depending on the nature of the device and the relevance of the risk to the patient. Randomized controlled trials (RCTs), the gold standard of evidence, have so far rarely been used in the development of digital medicine products, partly because the current classification of clinical trials does not fit the iterative nature of product design and because the cost of such trials is high relative to the perceived risk level of the product.

Although such digital products may collect a large amount of data in real time, and therefore new methods of assessing their efficacy and tolerability can be developed, as long as there is no consensus on such alternative methods, it seems inappropriate to invoke digital exceptionalism (Frederix et al. 2019). Today, only results from confirmatory RCTs can provide the clinical evidence needed to reassure all stakeholders of the therapeutic value of DTx, and overcome the lack of healthcare provider belief in digital health care.

4.4.8.2 Technology Assessment

Each new health technology, after regulatory approval, must undergo a technological assessment to determine its therapeutic value and place in therapy with the aim of informing and supporting decision-makers at different levels in decisions regarding purchase, reimbursement, and use. In the case of emerging health technologies, such as DTx, the need for a comprehensive and systematic multidisciplinary assessment of the welfare, economic, social, and ethical consequences determined by their adoption in health practice, becomes even more critical and relevant. Experience to date is limited and in Europe mainly concerns the assessments of the National Institute for Health and Care Excellence (NICE) of Deprexis (NICE 2018), a Digital Therapeutic for the treatment of depression, and of Sleepio (Darden et al. 2021) for the treatment of insomnia.

4.4.8.3 Reimbursement

In several European countries, where public health is the core of care, the reimbursement of a proven technology is a necessary, though not sufficient, condition for its adoption in medical practice.

The implementation in Germany, in 2020, of the new Digital Health Law [*Digitales Versorgungsgesetz* (DVG)] which provides for the reimbursement of DTx and other digital health technologies for German citizens covered by public health insurance, has the potential to represent a turning point for the entry of DTx into European medical practice (Lauer et al. 2021).

With Germany's DIGA Fast Track process set in place, several EU countries have implemented as Belgium (mHealthBelgium 2022), or are willing to implement as France (HealthcareITNews 2021), a similar assessment framework to evaluate digital health apps and to provide a direct access to the public reimbursement system. However, the majority of EU countries have not yet shown a notable interest in providing a standardized process for digital health services into their statutory health insurance system (Chawla 2022).

4.4.8.4 Data Safety and Privacy

As a result of the increased perception of the risks of misuse of online data (email, social media), health apps and in particular DTx must ensure that storage, use, and sharing fulfil the standards for handling patient health electronic data (Torous et al. 2019). Recommendations from an international panel of experts were recently proposed and concern: (a) agreed standards for data storage, use, and sharing are needed; (b) data storage, use, and sharing policies must be made transparent to users of the app; (c) if data are shared with external partners (e.g. researchers), the partner's storage, use, and sharing plans must be shared with the end user; (d) the end user must have the option to "opt out" of sharing his/her information; (e) any language regarding data storage, use, and sharing must be written at a maximum of a sixth grade reading level; (f) technical security reviews and data audits are necessary to guarantee that apps follow the standards they set out, and ensure that new vulnerabilities are quickly identified (Torous et al. 2019).

4.4.8.5 Information e Education

In the world of tomorrow, we must rely on digitally empowered and capable citizens, a digitally skilled workforce and digital experts (European Commission 2022). Raising public awareness on digital health technologies in general and on DTx in particular is a necessary condition to overcome the lack of patient motivation to use these technologies. At the same time, healthcare professionals need to become aware of the potential benefit to patients from the use of DTx in order to adopt these products into the care workflow.

Medicine and healthcare (at least at the beginning of the COVID-19 pandemic) are lagging behind other sectors in terms of digital transformation processes, and in many countries Human Capital and Digital Skills, as documented in the Digital Economy and Society Index, is still very low.

Physician and patient training represent a critical activity in this process, as well as the availability of expert patients in digital health technologies who could work as team members in the research and development of new DTx.

4.5 Conclusions

Digital Health and Medicine has the potential to change the way health systems are organized and financed, the type of health professionals needed, the role of those professionals and of patients, as well as the health services provided and the process of delivery (World Economic Forum 2019).

As a result, we expect a different healthcare system, which will move to a "consumer-centric" model, based on patient empowerment, self-management, shared decision-making, and goal orientation towards the achievement of life goals of individuals. These changes will allow citizens to have much more responsibility for managing their health and healthcare (De Maeseneer and Boeckxstaens 2012). As the grey areas around DTx are still large, the introduction, implementation, use, and funding of digital health technologies should be carefully evaluated and monitored. In the context of public healthcare, such evaluation and monitoring are necessary and must be performed in relation to the goals health systems pursue (Ricciardi et al. 2019).

DTx are not futuristic or future therapeutics, they are already available in a number of Countries worldwide. However, information, training, research, and a system of rules for their appropriate evaluation, reimbursement, and use are needed, in order to remove the barriers that still limit their success. The commitment of all health actors, starting with the institutions, is the necessary condition for these new health promises to be confirmed and delivered.

References

Alessa T, Hawley MS, Hock ES, de Witte L (2019) Smartphone apps to support self management of hypertension: review and content analysis. JMIR Mhealth Uhealth 7:e13645

Ambrose M, Seiler D, Barrett B et al (2020) The role of public standards in assuring quality of digital therapeutics. Source: https://qualitymatters.usp.org/sites/default/files/user-uploaded-files/USP_Digital_Therapeutics_Paper_2020-06-11.pdf Last consultation: 18/02/2022

American Psychiatric Association (APA) (2022) AppAdvisor. Source: https://www.psychiatry.org/psychiatrists/practice/mental-health-apps Last consultation: 18/02/2022

Baglioni C, Altena E, Bjorvatn B et al (2020) The European academy for cognitive behavioural therapy for Insomnia: an initiative of the European Insomnia Network to promote implementation and dissemination of treatment. J Sleep Res 9:e12967
Brittain D, D'Andrea P, Gruen E, et al (2022) A review of the unique drug development strategy of indacaterol acetate/glycopyrronium bromide/mometasone furoate: a first-in-class, once-daily, single-inhaler, fixed-dose combination treatment for asthma. Adv Ther. Online ahead of print
Chawla V (2022) Where is the money in digital health? The roadmap to digital health app reimbursement in Europe. Source: https://www.google.com/url?sa=t&rct=j&q=&esrc=s&source=web&cd=&ved=2ahUKEwji8syYxNz5AhW_hf0HHUqCCYoQFnoECA0QAQ&url=https%3A%2F%2Fresearch2guidance.com%2Fwhere-is-the-money-in-digital-health-the-roadmap-to-digital-health-app-reimbursement-in-europe%2F&usg=AOvVaw1Kb4MlHYIoCBa4uSRun4B1 Last consultation: 18/02/2022
Darden M, Espie CA, Carl JR et al (2021) Cost-effectiveness of digital cognitive behavioral therapy (Sleepio) for insomnia: a Markov simulation model in the United States. Sleep 44(4):zsaa223
De Maeseneer J, Boeckxstaens P (2012) James Mackenzie Lecture 2011: multimorbidity, goal-oriented care, and equity. Br J Gen Pract 62(600):522–524
European Commission (2022) Human capital and digital skills in the digital economy and society index. Source: https://digital-strategy.ec.europa.eu/en/policies/desi-human-capital Last consultation: 18/02/2022
European Union (2017) Regulation (EU) 2017/745 of the European Parliament and of the Council of 5 April 2017 on medical devices, amending Directive 2001/83/EC, Regulation (EC) No 178/2002 and Regulation (EC) No 1223/2009 and repealing Council Directives 90/385/EEC and 93/42/EEC. Source: https://eur-lex.europa.eu/eli/reg/2017/745/oj Last consultation: 18/02/2022
Frederix I, Caiani EG, Dendale P et al (2019) ESC e-Cardiology Working Group Position Paper: overcoming challenges in digital health implementation in cardiovascular medicine. Eur J Prev Cardiol 26(11):1166–1177
Gallen CL, Anguera JA, Gerdes MR et al (2021) Enhancing neural markers of attention in children with ADHD using a digital therapeutic. PLoS One 16(12):e0261981
Goldsack J (2019) Digital Health, Digital Medicine, Digital Therapeutics (DTx): What's the difference? DiMe. Source: https://bit.ly/33mqYNe Last consultation: 18/02/2022
Gussoni G (ed) (2021) Digital therapeutics: an opportunity for Italy, and beyond. Tendenze Nuove 4
HealthcareITnews (2021) France to enable rapid market access for digital therapeutics. Source: https://www.healthcareitnews.com/news/emea/france-enable-rapid-market-access-digital-therapeutics Last consultation: 18/02/2022
IQVIA (2021) Digital Health Trends 2021: innovation, evidence, regulation, and adoption. Source: https://bit.ly/3rtmk7V Last consultation: 18/02/2022
Khozin S, Coravos A (2019) Decentralized trials in the age of real-world evidence and inclusivity in clinical investigations. Clin Pharmacol Ther 106(1):25–27
Khusial RJ, Honkoop PJ, Omar Usmani O et al (2020) Effectiveness of myAirCoach: a mHealth self-management system in asthma. J Allergy Clin Immunol Pract 8:1972–1979
Kollins SH, Childress A, Heusser AC, Lutz J (2021) Effectiveness of a digital therapeutic as adjunct to treatment with medication in pediatric ADHD. NPJ Dig Med 4:58
Lancet T (2018) Editorial. Is digital medicine different? Lancet 392
Lauer W, Löbker W, Höfgen B (2021) Digital health applications (DiGA): assessment of reimbursability by means of the "DiGA Fast Track" procedure at the Federal Institute for Drugs and Medical Devices (BfArM). Bundesgesundheitsblatt Gesundheitsforschung Gesundheitsschutz 64(10):1232–1240
mHealthBelgium (2022) Home. Source: https://mhealthbelgium.be/ Last consultation: 18/02/2022
National Institute For Health And Care Excellence (NICE) (2018) Digital psychological therapy briefing: Deprexis for adults with depression. Source: https://www.nice.org.uk/media/default/about/what-we-do/nice-advice/iapt/iab-deprexis.pdf Last consultation: 18/02/2022

Nimri R, Nir J, Phillip M (2020) Insulin Pump Therapy. Am J Ther 27(1):e30–e41
Nkhoma DE, Soko CJ, Bowrin P et al (2021) Digital interventions self-management education for type 1 and 2 diabetes: a systematic review and meta-analysis. Comput Methods Prog Biomed 210
Patel NA, Butte AJ (2020) Characteristics and challenges of the clinical pipeline of digital therapeutics. NPJ Dig Med 159
Recchia G, Capuano DM, Mistri N, Verna R (2020) Digital therapeutics – What they are, what they will be. Acta Sci Med Sci 4(3):1–9
Ricciardi W, Pita Barros P, Bourek A et al (2019) Lehtonen L on behalf of the Expert Panel on Effective Ways of Investing in Health (EXPH). How to govern the digital transformation of health services. Eur J Pub Health 29(3):7–12
Ritterband LM, Thorndike FP, Ingersoll KS et al (2017) Effect of a web-based cognitive behavior therapy for insomnia intervention with 1-year follow-up. A randomized clinical trial. JAMA Psychiatry 74(1):68–75
Seron P, Oliveros MJ, Gutierrez-Arias R et al (2021) Effectiveness of telerehabilitation in physical therapy: a rapid overview. Phys Ther 101(6):pzab053
Torous J, Andersson G, Bertagnoli A et al (2019) Towards a consensus around standards for smartphone apps and digital mental health. World Psychiatry 18:1
World Economic Forum (2019) Building the healthcare system of the future. Source: https://reports.weforum.org/digital-transformation/building-the-healthcare-system-of-the-future/ Last consultation: 18/02/2022

5 Intelligence-Based Medicine: The Academic Perspective and Deep Humanism

Luca Tagliaferri, Loredana Dinapoli, Marika D'Oria, Alfredo Cesario, Maria Antonietta Gambacorta, Calogero Casà, and Vincenzo Valentini

Abstract

Artificial Intelligence (AI) and other statistical methods are providing new opportunities of data modelling for patient benefit. The number of studies on AI-driven predictions is increasingly growing, since hundreds of models have been developed for similar targeted populations and outcomes. Predictive models are used and tested in a wide range of realities to achieve Deep Medicine, and it is therefore difficult to map every product in biomedicine, bioinformatics, and robotic surgery. Although we are beginning to understand the chances provided by these solutions, attention is growing among the academic community to define validated methodologies and guidelines for AI systems development and use, to undergo the same level of inspection. This chapter explores some methodological aspects of academic research to identify possibilities and challenges that lie behind an Intelligent-Based Medicine.

Keywords

Intelligence-based medicine · Methodology · Deep medicine · Deep learning · Machine learning · Clinical practice · Personalized medicine

L. Tagliaferri · L. Dinapoli · M. A. Gambacorta · V. Valentini (✉)
UOC Radioterapia Oncologica, Dipartimento di Diagnostica per Immagini, Radioterapia Oncologica ed Ematologia, Fondazione Policlinico Universitario A. Gemelli IRCCS, Rome, Italy
e-mail: vincenzo.valentini@policlinicogemelli.it

M. D'Oria · A. Cesario
Open Innovation Unit, Direzione Scientifica, Fondazione Policlinico Universitario A. Gemelli IRCCS, Rome, Italy
e-mail: alfredo.cesario@policlinicogemelli.it

C. Casà
UOC Radioterapia Oncologica, Fatebenefratelli Isola Tiberina – Gemelli Isola, Rome, Italy

A. Cesario et al. (eds.), *Personalized Medicine Meets Artificial Intelligence*,
https://doi.org/10.1007/978-3-031-32614-1_5

5.1 Introduction

Artificial Intelligence (AI) and other statistical methods are providing new opportunities of data modelling for patient benefit. Although we are beginning to understand the chances provided by these solutions, attention is growing among the academic community to define validated methodologies and guidelines for AI systems development and use, to undergo the same level of inspection.

According to Vollmer and colleagues, research on AI must address four challenges to consolidate evidence and deliver health benefits: transparency, replicability, ethics, and effectiveness (TREE) (Vollmer et al. 2020). Transparency is fundamental to recognize the reliability of data sources and methodological trustworthiness of algorithms validation. Patients' data must follow a plain process of retrieving, storage, elaboration, analysis, and dissemination. Research methodology and results replicability are mandatory for evidence generation and scalability. Moreover, the ethics and algorethics (the ethics of algorithms) framework covers several issues, from preserving intellectual property to patients' rights, from securing data usage and interpretation to reach empirical evidence without neglecting innovation. Finally, the effectiveness of such algorithms have to be demonstrated, especially if they are modelled to reproduce, and therefore represent the patient (especially with multi-morbidity or co-morbidity conditions).

This chapter explores some methodological aspects of academic research to identify possibilities and challenges that lie behind an Intelligent-Based Medicine (Chang 2020).

5.2 Understanding the Methodological Pathway

Research methodology comprises specific procedures and techniques to identify, select, analyse, and interpret information on a specific phenomenon. Methodological rigour allows a critical evaluation of the study as well as enables the replicability of the study on the same phenomenon under similar conditions. For this purpose, research on AI-driven predictive models must give clear understanding of the methodological pathway to understand its implications. Research methodology varies for several aspects, for example the nature of the selected data source.

Observational and interventional studies can retrieve data from several sources (excel files, relational databases, omics containers, data warehouses, or real-world data), and each of them must be analysed through specific methods related to biostatistics and bioinformatics (Fig. 5.1), even if a different mention should be spent for big data analysis, since it requires proper bio-machine learning algorithms.

Confirmation of conformity relies on clinical data, which evaluation must follow a rigorous methodology (Vollmer et al. 2020). The process that goes from the research question to the use of predictive models in real life follows three steps represented in Fig. 5.2. Choosing the right methodology helps researchers find trustworthy results to support decisions when translated into medical practice.

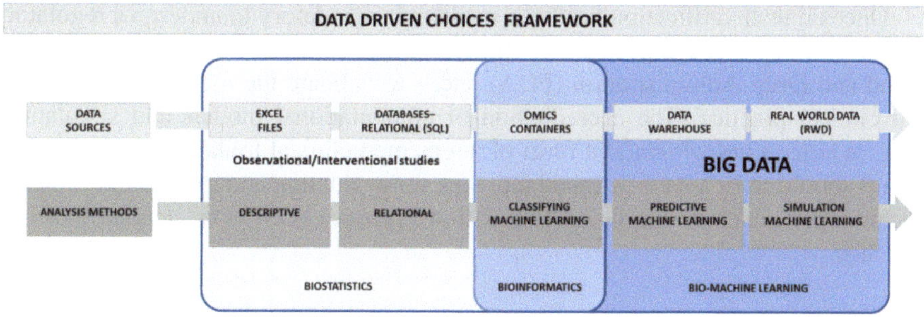

Fig. 5.1 The data-driven choices framework

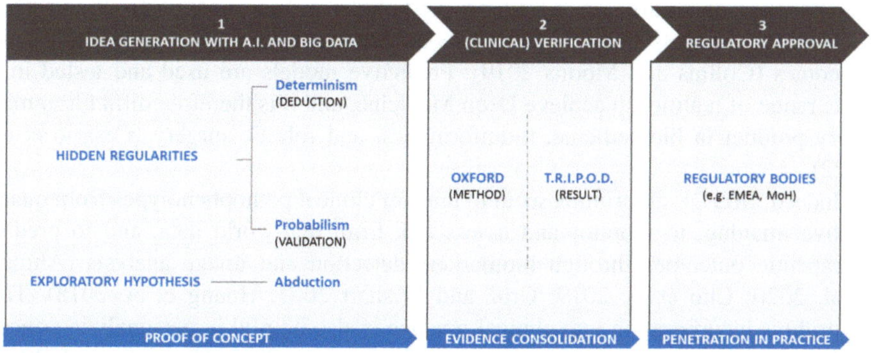

Fig. 5.2 The three-step process for predictive model validation

More specifically, researchers can investigate AI-driven predictive models for two main reasons:

1. To find hidden regularities among data towards a deductive/determinist or predictive/probabilistic approach
2. To explore new hypothesis towards an abductive approach

Deterministic and probabilistic models are both data-driven and knowledge-based approaches. When results are achieved, the proof of concept must undergo a clinical verification to consolidate evidence on their effectiveness. Algorithms must be validated and optimized through supervised trainings to sustain scientific replicability and credibility. To achieve this goal, two approaches can be followed:

1. The Oxford approach: it aims at investigating and replicating the research method through which results are achieved.
2. The TRIPOD-AI approach: it focuses on investigating and replicating the results.

Once clinical verification has been made, it is mandatory to undergo a regulatory approval from Notifying Bodies—the European Medicines Agency (EMA), the Food and Drug Administration (FDA), etc.—to validate the model before using it for clinical practice. The accreditation of computational models and simulations (e.g. *in silico clinical trials* [a form of augmented clinical trials] for drug development) should fit in a well-defined landscape for evaluation and assessment of health technologies, such as the Good Simulation Practices (GSPs) as standards of best practice.

5.3 Current Deep Medicine Interventions

The number of studies on AI-driven predictions is increasingly growing, since hundreds of models have been developed for similar targeted populations and outcomes (Collins and Moons 2019). Predictive models are used and tested in a wide range of realities to achieve Deep Medicine, and it is therefore difficult to map every product in biomedicine, bioinformatics, and robotic surgery (Cesario et al. 2022).

Indeed, ML/DL algorithms seem to unravel clinical pathophenotypes from quantitative imaging, to monitor and assess risk from real-world data, and to predict therapeutic outcomes through biomarkers detection and image analysis (Ahmed et al. 2020; Cho et al. 2019; Cruz and Wishart 2007; Huang et al. 2018). The algorithms incorporation into clinical practice is also helpful in personalizing oncologic patient management and follow-up (Esteva et al. 2017; Haenssle et al. 2018; Hosny et al. 2018; Langlotz et al. 2019). The potential of ML/DL is also indicated in gastrointestinal endoscopy for early disease detection (Min et al. 2019), and some real-world applications are applied in ophthalmology (e.g. to detect diabetic retinopathy) (Gulshan et al. 2016). Most AI solutions are related to radiology and biomedical imaging, for example to interpret breast cancer or cardiovascular imaging, but also applied to neurological diseases (e.g. the Alzheimer's disease) (O'Bryant et al. 2010). On the other hand, public health surveillance can benefit from AI tools for detection, monitoring, and control of local and global epidemic (Tayarani 2020). Future research should deepen understanding on user acceptance and interpretation of AI models, especially "virtual patients" that should "represent" a sample of a wider ontology and not the "whole" ecosystem of the person.

This scenario challenges academic research and clinical practice; hence, it is mandatory to develop recognized standards for integration of AI systems in medicine (Xing et al. 2020). To this aim, the CONSORT-AI and guidelines SPIRIT-AI represent the first international standards for AI-supported clinical trials and ensure transparent reporting of protocols whilst potentially improve their design and delivery (Ibrahim et al. 2021).

5.3.1 Patient Profiling and Deep Humanism

The preliminary step for developing AI-driven predictive models is patient profiling. Patient profiling gives a holistic view of the person and his/her preferences, to forecast tailored treatment and supportive therapies, allowing real-time monitoring of the patient. Despite the still controversial aspect on the ethical usage of patient profiling for misjudgement and mistreatment of their information, to which we should pay careful attention and maximum respect, some validated clinical applications are available in our context, such as:

- The creation of large databases for large multicentric databases analysis (data mining) or modelling (distributed learning) (Tagliaferri et al. 2016, 2018a, b; Lancellotta et al. 2020).
- The analysis that combines administrative data and clinical data to identify critical aspect within clinical process (process mining) (Lenkowicz et al. 2018; Gatta et al. 2018).
- The quantitative analysis of radiological images to generate new predictive models or clinical decision support systems (Gatta et al. 2019; Cusumano et al. 2020a, b, 2021; Soror et al. 2020; Chiloiro et al. 2020) even using AI (Fionda et al. 2020).

Patient profiling can be obtained by integrating clinical data with Internet-of-Medical Things and artistic stimuli. This integration lies at the basis of Deep Humanism considered as an opportunity to understand emotional preferences of the patient (especially those with oncological diseases). An example of Deep Humanism in medicine is the “Art4ART” project of the Advanced Radiation Therapy (ART) department at Fondazione Policlinico Universitario A. Gemelli IRCCS (Rome, Italy). The goal of the project is to empower patients by offering a clinical experience which connects artistic narratives (cinema, music, painting, nature, literature, and design) and the spiritual dimensions of human beings (friendship, love, passion, faith, etc.).

Artistic stimuli are strategically placed in several department places, aiming at welcoming the patient in a more familiar context during her/his journey. In this scenario, AI can be helpful:

- To customize artistic stimuli according to models of patients’ preferences
- To tailor patients’ communication and education by focusing on their preferences related to their mood, disease phase, and treatment, to enable their resilience during the oncological journey

The pathway to use artistic products to patients according to their humanistic and clinical profiling is represented in Fig. 5.3.

The impact of integrating digital technologies with artistic inputs can positively affect psychological, clinical, organizational, and territorial context. Psychological effects are measured with validated tools in order to assess: patient distress and

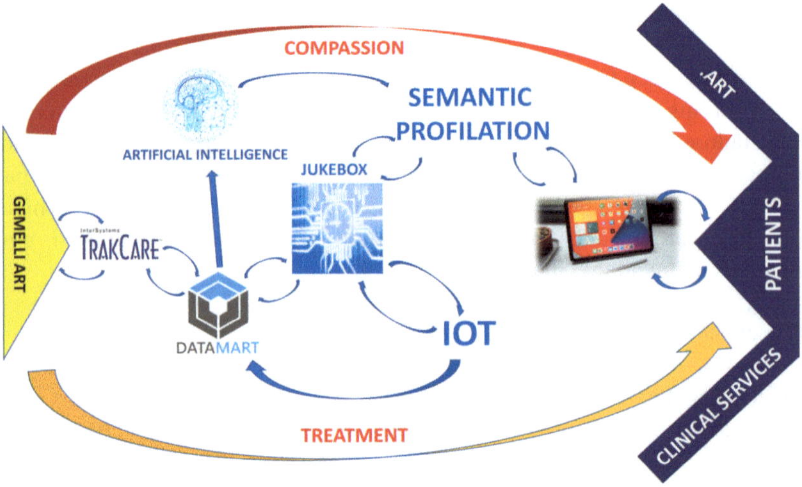

Fig. 5.3 Patients accessing Gemelli ART's clinical services can interact with digital devices to receive personalized artistic stimuli (music, images, etc.) to ameliorate their illness experience. The interaction generates patient-specific information that, in combination with data from wearables to understand their lifestyle (Internet-of-Things, IoT), and clinical data generated by the services of the department (uploaded in the organizational IT system (TrakCare) that populates a DataMart safe-by-design), allows a semantic profilation. All information is collected in a "Jukebox" that stores several artistic features (music, images, etc.), which are used to improve patients' Quality of Life according to his/her humanistic and clinical profile. Artistic stimuli are intertwined in a major care experience convened towards compassion and treatments

anxiety/depression (Holtzman et al. 2018; Mackenzie et al. 2013), effects of amusement and awe (Allen 2018), illness experience (Al-Rashdan et al. 2021), and functional coping styles (Morris et al. 2018). Consequently, patient profiling also leads clinicians to understand how therapeutics acts are perceived from a patient-centred perspective. The concept of *individualized* treatment is thus evolving towards that of *personalized* care: digital health not only helps in choosing the best treatment for that specific patient, it also contributes to a better patient welcoming and, through patient profiling, allows to identify eventually pattern of choice that should correlate with better outcomes, such as lower toxicity, improved management of certain symptoms moving from entertainment to an integral part of the treatment.

5.4 Safety in Research and Development

The regulatory and scientific framework guides research to comply with standards while incentivizing innovation. Research & Development (R&D) must ensure privacy-preserving behaviours and procedures, along with cybersecurity and data protection. In a practical fashion, predictive modelling is achieved through the combination of clinical and experimental information with data from patients having similar characteristics. The scientific community agrees on building a safe ecosystem between disciplines and actors that share the same assumptions (ethical, at least) to preserve patient information, while validating secure standards and methodologies, especially in multi-stakeholder collaborations (e.g. multicentric studies) and algorithms training on external databases. Predictive models' generation should pursue "quality by design", an approach that regulates the quality of the product towards statistical, analytical, and risk-management methodologies (EMA 2022).

5.5 Outlook for an Intelligence-Based Medicine

If we want to manage R&D, some improvements are suggested to update medical education. The awareness on AI potential has increased, but many physician still feel uncomfortable with AI-driven research because of a limited understanding of technical language and methodologies. Students, clinicians, and faculties should be educated to acquire an AI-literacy since Intelligence-Based Medicine will progressively blend the actual knowledge on human complexity with stratified heterogeneous data (omics, RWD, etc.) on diseases patterns. Educational topics should cover not only technical jargon, basic knowledge of statistics (e.g. sample size, overfitting, underfitting, cross validation), and a broad knowledge of the metrics and interpretation of ML/DL algorithms (Handelman et al. 2019); a dedicated digital health education needs to be closely integrated with clinical disciplines. In fact, it is reported in the literature that there is a knowledge gap that needs to be bridged in order to safely apply new technologies in healthcare (Casà et al. 2021; Khamisy-Farah et al. 2021).

References

Ahmed Z, Khalid M, Zeeshan S, Dong X (2020) Artificial Intelligence with multi-functional machine learning platform development for better healthcare and precision medicine. Database 2020:1–35

Allen S (2018) The science of awe. John Templeton Foundation, Berkeley (CA)

Al-Rashdan A, Watson L, Yannitsos D et al (2021) Comparison of patient-reported experience of patients receiving radiotherapy measured by two validated surveys. Curr Oncol 28(3): 2180–2189

Casà C, Marotta C, Di Pumpo M et al (2021) COVID-19 and digital competencies among young physicians: are we (really) ready for the new era? A national survey of the Italian Young Medical Doctors Association. Ann Ist Super Sanità 57(1):1–6. https://doi.org/10.4415/ANN_21_01_01

Cesario A, D'Oria M, Valentini V (2022) Artificial Intelligence and AI-guided robotics for personalized precision medicine. In: Riva G, Marchetti A (eds) Humane Robotics. Vita e Pensiero, Milano

Chang AC (2020) Intelligence-based medicine. Data science, artificial intelligence, and human cognition in clinical medicine and healthcare. Academic Press – Elsevier, London

Chiloiro G, Rodriguez-Carnero P, Lenkowicz J et al (2020) Delta radiomics can predict distant metastasis in locally advanced rectal cancer: the challenge to personalize the cure. Front Oncol 10:595012

Cho G, Yim J, Choi Y et al (2019) Review of machine learning algorithms for diagnosing mental illness. Psychiatry Investig 16:262–269

Collins GS, Moons KGM (2019) Reporting of artificial intelligence prediction models. Lancet 393(101818):1577–1579

Cruz JA, Wishart DS (2007) Applications of machine learning in cancer prediction and prognosis. Cancer Inform 2:59–77

Cusumano D, Catucci F, Romano A et al (2020a) Evaluation of an Early Regression Index (ERITCP) as predictor of pathological complete response in cervical cancer: a pilot-study. Appl Sci 10:8001. https://doi.org/10.3390/app10228001

Cusumano D, Meijer G, Lenkowicz J et al (2020b) A field strength independent MR radiomics model to predict pathological complete response in locally advanced rectal cancer. Radiol Med [Internet]. Source: http://link.springer.com/10.1007/s11547-020-01266-z Last consultation: 05/08/2022

Cusumano D, Boldrini L, Yadav P et al (2021) Delta radiomics analysis for local control prediction in pancreatic cancer patients treated using magnetic resonance guided radiotherapy. Diagnostics (Basel) 11(1)

EMA (2022) Quality by design. Source: https://www.ema.europa.eu/en/human-regulatory/research-development/quality-design Last consultation: 23/08/2022

Esteva A, Kuprel B, Novoa RA et al (2017) Dermatologist-level classification of skin cancer with deep neural networks. Nature 542:115–118

Fionda B, Boldrini L, D'Aviero A et al (2020) Artificial intelligence (AI) and interventional radiotherapy (brachytherapy): state of art and future perspectives. J Contemp Brachyther 12: 497–500. https://doi.org/10.5114/jcb.2020.100384

Gatta R, Vallati M, Lenkowicz J et al (2018) A framework for event log generation and knowledge representation for process mining in healthcare. In 30th IEEE International Conference on Tools with Artificial Intelligence (ICTAI 2018) [Internet]. IEEE, 647–54

Gatta R, Vallati M, Dinapoli N et al (2019) Towards a modular decision support system for radiomics: a case study on rectal cancer. Artif Intell Med 96:145–153

Gulshan V, Peng L, Coram M et al (2016) Development and validation of a deep learning algorithm for detection of diabetic retinopathy in retinal fundus photographs. JAMA 316:2402–2410

Haenssle HA, Fink C, Schneiderbauer R et al (2018) Man against machine: diagnostic performance of a deep learning convolutional neural network for dermoscopic melanoma recognition in comparison to 58 dermatologists. Ann Oncol 29:1836–1842

Handelman GS, Kok HK, Chandra RV et al (2019) Peering into the black box of artificial intelligence: evaluation metrics of machine learning methods. AJR Am J Roentgenol 212(1): 38–43. https://doi.org/10.2214/AJR.18.20224

Holtzman AL, Pereira DB, Yeung AR (2018) Implementation of depression and anxiety screening in patients undergoing radiotherapy. BMJ Open Qual 7:e000034. https://doi.org/10.1136/bmjoq-2017-000034

Hosny A, Parmar C, Quackenbush J et al (2018) Artificial intelligence in radiology. Nat Rev Cancer 18:500–510

Huang S, Cai N, Pacheco PP et al (2018) Applications of Support Vector Machine (SVM) learning in cancer genomics. Cancer Genomics Proteomics 15(1):41–51

Ibrahim H, Liu X, Rivera SC et al (2021) Reporting guidelines for clinical trials of artificial intelligence interventions: the SPIRIT-AI and CONSORT-AI guidelines. Trials 22(1):11. https://doi.org/10.1186/s13063-020-04951-6

Khamisy-Farah R, Gilbey P, Furstenau LB et al (2021) Big data for biomedical education with a focus on the COVID-19 era: an integrative review of the literature. Int J Environ Res Public Health 18(17):8989. https://doi.org/10.3390/ijerph18178989

Lancellotta V, Guinot JL, Fionda B et al (2020) SKIN-COBRA (Consortium for Brachytherapy data Analysis) ontology: the first step towards interdisciplinary standardized data collection for personalized oncology in skin cancer. J Contemp Brachyther 12:105–110. https://doi.org/10.5114/jcb.2020.94579

Langlotz CP, Allen B, Erickson BJ et al (2019) A roadmap for foundational research on artificial intelligence in medical imaging from the 2018 NIH/RSNA/ACR/the Academy Workshop. Radiology 291:781–791

Lenkowicz J, Gatta R, Masciocchi C et al (2018) Assessing the conformity to clinical guidelines in oncology: an example for the multidisciplinary management of locally advanced colorectal cancer treatment. Manag Decis 56:2172–2186

Mackenzie LJ, Carey ML, Sanson-Fisher RW, Este CA (2013) Psychological distress in cancer patients undergoing radiation therapy treatment. Support Care Cancer 21:1043–1051. https://doi.org/10.1007/s00520-012-1624-3

Min JK, Kwak MS, Cha JM (2019) Overview deep learning in gastrointestinal endoscopy. Gut Liver 13(4):388–393

Morris N, Moghaddam N, Tickle A, Biswas S (2018) The relationship between coping style and psychological distress in people with head and neck cancer: a systematic review. Psycho-Oncology 27(3):734–747. https://doi.org/10.1002/pon.4509

O'Bryant SE, Xiao G, Barber R et al (2010) A serum protein-based algorithm for the detection of Alzheimer disease. Arch Neurol 67:1077–1081

Soror T, Lancellotta V, Kovács G et al (2020) kOBCS©: a novel software calculator program of the Objective Breast Cosmesis Scale (OBCS). Breast Cancer 27(2):179–185

Tagliaferri L, Kovács G, Autorino R et al (2016) ENT COBRA (Consortium for Brachytherapy Data Analysis): interdisciplinary standardized data collection system for head and neck patients treated with interventional radiotherapy (brachytherapy). J Contemp Brachyther 8:336–343. https://doi.org/10.5114/jcb.2016.61958

Tagliaferri L, Gobitti C, Colloca GF et al (2018a) A new standardized data collection system for interdisciplinary thyroid cancer management: Thyroid COBRA. Eur J Intern Med 53:73–78. https://doi.org/10.1016/j.ejim.2018.02.012

Tagliaferri L, Budrukkar A, Lenkowicz JB et al (2018b) ENT COBRA ONTOLOGY: the covariates classification system proposed by the Head & Neck and Skin GEC-ESTRO Working Group for interdisciplinary standardized data collection in head and neck patient cohorts treated with interventional radiotherapy (brachytherapy). J Contemp Brachyther 10:260–266. https://doi.org/10.5114/jcb.2018.76982

Tayarani M-HN (2020) Applications of artificial intelligence in battling against COVID-19: a literature review. Chaos Solitons Fractals 142:110338

Vollmer S, Mateen BA, Bohner G et al (2020) Machine learning and artificial intelligence research for patient benefit: 20 critical questions on transparency, replicability, ethics, and effectiveness. BMJ 368:l6927. https://doi.org/10.1136/bmj.l6927

Xing L, Giger M, Min JK (eds) (2020) Artificial intelligence in medicine. Technical Basis and Clinical Applications. Elsevier

Part II

Consolidated Evidence

The Role of Big Data and Artificial Intelligence in Clinical Research and Digital Therapeutics

6

Sofoklis Kyriazakos, Aristidemos Pnevmatikakis, Harm op den Akker, and Konstantina Kostopoulou

Abstract

Healthcare is among the pioneering industries in the velocity of generation of big data, from the time that Electronic Health Records, Internet-of-Things, and electronic medical devices have been introduced. Researchers around the world are curating large volumes of data and applying Artificial Intelligent (AI) algorithms that give meaning to medical conditions, as well as prediction of clinical outcomes. The clinical research domain, which has been a laggard in digital transformation, is running today at high speed, utilizing electronic solutions that enable capturing of Real-World Data (RWD) in decentralized virtual clinical studies that aim to shorten the life cycle of drug development and the associated costs. At the same time, the domain of Digital Therapeutics (DTx) has been recently established, with various electronic solutions that process RWD to provide digital interventions that improve health-related endpoints. In this chapter, we will review how clinical research and DTx domains have been accelerated by the existence of Big data and AI algorithms and we will describe smart services that are expected to further boost the healthcare industry.

Keywords

Artificial Intelligence · Big data · Clinical research · Real-World Data · Digital therapeutics

S. Kyriazakos (✉) · A. Pnevmatikakis · H. op den Akker · K. Kostopoulou
Innovation Sprint, Brussels, Belgium

BTECH, Aarhus University, Herning, Denmark
e-mail: skyriazakos@innovationsprint.eu

A. Cesario et al. (eds.), *Personalized Medicine Meets Artificial Intelligence*,
https://doi.org/10.1007/978-3-031-32614-1_6

6.1 Introduction

Clinical research studies are demanding in terms of time, resources and cost, while actual outcomes are usually different from the expected ones and often not reproducible (Henegan et al. 2017; Nivel et al. 2018). The outcomes' difference is often caused by parameters, such as adherence and compliance to the trial protocol, patient dropout rate, and the recording of adverse effects (Fogel 2018). As a result of the differences in outcomes, drug profit margins may decrease due to the intervention results, the marketed solution loses competitiveness, and foremost the overall benefits to the citizens are reduced.

Further to that, hard facts about the cost and development duration indicate the size of the problem, which can be described with these numbers: $1.5–2B is the average R&D expenditure per new drug, with 50% of the cost accounting for the clinical studies; 10–15 years is the duration of development of a new drug (Harrer et al. 2019). Another parameter in drug development and clinical research is the fact that lifestyle is not considered, even though studies say it can be up to 40% health determinant.

Digital transformation is the vehicle to change these facts, while an improvement of the above numbers is already observed, whenever big data technologies and AI are applied. In this challenging environment, the regulatory framework is a mandatory component for safety of patients, data integrity, and credibility of results, while at the same time it is often creating barriers. Especially in the DTx domain, where electronic applications could offer effective digital interventions, the regulatory authorities demand solutions that are certified against the potential risks they might be linked to and often reject AI approaches; at least those that are not well justified.

The rest of the chapter is organized around four sections. In Sect. 6.2, Big data and Real-World Data collection mechanisms are presented, followed by the outcomes of processing with AI algorithms for both clinical research and DTx domains in Sect. 6.3. Next, in Sect. 6.4, we present the challenges of patients' engagement, as a major factor for the adoption of digital technologies by the patients and finally, we sum up with the conclusions in Sect. 6.5.

6.2 Big Data and RWD Collection

Data is the new medicine (Sinipuru 2021). When shared, reused in a privacy-respecting way and maintaining the control to the people providing it, data can improve patient outcomes, foster research, and accelerate deployment of novel health services (Hämäläinen et al. 2020).

Health data are readily associated with clinical tests performed invasively on samples taken from our bodies, or non-invasively using modern sensing techniques. Such is the clinical data obtained in a clinical setting that cover the lifetime of subjects in a study, however, not forming the complete spectrum of health data.

Hippocrates (approx. 460–370 BC), the father of medicine, believed that disease was not a punishment inflicted by the Gods but rather the product of environmental factors, diet, and living habits, a fact that is well-established today (Grey 2017). Our living habits can be enumerated using data attributes about our lifestyle, obtained in our natural environment, outside of the clinical setting. This data is termed "Real-World Data" (RWD) and although it does not provide a detailed understanding of the clinical status of a person, its ubiquitous collection allows a dense description of peoples' lifestyle. RWD are formally defined by the FDA (2017) as:

> data related to patient health status and/or the delivery of health care routinely collected from EHRs, claims and billing data, data from product and disease registries, patient-generated data including home-use settings, and data gathered from other sources that can inform on health status, such as mobile devices.

While clinical data is generated sparingly, during testing or hospitalization sessions, RWD is generated continuously, throughout the days, weeks, and months of our lifetime. It would thus be good practice to actually collect RWD in a continuous mode, rather than the sporadic mode in which clinical data are collected.

6.2.1 Attributes of RWD

RWD comprises attributes that enumerate different important aspects of the way we live our lives. The attributes are grouped in the physiological, psychological, social, and environmental categories discussed next.

The physiological attributes of RWD have to do with the human body, its activities and adverse events. They are mostly measured using activity trackers and/or smartphones but are also reported (Kyriazakos et al. 2021). The measured attributes related to activity are *steps and distance walked*, *elevation* (expressed as *floors climbed*), energy burned, time spent in different activity intensity zones (e.g., mild, moderate, and high intensity physical activity, as it is formally defined as a function of age), and auto-detected exercise activities (walking, running, cycling, etc.), as well as their distribution in the day. The location services of smartphones facilitate the detection of indoors or outdoors presence. Composite physical activity measurements can also be obtained via physical tests like the six-minute walk test (6MWT), the frailty test, or games specifically designed to measure muscular responses (tapping on a mobile phone screen for Parkinson's disease or performing other exercises while monitored and analyzed by depth cameras to measure features important in stroke or accident rehabilitation). All these tests are scripted and hence can be measured using sensors and audio-visual instructions to the people on their smartphone. Measurements of attributes related to the heart include the heart rate variability, the time spent in different heart rate zones, and the resting heart rate. Measurement of sleep related attributes include the time spent in the different sleep stages (awake in bed, light, REM, deep sleep), the sleep duration and quality.

Other physiological attributes can be self-reported by the participant. Symptoms of interest in general or to specific therapeutic areas (e.g., headache, body temperature, blood pressure, pains, diarrhea, fatigue, nausea), including their intensity can be collected. Weight should be regularly reported, and less so height (especially at younger or older ages). Nutrition is paramount, starting at a higher level with the consumption of food categories of interest, but more detailed analysis can also be used when available. Water, coffee, tea, refreshments, and alcohol intake can be reported. Finally, the menstrual cycle can also be of importance.

The psychological RWD attributes refer to the emotions of the study participants (Kyriazakos et al. 2021). They are mostly reported, either as simple emotional state self-assessment, or, when deemed necessary, using standardized reports from professional therapists monitoring the patients. Measurements can also be used to indirectly capture psychological aspects. Video recordings of the face can be analyzed by computer vision algorithms for emotion recognition. Speech sentiment analysis of voice audio and text analysis of the social media posts also yield emotion measurements. An indication of the psychological state can be given by the places visited (which, how diverse they are). Finally, aspects like the weather or spending unusual time commuting can have some importance.

The social RWD attributes enumerate study participants' social life. Such information can be measured indirectly using attributes about the usage of the phone (diversity, duration, frequency of calls) and social media (diversity, number, frequency of interactions). More direct information can be reported using questionnaires on activities with friends, family, or co-workers, or can be obtained in conversation with a digital virtual coach.

The environmental RWD attributes attempt to enumerate the environment the study participants live in. They include reported environmental indicators for the assessment of the quality of life, like the 11 attributes of the OECD better life index (OECD 2021). Precise measurements of living or working environment quality can be obtained by integrating relevant commercial devices (e.g., for air quality analysis), or by integrating with data services that report, e.g., Air Quality Index at specific locations.

6.2.2 Collecting RWD

RWD are collected outside the clinical setting, directly from the study participants using ubiquitous, easy to operate devices, or simply by asking people about the necessary information.

RWD measurements involve devices that are commonly used by people. An important source of physiological information is the activity tracker. These are consumer devices gaining popularity amongst health- and wellness aware people. Study participants can already own one, easily get a cheap one, or the studies can distribute them for free as a participation incentive, especially for longer studies. Other devices can be scales and ubiquitous medical devices (e.g., thermometers, blood pressure monitors, SPO2 monitors). Even more specialized medical devices can be included here, of the type used at home not by the general population, but by certain patient categories.

No matter the devices, there are two modes of measurement collection: the automatic and the manual. Automatic measurement collection refers to having the measurement device integrated with the data collection system, the measurements flowing from the device into the RWD collection system in an unattended manner. Manual measurement collection refers to having the study participant reading out the measurement from the device and manually reporting the measurement to the RWD collection system.

The automatic is clearly the preferred option, the only one when the measurements have high volume and/or frequency. It minimizes study participants' burden and data entry errors. It is of this option only that "measurement from devices" refers to. Devices can be integrated with RWD capturing systems in two ways: 1) by employing the devices manufacturers' provided API, or 2) through SDK. To understand the difference, it is important to understand the flow of information from the device.

All devices have a short-range communication capability, which is almost always Bluetooth Low Energy (BLE). This necessitates the use of another device that collects the information, called a controller, base station, edge node or bridge depending on the manufacturer. In case of domotics devices being fixed in some location, this controller is a fixed device, located somewhere in the home. In the case of wearables, the most natural choice is to use the mobile phone that is mostly around the study participant as a controller, in which case the controller is some software installed on the phone. The controller receives the information from the device(s) it is paired with using the BLE protocol of its communication module. It then transmits this information to the servers of the device manufacturers using WLAN. The study participants control the devices using SW on their mobile phones and/or on the web. They also view (and control) their data using the same apps. The process is depicted in the upper section of Fig. 6.1, the third-party device section.

Since the study participants need to be more in control of their data and the advent of GDPR, device manufacturers typically offer means to the study participants to get their data, not just view it. Data exports in files have long been available, but recently there are more automated options, allowing the study participants to get their data in online ways. APIs and/or SDKs are being offered so that requests are forwarded by third-party SW systems to get the data. The study participants only need to authorize these third-party SW systems to collect the data on their behalf.

When a device manufacturer offers an API to get data, then the data is captured from the cloud platform of the integrated device. The API offers endpoints to be used by an authenticated entity to get data they are authorized for. When the entity is a data collection system, the API endpoints are called by the data collection cloud platform. Depending on the manufacturer, the API can passively wait to be utilized to offer the requested data, or can notify the data collection system of the availability of new data to be collected. The data loop back to the study participant is closed by one of the tasks of the mobile app of the data collection system: It utilizes another

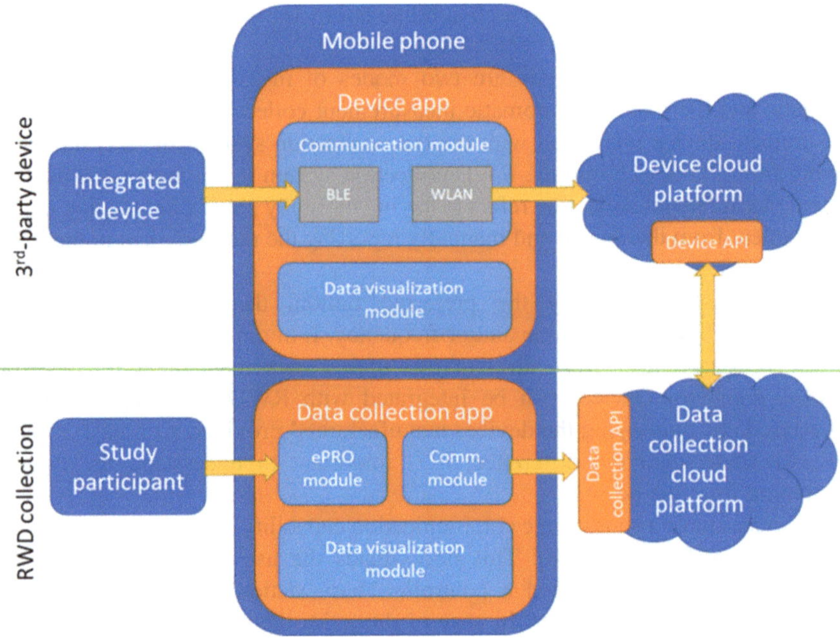

Fig. 6.1 RWD collection via the device API

API, that of the data collection system, to get the integrated device data and visualize it alongside the rest of the secondary data it collects.

While the usage of a third-party API is very easy to support by the data collection system, the downside of the API approach is that many SW entities are involved (the two apps and the two cloud platforms) and the round trip of the data that travels from the device to the mobile phone, to two different cloud platforms back to the mobile phone, is very long. The device integration via the provided API is presented in Fig. 6.1.

When a device manufacturer offers an SDK, then the data collection app receives the data from a local source, without involving the device cloud platform. The SDK offers direct access to the device, effectively sending the collected data to the data collection app of the mobile phone, as depicted in Fig. 6.2.

A variant of the SDK access is when the devices being integrated are the actual sensors on the mobile phone. In this case, the SDK used is the phone's SDK, which is always well known and documented. On Android devices a data collection service is programmed within the data collection app. On iPhone devices, the Apple Health Kit is used to access the data from Apple Health, the app that collects all health and wellness sensor data of the phone or other integrated devices. Similar solutions to Apple Health exist from all the big Android players (e.g., Samsung Health) but the Android manufacturer segmentation does not allow a data collection system to interface with one such app and get data from a significant portion of the possible study participants. Apart from Android manufacturer segmentation, another

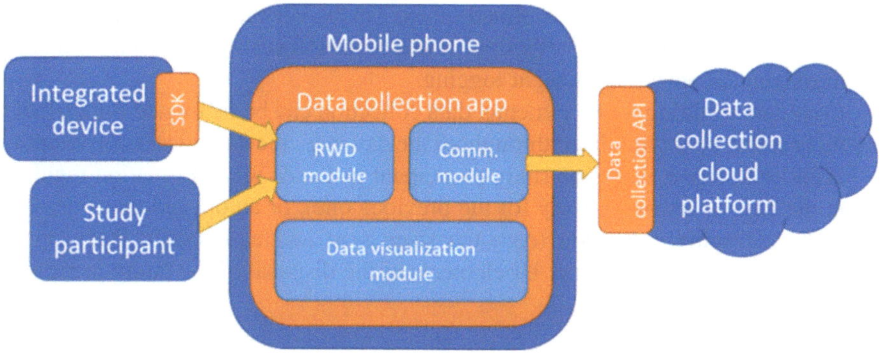

Fig. 6.2 RWD collection by direct access to the device via its SDK

usability issue arises. For the mobile phone sensors to be used, the mobile phone must be used as a wearable, always in the pocket of the study participant, which is clearly not feasible for most people. Additionally, the location of measuring (i.e., the pocket) is of influence in the accelerometer-based estimation of a person's energy expenditure throughout the day. When using the mobile phone as a measurement device, this location of wearing is not always known, and not always the same. Wrist-worn devices—especially when it is known whether the user is wearing it on the dominant or non-dominant hand, and therefore preferred.

Obviously, the easiest integration method is the API, but the most open to the needs of the data collection system is the SDK that allows direct access to a device. Unfortunately, the device measurement method is not for the data collection system to select. Almost all device manufacturers offer API access to the collected data (the manufacturer maintains ownership of the data), while only a handful offer an SDK, and those that do usually represent more experimental and less commercial devices. Also, API usage is usually free, while SDK access is reserved for very large clients of the device manufacturers, or comes with a very high cost. Finally, device manufacturers can apply usage constraints in both modes of access, limiting the data attributes offered, the temporal granularity, the number of times data can be requested, or the volume of transferred data. All these aspects need to be taken into account when selecting the device to integrate.

Automated measurements cannot cover all aspects of RWD. Study participants need to report on outcomes and their experiences, quantifying measures about them: the patient-reported outcome measures (PROMs) and patient-reported experience measures (PREMs). Traditionally the collection of the reports had been a manual process involving pen and paper, but now ePRO systems are facilitating this process in three important ways:

- Scheduled questionnaires can be addressed to study participants and the answers can be collected and processed much easier with the online ePRO tools.
- Spontaneous answers can be included with study participants reporting on outcomes and experiences at the moment they happen, without needing to collect

them and report them at predefined times. This is very important for event-type outcomes like a symptom, affliction, or discomfort. It is also important for manual measurement entry, like weight logging.
- Continuous editing for incremental data collection can be facilitated for attributes that need to be accumulated throughout the day (e.g., the daily water consumption is easier to report when the study participant just adds this information regularly in the day, or retrospectively for the previous one), or automated measurements that need some correction (e.g., the sleep start and end times). Both can be achieved using widgets, i.e. UI elements at the disposal of the study participant for continuous data entry.

6.2.3 Healthentia for RWD Collection and Management

The data collection is facilitated by Healthentia, an e-clinical platform. The platform provides secure, persistent data storage and role-based, GDPR-compliant access. It collects the data from the mobile applications of all study participants, facilitating smart services such as risk assessment, and providing both original and processed information to the mobile and portal applications for visualization. The high-level architecture of the platform is shown in Fig. 6.3. This is a layered architecture, comprising the API, data management, core functionalities, study management, and services layers.

The Healthentia API layer provides the means to connect Healthentia with the outside world. Data importing (IoT integration) and exporting (eCRF integration), as well as internal communication with the portal and mobile apps is facilitated through it. The low-level operations on the data are hosted in the data management layer. The data handling functionalities are utilized by the API I/O endpoints. The smart data generation functionalities drive synthetic data generation. The Healthentia core layer comprises high-order functionalities on top of the data, like role-based control, participant management, participants' reports management, and ML functionalities. The study layer allows managing of the studies, the entities in which HCP, participants and their data are organized. They can be formal clinical studies, or informal ones managed by pharmaceutical companies, hospitals, or research centers. Finally, the services layer implements the necessary functionalities of the web portal and the mobile application. These include dashboard services in both portal and mobile apps, e-diary for reporting by study participants and other services like treatment adherence, teleconsultation, and smart services.

The Healthentia mobile application (Fig. 6.4) enables data collection at the study participant end. Measurements are obtained from IoT devices, third-party mobile services, or a proprietary sensing service. Study participants' reports are obtained via answering questionnaires that are either regularly pushed to the study participants' phones or are accessed on demand by the study participants themselves. Both the measured and reported data are displayed to the study participants, together with any insights offered by the smart services of the platform.

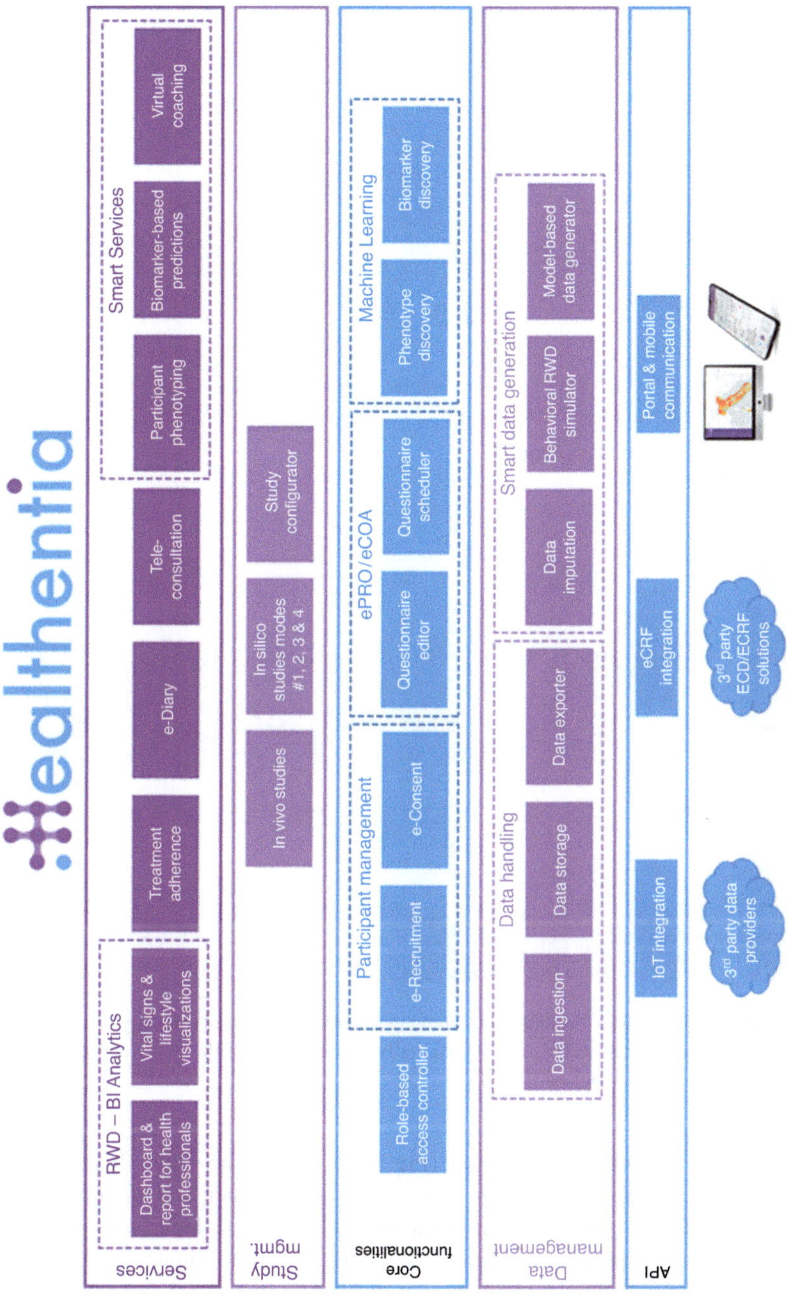

Fig. 6.3 Healthentia high-level architecture

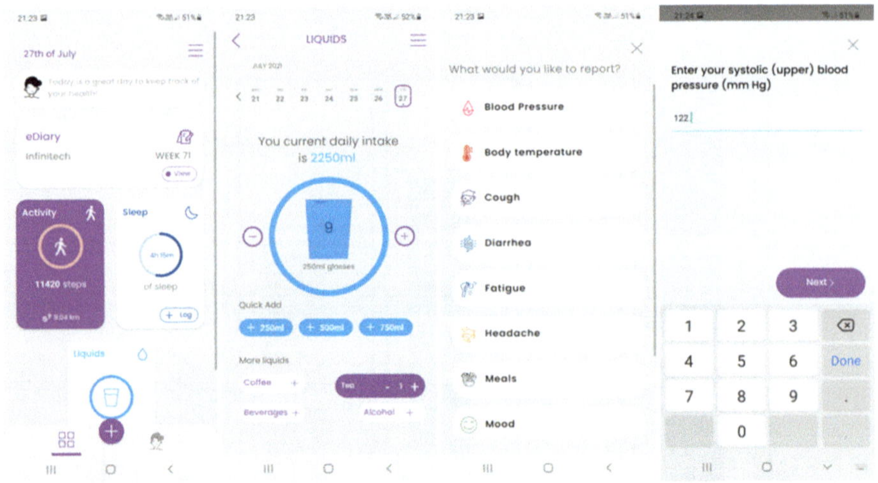

Fig. 6.4 Healthentia mobile application

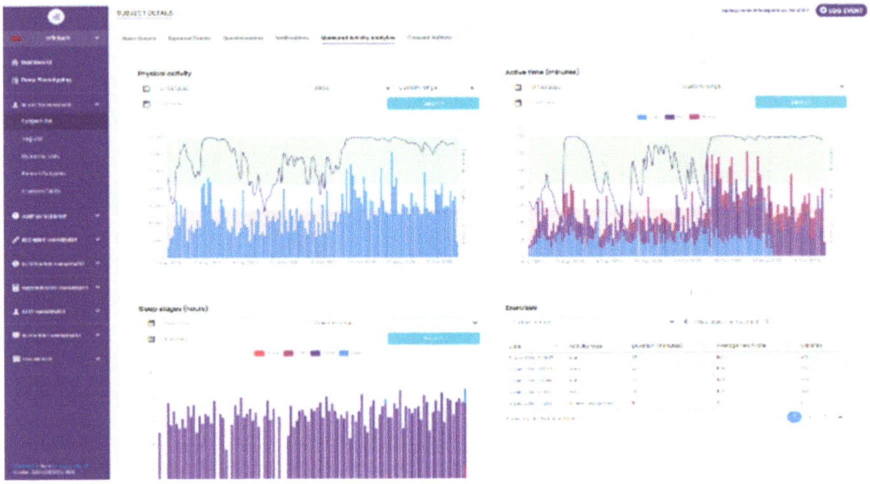

Fig. 6.5 Healthentia portal application—viewing measurements and creating questionnaires. Adapted from (Pnevmatikakis et al. 2021), originally published under the terms and conditions of the Creative Commons Attribution (CC BY)

The Healthentia portal application (Fig. 6.5) targets the HCP. It provides an overview of the study participants of each clinical partner and details for each study participant. Both overview and details include analytics based on the collected data and the risk assessment insights. It also facilitates managing the studies, e.g. provides a questionnaire management system to determine the types of self-assessments and reports provided by the participants.

6.3 AI in Clinical Research and Digital Therapeutics

The Healthentia DTx functionalities are provided by the smart services shown in Fig. 6.6. The depicted user-interface components (the Mobile App and the Web Portal) have already been discussed in Sect. 6.2.3. The central component—the User Data Service—handles storage of and access to all collected RWD. The main digital therapeutic functionalities are provided by the five surrounding services: Real-World Data Collection Service (already discussed in Sect. 6.2.3), Learning- and Intelligent Decision Services, Clinical Pathway Services, and the Virtual Coaching Services.

6.3.1 AI for Patients' Selection and Adherence

Patient recruitment accounts on average for one third of the clinical study duration, therefore, it is highly important to identify the most appropriate candidates in the cohort. Harrer (2015) demonstrated that the assessment of suitability of subjects can be assisted by using patient-specific diagnostic genome-to-exposome profiling for determining whether biomarkers are sufficiently strongly represented in the patient profile. In addition, AI can be used to enhance patient cohort selection through one or more of the following means identified in (FDA 2019). The first recommendation of the FDA is to reduce the population heterogeneity. Furthermore, clinical investigators using AI mechanics for subjects' recruitment can choose patients that are more likely to have measurable clinical endpoints, the so-called *prognostic enrichment*. Finally, clinical investigators can identify through AI a study population that is more capable of responding to a treatment, the so-called *predictive enrichment*.

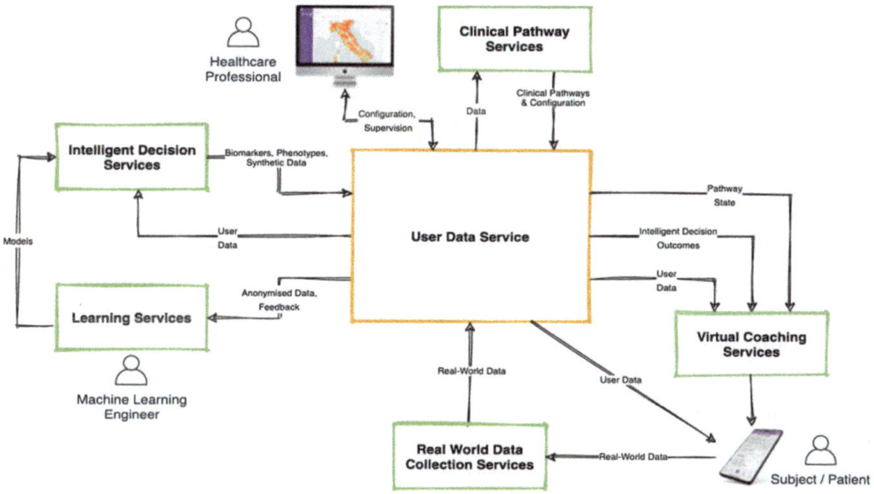

Fig. 6.6 Healthentia smart services for DTx

The dropout rate across clinical trials varies based on the pathology and other study parameters, however 30% or more is a very likely scenario for many of the cases. In general, the patients' drop out is linked with lack of adherence and compliance to trial protocol. A linear increase of the non-adherence rate results in an exponential increase in additional patients required to be recruited to maintain the statistical power of the outcomes (Harrer et al. 2019). Processing of RWD from various sources, including IoTs and wearables, as well as legacy big data with AI can result in services for the PIs to improve patients' adherence. Such services include: (a) Patients' diary; (b) Medication reminders; (c) Virtual Coaching; and (d) Risk stratification with automatic alerts.

6.3.2 Discovering Digital Composite Biomarkers and Phenotypes

Healthentia DTx considers RWD as the primary source of patient information, and it uncovers second-order information through close collaboration between its learning and intelligent decision services, as described in (op den Akker et al. 2021). The system employs signal processing to obtain the simpler forms of second-order information, such as long- and short-time averages and trends of the collected data. More complex processes using machine learning are also utilized to extract second-order information. Firstly, the system predicts important clinical outcomes from composite biomarkers. Then, it analyzes the outcome decisions to model the attributes of patients that have the most influence on the decisions. Finally, it groups patients into phenotypes based on their unique characteristics.

Biomarkers are quantities that describe a clinical outcome or a stage of disease (Coravos et al. 2019), and they can be direct, indirect, or composite. Direct biomarkers are values of a single quantity that allow for the diagnosis of a disease outcome or stage, and they are typically measured in a clinical setting. Indirect biomarkers, on the other hand, are values of quantities that are indirectly associated with a factor or product of the disease and are highly correlated with it. Indirect biomarkers can be obtained using ubiquitous devices in everyday life settings. Indirect biomarkers can also be combined into composite biomarkers (Kovalchick et al. 2017), which are non-linear combinations of multiple quantities into a single value that characterizes an outcome. Composite biomarkers are better at correlating with outcomes than single quantities alone. In some cases, the combination can be done analytically, such as with the Body Mass Index (Garrow and Webster 1985), which is a simple composite biomarker used to measure obesity. Body mass and height are nonlinearly combined in a simple equation to yield the composite biomarker. Finally, biomarkers can be digital if they are collected using sensors and software or hardware computational tools.

Analytic combinations are the exception when it comes to composite biomarkers. In most cases, predictive models are used to perform non-linear combinations. These models are learnt using machine learning algorithms (Geron 2019), which can be classifiers or regressors. Classifiers are used when biomarker values are finite and discrete, while regressors are used when they are continuous. For example, Random

Forests (Breiman 2001) can be employed to learn a predictive model forming a binary classifier that predicts whether the systolic blood pressure of patients is expected to improve or worsen.

The process of learning the models is done offline and involves various experts. Initially, domain experts are consulted to determine the outcomes of interest and the direct or indirect biomarkers that are expected to affect them. These biomarkers are used as attributes to be combined nonlinearly by the predictive model. The attributes' input vector is formed, and the outcomes to be predicted from the output vector. When faced with multiple outcomes, then a machine learning engineer can opt either for one predictive model with multiple outputs, or for multiple models, each with a single output. The process of model learning involves collecting data through the user data services and processing it. The processing aims at the correction of measurement errors, the imputation of missing values, and the anonymization (Jaidan et al. 2019). The processed and anonymized data are then split into three sets—training, validation, and testing sets. Candidate models are trained using the training set, and their performance is evaluated by tuning their hyperparameters with the validation set. The final model is trained using the optimal parameters with the combination of the training and validation sets, while the performance is evaluated with the testing set. Finally, the resulting model is stored using the user data services.

The Healthentia DTx's Intelligent Decision Services use the learned models to generate composite biomarkers. These services include timed components that periodically request new data from the user data services for the patients, enumerate all the attributes of the predictive models, and evaluate the composite biomarkers. The resulting composite biomarkers are then stored as second-order data for the patients using the user data services. It is important to note that unlike model learning, the data used to evaluate the composite biomarkers is not anonymized but is processed for measurement errors and missing values imputation.

After the composite biomarkers are evaluated, they are analyzed to identify the most important single quantities, which are the predictive model attributes that have the greatest influence on the predicted outcome. This analysis is done using *SHapley Additive exPlanations* (SHAP) analysis (Lundberg et al. 2020), which for the given decision and patient, assigns a SHAP coefficient to each model attribute value indicating its significane in this decision.

The Healthentia DTx utilizes the SHAP coefficients obtained from the analysis of composite biomarkers in its Learning Services. By accumulating the SHAP coefficients across patients, the system can identify attributes that have minimal impact on the overall decisions. This allows for the model complexity to be reduced without sacrificing performance, or even improving it, especially when working with small training sets. The individual SHAP coefficients, on the other hand, are used in an online learning process facilitated by the Learning Services. The Healthentia DTx generates models for each attribute and patient, providing insight into the attribute's significance for a particular outcome. These models are updated continuously as new decisions are made, and the SHAP coefficients are analyzed to provide ongoing optimization of the learning process.

Phenotypes are models that describe the appearance of groups of patients who share similar characteristics. These characteristics are identified by analyzing the most impactful attributes and second-order data, such as time averages, trends, and composite biomarkers. These attributes form the phenotype clustering space, and their values form the data vectors that are used to cluster patients. Clustering is performed using unsupervised machine learning algorithms, such as k-means or hierarchical clustering (Theodoridis and Koutroumbas 2008), to uncover the underlying structure in the data. Each patient's snapshot is then assigned to a cluster, and all of their data attributes are considered to belong to that cluster. Once a patient's snapshot is clustered, certain attributes are selected as representative of the patient within the particular therapeutic area, forming the phenotype modeling space. Gaussian Mixture Models are then used to describe each cluster based on the attributes in the phenotype modeling space. These models are trained using the Expectation-Maximization algorithm (Theodoridis and Koutroumbas 2008). The aim of creating phenotypes is to group patients based on their characteristics, which can help clinicians better understand the disease and develop more personalized treatment plans.

The intelligent decision services use the learnt phenotype models for patient phenotyping, which involves assigning each patient snapshot in time to one of the phenotype models. This allows each patient to be considered as exhibiting a particular phenotype for a time interval, with the weight of that phenotype being updated every time a new snapshot of the patient is clustered. If the weight of a particular phenotype exceeds those of all others, the patient is considered to exhibit that phenotype for the time interval. While transitions to different phenotypes are not frequent, they are considered important and are monitored, as they often correspond to a transition to a new disease stage or new behavioral habits.

6.3.3 Virtual Coaching for DTx

Virtual Coaching in Healthentia is the set of services that can aid the end-users—whether they are patients or citizens—in setting and reaching their own health-related goals. The Healthentia Virtual Coaching platform is a set of technological tools that allows clinicians to define how and when to deliver coaching content to patients. In the Healthentia approach, we focus on virtual coaching as delivered through conversational agents that interact with end-users using natural language. Such conversational agents (or virtual agents, or virtual coaches) are receiving growing interest in the healthcare domain (e.g., Provoost et al. 2017; Laranjo et al. 2018); however, when applied to the field of chronic conditions, the literature is still scarce (Schachner et al. 2020). We first shortly discuss the principles of our virtual coaching approach, and then present the technology framework on which the virtual coaching platform is built.

The ultimate aim of virtual coaching in our supported use cases is to accompany patients throughout the use of Healthentia, and to assist them in making positive changes in their health-related behavior. Behavior change is difficult, especially

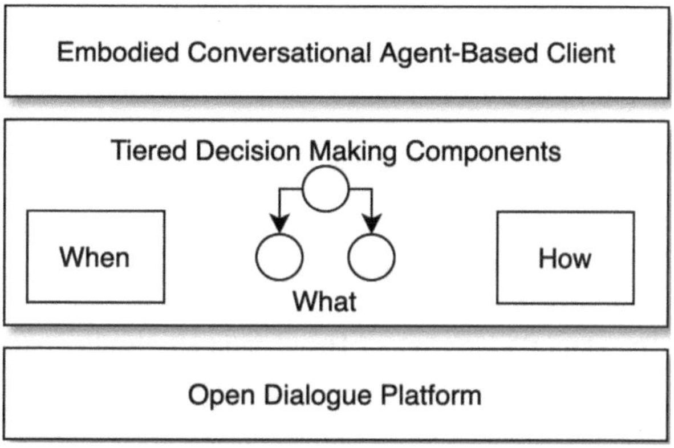

Fig. 6.7 High-level approach to Virtual Coaching in Healthentia, based on the principles of using natural language, tailoring and explainable AI

when it comes to internalizing change in the long-term (Bouton 2014). As such, we believe that supporting an individual in achieving this change cannot be achieved by simple nudges, reminders, or visualizations of performance alone. The support that is needed is complex and deep, and thus the medium of delivery should support this rich type of interaction. This is our first principle, and the core motivator to focus on virtual coaching based on natural language conversations between the coach and user. A *conversation*, especially when combined with visual aids such as graphs, is a very rich medium for conveying information from a computer system to an end-user.

Every individual end-user is different, and this includes the individual's preferences when it comes to receiving health-related coaching. Tailoring is the mechanism of adjusting the timing, the intention, the content, or the representation of communication to the user of an eHealth system (op den Akker et al. 2014). Within an eHealth application, the system's communication may be tailored to the user based on information from the user's profile. In Healthentia, the user's profile is rich due to the focus on real-world data collection. We thus have motive and opportunity to tailor our virtual coaching to the end-users. Tailoring is the second principle underlying our virtual coaching approach, meaning that every conversation with the user should be carefully adjusted to the individual and his or her context.

When building systems that automatically tailor elements of conversation to the user, such as *what to say* or *how to say it*, or even *when to say it*, certain decision-making components must be integrated into the system. These decision-making components are artificially intelligent: they operate in a fluid environment (the user and his/her context) and reason about certain actions to take (e.g., starting a conversation about healthy eating). Such decisions, although not immediately dictating life or death, are impacting someone's life and health. The final design principle is that it must be understandable at all times, why the virtual coach is delivers specific messages, by using expertly designed scripted dialogues as a basis, and Explainable AI (Goebel et al. 2018) as a guiding principle for the development of decision-making components.

Combined, our high-level virtual coaching approach is depicted in Fig. 6.7. The basis is an open dialogue platform that allows us to use expert knowledge to script coaching dialogue content for the various use cases served by the Healthentia platform. Then, a tiered approach to automated decision-making is applied, whereby the decision on when to say what and how to deliver it to the user is split up into a logical sequence of simpler decisions. Finally, an embodied conversational agent is used to deliver the coaching dialogue to the end-users of the platform.

6.3.4 Clinical Pathways

The clinical pathways of Healthentia encapsulate expert knowledge on the digital therapy plan to be offered to patients. This knowledge is turned into a set of predefined rules enabling actions and transitions to other branches of the pathway. A patient enters the pathway for their pathology at any of the defined entry nodes, their treatment being orchestrated by the pathway and their progress along it, until they reach the exit node.

Healthentia employs an offline visual pathway modeler and a pathway execution service. The modeler is offline, in the sense that it is used at the design phase of the digital therapy plan. It is implemented around Business Process Modeling Notation (BPMN), offering the visual tools to define pathways for different chronic diseases. The modeler is built having the healthcare professionals as its intended users. Their expertise, clinical practice, and medical knowledge will be encapsulated into the pathways they design for different pathologies.

Patients are initially assigned to any of the start states in the pathway, whereupon the pathway execution service maintains their state. The pathway execution service operates during the digital therapy execution phase, i.e., is an online service. It is implemented around Camunda that updates patient states and acts upon them.

6.4 Patient Engagement

Patient engagement in the last few years is becoming a key metric in the field of Life Sciences and Healthcare. It improves participation in clinical studies and ensures cost effectiveness. In the standard clinical care evidence also shows that when patients are actively engaged in their health management by gathering information of their symptoms, disease evolution, and medication adherence, it can lead to better patient's outcomes, faster recovery, reduced hospitalizations or readmissions, and as an extent a lower cost of care in general.

In the case of Healthentia, its role is to accompany patients in their disease journey whether they are part of a clinical study, or part of a clinical care program. It allows the possibility to gather lifestyle and disease related information and offers valuable insights to clinical teams and study sponsors. In patient programs, different strategies are used to address patient engagement, as follows:

- *Identify & Group patients*. In order to identify patients at higher risk or with special needs we use tagging, alerts, and clustering mechanisms. By understanding the different needs in terms of care, risks, or communication, the clinical team can better address them.

- *Patient Education*. It is essential for patients to understand their disease and therapy in order to be engaged, comply with a care plan, and see improvements in their outcomes. When clinicians educate their patients and give them personalized content and guidelines then they are more likely to see how their engagement efforts will result into their benefit.
- *Co-design & Feedback*. To ensure patient engagement in a program or study it is essential in the design process to include patients and share decision-making. It is a co-design process usually with focus groups with the participation of all stakeholders in collecting user requirements and objectives in order to address them with suitable features and processes. Thus, patients feel co-ownership and are more motivated to comply and advocate to others. An important aspect of co-design is the user evaluation for technology acceptance and user experience by patients, using established protocols, such as TAM+ and UEQ. The Technology Acceptance Model (TAM+) questionnaire consists of 34 items, divided into 7 domains: enjoyment, aesthetics, control, trust in technology, perceived usefulness, ease of use, and intention to use (Davis 1989). The User Experience Questionnaire (UEQ) is used to analyze the system's attractiveness, perspicuity, efficiency, dependability, stimulation, and novelty (Laugwitz et al. 2008). Attractiveness is defined as a pure valence dimension. Perspicuity, Efficiency, and Dependability are pragmatic quality aspects which are goal-directed, while Stimulation and Novelty are hedonic quality aspects that are not goal-directed. UEQ offers a benchmark to classify a product into 5 categories of the 6 scales: excellent, good, above average, below average, and bad (UEQ).
- *Continuity of Care & Accessibility*. Healthentia as a virtual solution enables patient engagement as it helps a clinical team create personalized care plans but also offers a continuity of care outside the hospital by collecting and monitoring patient data in real-time. At the same time patients from the comfort of their home can have access to a care team and a personalized and consistent care plan and support throughout their disease journey, which can lead to better outcomes for healthcare organizations and patients alike.

6.5 Conclusions

This chapter described the use of big data and AI technologies to improve the current status in clinical studies and DTx. Both big data and AI technologies are widely used in several industries and especially those that are not highly regulated can demonstrate big impact on key performance indicators, as well as multiplying the benefits for the end-user. Healthcare is the perfect industry to demonstrate improvement by applying such technologies, also including the social dimension due to the patient-centric focus; however, the sensitivity of collected data and the regulatory framework have delayed these outcomes for around a decade, compared to other industries. Nevertheless, the natural evolution of the penetration of such technologies in the healthcare domain and other events, such as the COVID-19 pandemic that accelerated their application, have created new horizons and are expected to contribute to the healthcare system, by improving the efficacy of drugs, the R&D duration and expenditure, as well as enabling the use of digital ingredients in the form of DTx applications.

References

op den Akker H, Jones VM, Hermens HJ (2014) Tailoring real-time physical activity coaching systems: a literature survey and model. User Model User-Adap Inter 24(5):351–392. https://doi.org/10.1007/s11257-014-9146-y

op den Akker H, Cabrita M, Pnevmatikakis A (2021) Digital therapeutics: virtual coaching powered by artificial intelligence on real-world data. Front Comput Sci 3

Bouton ME (2014) Why behavior change is difficult to sustain. Prev Med 68:29–36. https://doi.org/10.1016/j.ypmed.2014.06.010

Breiman L (2001) Random forests. Mach Learn 45:5–32

Coravos A, Khozin S, Mandl KD (2019) Developing and adopting safe and effective digital biomarkers to improve patient outcomes. NPJ Digital Med 2(14)

Davis FD (1989) Perceived usefulness, perceived ease of use, and user acceptance of information technology. MIS Q 13(3):319–340. https://doi.org/10.2307/249008

FDA (2017) Use of real-world evidence to support regulatory decision-making for medical devices: guidance for industry and Food and Drug Administration staff. Source: https://www.fda.gov/downloads/medicaldevices/deviceregulationandguidance/guidancedocuments/ucm513027.pdf Last consultation: 23/08/2022

FDA (2019) Enrichment strategies for clinical trials to support determination of effectiveness of human drugs and biological products. Guidance for industry. Source: https://www.fda.gov/regulatory-information/search-fda-guidance-documents/enrichment-strategies-clinical-trials-support-approval-human-drugs-and-biological-products Last consultation: 23/08/2022

Fogel DB (2018) Factors associated with clinical trials that fail and opportunities for improving the likelihood of success: a review. Contemp Clin Trials Commun 11:156

Garrow JS, Webster J (1985) Quetelet's index (W/H2) as a measure of fatness. Int J Obes 9(2): 147–153

Geron A (2019) Hands-on machine learning with Scikit-Learn, Keras and TensorFlow: concepts, tools, and techniques to build intelligent systems. O'Reilly

Goebel R, Chander A, Holzinger K et al (2018) Explainable AI: the new 42? In: Holzinger A, Kieseberg P, Tjoa A, Weippl E (eds) Machine learning and knowledge extraction, lecture notes in computer science. Springer, Cham, p 11015. https://doi.org/10.1007/978-3-319-99740-7_21

Grey M (2017) Lifestyle determinants of health: isn't it all about genes and environment? Nurs Outl 65(5):501–515

Hämäläinen H, Malkamäki S, Räsänen I, et al (2020) Towards trustworthy health data ecosystems. Sitra. Source: https://www.sitra.fi/en/publications/towards-trustworthy-health-data-ecosystems/ Last consultation: 23/08/2022

Harrer S (2015) Measuring life: sensors and analytics for precision medicine. Microtechnologies for the New Millennium https://doi.org/10.1117/12.2178956

Harrer S, Shah P, Antony B, Hu J (2019) Artificial intelligence for clinical trial design. Trends Pharmacol Sci 40(8):577–591. https://doi.org/10.1016/j.tips.2019.05.005

Henegan C, Goldacre B, Mahtani KR (2017) Why clinical trials outcomes fail to translate into benefits for patients. Trials 18:22

Jaidan DN, Carrere M, Chemli Z, Poisvert R (2019) Data anonymization for privacy aware machine learning. In Nicosia G, Pardalos P, Umeton R, Giuffrida G, Sciacca V (eds), Machine Learning, Optimization, and Data Science. LOD 2019. Lect Notes Comput Sci, 11943:725–37

Kovalchick C, Sirkar R, Regele OB et al (2017) Can composite digital monitoring biomarkers come of age? A framework for utilization. J Clin Transl Sci 1(6):373–380

Kyriazakos S, Pnevmatikakis A, Cesario A et al (2021) Discovering composite lifestyle biomarkers with artificial intelligence from clinical studies to enable smart eHealth and digital therapeutic services. Front Digit Health 3:648190

Laranjo L, Dunn AG, Tong HL et al (2018) Conversational agents in healthcare: a systematic review. JAIMA 25(9):1248–1258

Laugwitz B, Held T, Schrepp M (2008) Construction and evaluation of a user experience questionnaire. In: Holzinger A (ed) HCI and Usability for Education and Work, USAB 2008, Lecture Notes in Computer Science, vol 5298. Springer, Berlin, Heidelberg. https://doi.org/10.1007/978-3-540-89350-9_6

Lundberg SM, Erion G, Chen H et al (2020) From local explanations to global understanding with explainable AI for trees. Nat Mach Intell 2:56–67

Nivel DJ, McCormick TJ, Strause SE et al (2018) Reproducibility of clinical research in critical care: a scoping review. BMC Med 16:26

OECD (2021) Better Life Index. Source: http://www.oecdbetterlifeindex.org Last consultation: 23/08/2022

Pnevmatikakis A, Kanavos S, Matikas G et al (2021) Risk assessment for personalized health insurance based on real-world data. Risks 9

Provoost S, Lau HM, Ruwaard J, Riper H (2017) Embodied conversational agents in clinical psychology: a scoping review. J Med Internet Res 9(19):e151

Schachner T, Keller R, Wangenheim F (2020) Artificial intelligence-based conversational agents for chronic conditions: systematic literature review. J Med Internet Res 22(9):e20701

Sinipuru J (2021) “Statement & vision”, Next Generation IoT thematic workshop, New concepts in smart health and care. Source: https://www.ngiot.eu/event/ngiot-thematic-workshop-health-and-care/ Last consultation: 23/08/2022

Theodoridis S, Koutroumbas K (2008) Pattern recognition. Academic Press Inc, Orlando (FL)

Systems Pharmacology for Immunotherapy: A Network Analysis of Signaling Pathways in Multiple Sclerosis for Combination Therapy Development

7

Pablo Villoslada, Keith Kennedy, and Jordi Garcia-Ojalvo

Abstract

Developing combination therapies aimed to target different biological processes is a sought-after goal for treating complex diseases. However, there are drawbacks due to the number of combinations of different targets, doses, therapeutic regimens, and individual patient responses, making this a severe combinatorial problem. In the case of immunotherapies, this is further hampered due to the complexity of immune responses and significant individual variability. In Multiple Sclerosis (MS), for instance, physicians have considerable difficulties selecting the proper treatment for each patient, even though there are more than 20 immunotherapies available. Consequently, the disease is far from being controlled. Systems Pharmacology, specifically network modeling of signaling pathways, provides a framework to address such complexity at both the research and clinical levels. Network analysis can help us search for new targets and combination therapies to achieve better control of MS. By examining signaling networks in immune cells combining in vitro assays, proteomic analysis, and logic network modeling, network models can predict response to therapy at the individual patient level. A druggability algorithm identifies the network nodes (proteins) that will qualify as therapeutic targets in combination with approved drugs. These efforts are showing that network pharmacology can be a valuable

P. Villoslada (✉)
Dynamical Systems Biology Group, Department of Experimental and Health Sciences, Pompeu Fabra University, Barcelona, Spain

Department of Psychiatry, Stanford University, Stanford, USA

Institut d'Investigacions Biomediques August Pi Sunyer, Barcelona, Spain
e-mail: pablo.villoslada@upf.edu

K. Kennedy · J. Garcia-Ojalvo
Dynamical Systems Biology Group, Department of Experimental and Health Sciences, Pompeu Fabra University, Barcelona, Spain

A. Cesario et al. (eds.), *Personalized Medicine Meets Artificial Intelligence*,
https://doi.org/10.1007/978-3-031-32614-1_7

tool for supporting drug discovery efforts, and for guiding clinicians in testing the most promising drug combinations in immunotherapy.

Keywords

Systems pharmacology · Network pharmacology · Immunotherapy · Combination therapy · Multiple sclerosis · Personalized medicine

7.1 Introduction

Complex diseases involve the participation of many biological pathways leading to similar cell phenotypes that generate tissue damage, but with significant inter-patient heterogeneity (Villoslada et al. 2009). From a therapeutic perspective, this implies that the traditional model successfully used in the twentieth century of "one gene/protein—one disease—one drug" is not valid for curing or controlling most of these diseases. A possible solution to overcome this limitation is systems pharmacology, which aims to provide a new, dynamic method for drug discovery and therapeutics (Butcher 2005; Pujol et al. 2010; Trame et al. 2016).

Complex diseases often operate by perturbing the signal transduction machinery among pathways and cells. Hence, the treatment of complex diseases such as cancer, cardiovascular, immunological, or brain disorders focuses on modifying sequences of molecular reactions (pathways) to stop the progression of the disease. As a result, kinases (enzymes responsible for the phosphorylation of proteins transmitting the information from membrane receptors to the nucleus) have become a primary focus of many illnesses (Owens 2007). However, using therapies acting upon single molecular targets to treat complex diseases has seen modest success. There is hope that using multiple treatments that target simultaneously various molecules and/or pathways will allow more precise manipulation of the underlying mechanisms of the disease (Jia et al. 2009; Cully 2015). It has been nevertheless difficult to date to determine how the combination of multiple therapies will modify a complex signaling network, avoiding unwanted secondary effects such as drug-induced interference between the pathways (Bozic et al. 2013; Klinger et al. 2013).

As an additional source of complexity, patients differ in the characteristics of their disease, regarding both the symptoms manifested and their underlying molecular signaling networks. This makes the efficacy of the treatment difficult to predict in a clinical setting. Finally, the combinatorial nature of composite therapies in terms of the number of targets, drugs, doses, and therapeutic regimen implies many experiments and associated costs, preventing a complete analysis for all possible alternatives. Altogether, the field of combination therapy has not reached its full potential, and modeling-based studies may be the key to characterize the effect of drug combinations at the molecular level, allowing prediction of both efficacy and reduction of off-target effects (Lee et al. 2012; Bozic et al. 2013; Korkut et al. 2015).

Systems Biology, by considering signaling networks instead of collections of parallel pathways and applying modeling techniques to discover proper combinations of treatments, may provide a new way to approach this question (Jia et al. 2009; Villoslada et al. 2009). Understanding the biochemical mechanisms underlying the network allows predicting how cells respond to stimuli, either naturally from the environment or due to drug effects. Much work has already been done to identify the signaling cascades among cells, and systems biology can bolster how exactly they interact (Kholodenko et al. 2010; Kolch et al. 2015; Davis et al. 2017). Furthermore, pathway changes can be associated with disease outcomes. Ultimately, the efficacy and safety of combination therapies can be expected to decrease the number of drugs to be tested in vivo and clinical trials. Mathematical modeling of signaling networks has been used to identify targets for therapies and specific molecular aspects of the disease, such as cell surface receptors or intracellular molecules, by training models to monitor responses of key pathway components in combinatorial in vitro assays using inhibitors and activators (Palacios et al. 2007; Klinger et al. 2013; Korkut et al. 2015). Within that context, logic models are a helpful tool to determine the makeup of the signaling network underlying a disease and how it would respond to a given drug. Computationally, these types of models are useful because they can describe large networks with a low number of parameters (Bernardo-Faura et al. 2014; Flobak et al., 2015; Eduati et al. 2017; Silverbush et al. 2017).

7.2 Systems Pharmacology for Combination Therapy in Multiple Sclerosis

MS is an autoimmune disease in which the immune system is chronically activated, damaging the central nervous system (CNS) (Ransohoff et al. 2015). The deregulation of the immune system in MS manifests in changes in the activity of blood lymphocytes and monocytes (Ransohoff et al. 2015) and through its association with genetic polymorphisms of immune genes (Sawcer et al. 2011). In addition, several immune-related pathways have been identified as dysregulated in MS, including NfKB, MAPK, and others (Fig. 7.1) (Kotelnikova et al. 2015).

At present, there are more than twenty FDA-approved immunomodulatory drugs for MS, with many others being currently tested in trials. These drugs mainly act by controlling, to a certain degree, the levels of inflammation in patients with MS. However, these treatments can only control the disease to a certain extent. Sometimes treatments are not effective enough, some cause further problems to the immune system, and there is a need for those that can protect or even regenerate neural tissue (Villoslada and Steinman 2020). Combination therapies may surpass these limitations but predicting how patients would react to a specific combination of therapies remains an unresolved challenge (Conway and Cohen 2010; Kieseier and Stuve 2011; Milo and Panitch 2011; Kotelnikova et al. 2015).

Systems pharmacology can be used to predict new combination therapies using signaling networks in primary immune cells obtained from the blood for MS patients, based on the differences between treated MS patients and controls (Bernardo-Faura et al. 2021). The first step of the approach is to build a network

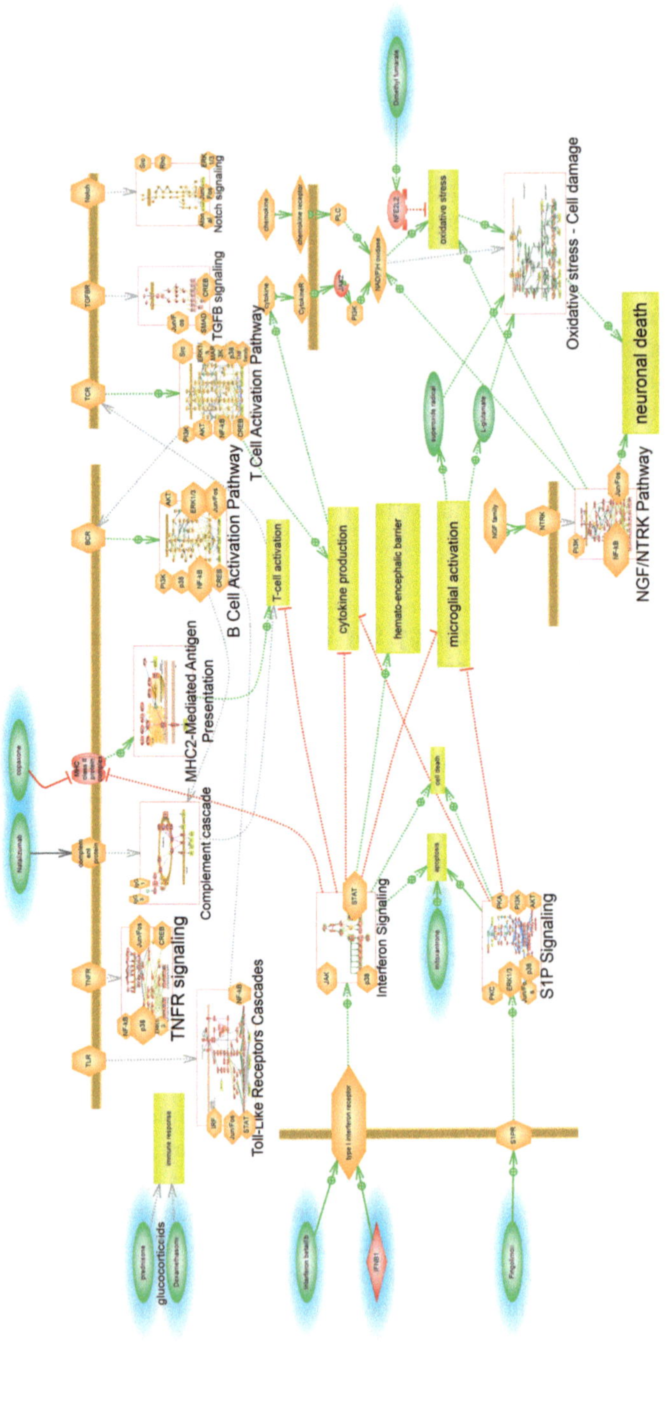

Fig. 7.1 Signaling pathways associated with multiple sclerosis [MS]. The genes and proteins related to MS are displayed in orange, drugs are in green, and the main MS pathways targeted by therapies are in yellow (reproduced with permission from Kotelnikova et al. 2015)

of signaling interactions based on existing literature and databases (Saez-Rodriguez et al. 2009), including the pathways involved in immune and MS signaling (Kotelnikova et al. 2015). Next, Boolean logic models are applied to the signaling network and trained to represent the kinase de/phosphorylation levels in peripheral blood mononuclear cells (PBMCs) of healthy controls and MS patients, after they are perturbed with ligands and drugs. The kinase interactions that the drugs failed to revert to a healthy-like activity level in the *ex vivo* assays become candidates to be targeted by a second drug, with the goal of developing a personalized combination therapy. To guide target identification and combination selection, a score of co-druggability of signaling interactions can be defined according to quantitative differences in network topology among healthy controls and untreated and treated MS patients guides. This network-based approach can predict combination therapies based on current immunomodulatory drugs.

7.2.1 Modeling Signaling Pathways from Ex Vivo Proteomic Assays in MS

A literature-based search is performed as the first step to build the MS-related immune signaling network that connects molecular species based on their biochemical interactions, such as kinase phosphorylation. The network is constructed using omics data, functional assays, and target pathways of MS therapies (Kotelnikova et al. 2015), including interferon response, B and T cell receptor signaling, cellular survival and apoptosis, innate immunity, and multi-drug response genes (Fig. 7.2).

Once the network is built, experimental data from ex vivo phosphoproteomic assays in PBMCs is used to constrain it to the specific case of patients with MS using Boolean logic models (Poussin et al. 2014). Since the outputs of the Boolean gates are binary, the data from the assays need to be normalized. This rigorous method has two aims: first, to transform the raw phosphorylation levels to values between 0 and 1 using a non-linear mapping (Saez-Rodriguez et al. 2009), and second to remove the measurements which did not undergo a significant change in phosphorylation when perturbed. The modeling tool CellNOpt is used to optimize the Boolean model, selecting the model which best reproduces the assay data while avoiding large models with unnecessary edges (Terfve et al. 2012). This optimization is repeated ten times for each patient to ensure that the model is robust. The final model is selected to have the median value (within the relative tolerance of the best solution) for each reaction between proteins.

Individual models for each subgroup of subjects (untreated MS patients, MS patients with treated with each drug, and healthy controls) are merged using the mean of each interaction per subgroup to characterize MS signaling. To confirm that the modeling approach can capture the effect on signal transduction of MS drugs, the Jaccard similarity index is calculated to assess both intra- and inter-patient model similarity, by quantifying the number of interactions in pairs of networks that are equal over the total interactions, after stratification of patients by treatment subgroups. The pathways found to be active for each patient group uncovered the

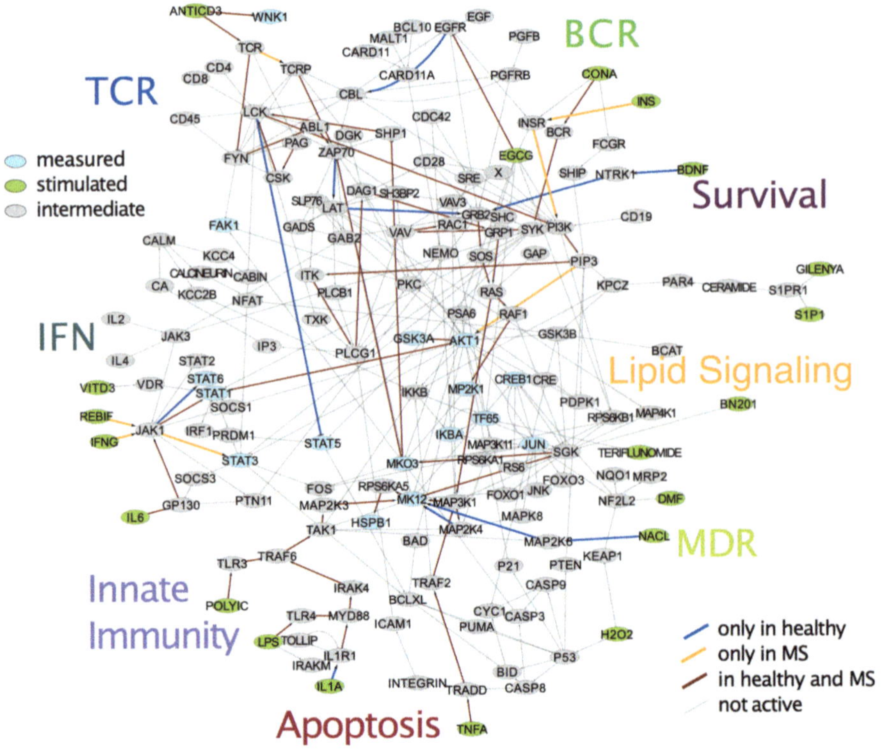

Fig. 7.2 Logic modeling identifies MS-specific signaling pathways. The background network shows the literature-based signaling network for MS. Experimentally activated pathways between healthy controls [HC] and untreated MS patients (MS) are highlighted: controls (blue), untreated MS patients (orange), and both (brown). Gray: Inactive interactions from the MS, immune- and treatment-related network (reproduced with permission from Bernardo-Faura et al. 2021)

signaling activity in PBMCs from MS patients, providing a map of interactions (directed graph of activation or inhibition) for each subgroup (Fig. 7.2). The MS network shows the activation of several pathways in PBMCs after stimulation both in patients and controls, namely T cell receptor (TCR), STAT-JAK, PI3K, TLR3, NTRK1, TRAD, or MKO3 pathways. This method can identify previously described pathways through the ex vivo analysis of human PBMCs and propose new ones.

7.3 Network Topology-Based Prediction of Targeted Combination Therapy

To predict novel combination therapies for MS, the therapeutic goal is to use a combination of treatments to alter the kinase reactions within the signaling network in MS patients, to bring it closer to that of the healthy controls. The combinations are first defined using an approved MS drug as the base to determine the topology of its

Healthy	Disease untreated	Disease drug	Druggability score	Interaction modulated by drug?	Drug induced healthy-like interaction?
0	0	0	0	No	Yes
0	0	1	-1	Yes	No
0	1	0	-1	Yes	Yes
0	1	1	0	No	No
1	0	0	0	No	No
1	0	1	1	Yes	Yes
1	1	0	-1	Yes	No
1	1	1	0	No	Yes

Yes (green)	Yes	Yes	-1. No Druggable: drug is effective in the interaction
No (orange)	Yes	No	-1. Druggable: drug do not induce healthy-like interaction
	No	No	0. Druggable: drug is not targeting the interaction

Fig. 7.3 Druggability algorithm. The algorithm considers all combinatorial options to define if a given interaction of the signaling network models is a candidate to be modulated by a drug based on its differential activation between healthy controls, untreated MS patients, and MS-drug specific models. For example, interaction with a negative co-druggability score indicated a treatment effect that produced signaling activity different from that found in healthy patients and was selected as co-druggable. Conversely, a co-druggability score of 0 indicated that there was no effect due to drug treatment. From those cases, the interactions in which signaling activity differed between healthy donors and treated patients were also selected as co-druggable, i.e., those where the drug alone failed to revert signaling to a healthy state. The double or single red squares identify Co-druggable interactions (reproduced with permission from Bernardo-Faura et al. 2021)

respective signaling network. Although these drugs are indeed effective in treating MS, their associated signaling networks do not fully reflect those of the healthy controls. The goal then is to determine the kinase reactions (influenced by the ongoing therapy) that are crucial to change the network into one that more closely resembles a healthy state. To that end, the algorithm searches for a co-druggable interaction, which is when the selected treatment could not return to a healthy-like state. These should be the model interactions with a signaling value more distant between healthy controls and MS drug models than between healthy and MS models. To identify the co-druggable interactions, the quantitative network interaction scores of untreated MS patients are subtracted from those from healthy controls, and similarly between the healthy controls and the model for each MS drug (Fig. 7.3).

This provides a score for each interaction that tells whether it is a candidate to be considered co-druggable. These co-druggable interactions are both those that are different from healthy controls due to both the disease and unintended effects from the treatment (Fig. 7.4a). In the last step, the co-druggable interactions are integrated into the signaling network assessed for each subgroup (Fig. 7.4b). Mapping the co-druggability scores in this way allows additional therapies to be suggested due to the makeup of the networks for the individual treatments. Finally, a graph search algorithm is used to determine the co-druggable interactions that can be activated by the given in vitro stimuli used in the study. For example, Fig. 7.4c shows the

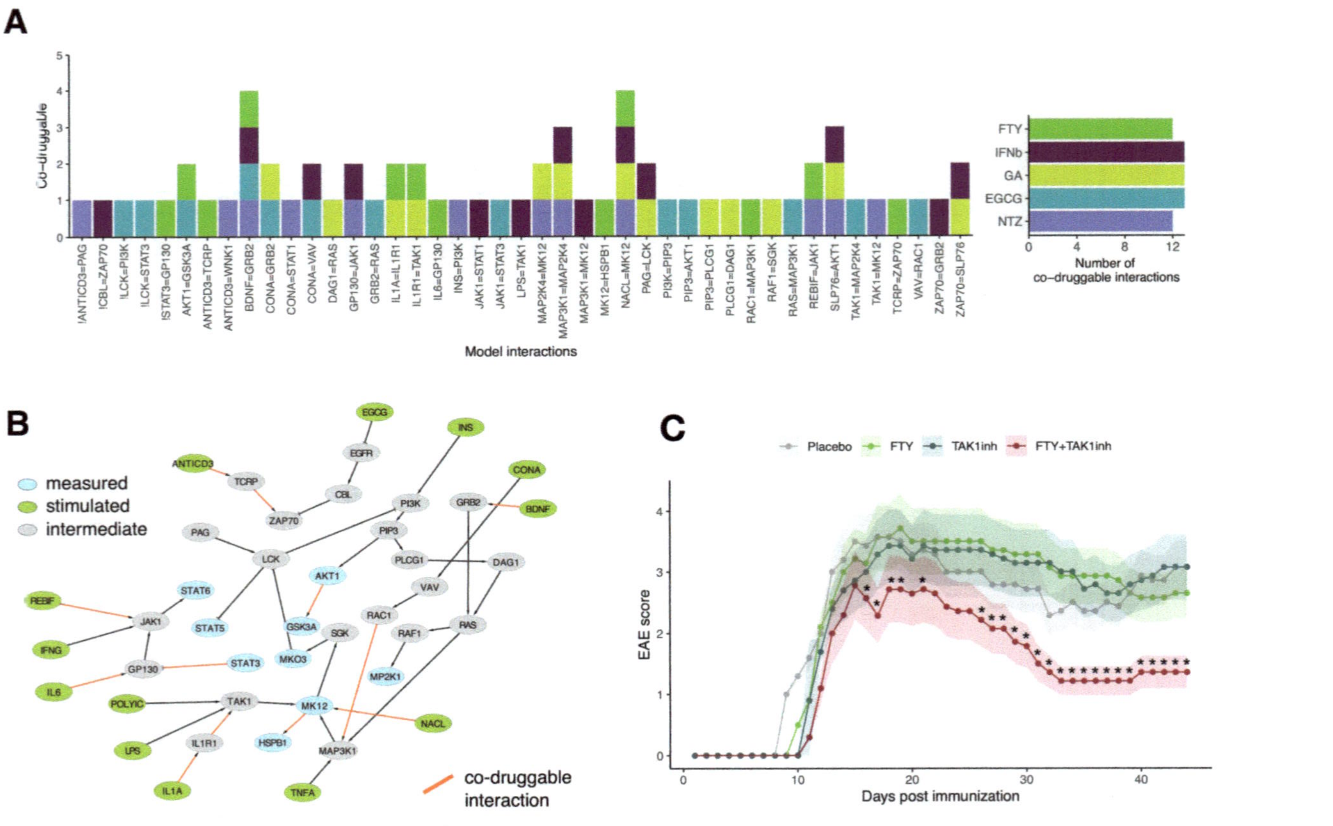

Fig. 7.4 Combinatorial immunotherapies prediction. (**a**) All predicted co-druggable interactions of the MS drugs models. Based on the subgroup models, the co-druggability of all network interactions (*X*-axis) is assessed using the co-druggability score. Those identified as co-druggable are shown. For each interaction

(*X*-axis), the number of drugs (*Y*-axis) is shown, in which it was found to be co-druggable using the co-druggability criteria. (**b**) FTY network co-druggability: the case of FTY network co-druggability is shown as an example (red line: interactions predicted to be co-druggable). (**c**) In vivo validation of the combination FTY + TAK1-inhibitor in the EAE model. The graph shows the mean and the standard error of the clinical score for each group ($n = 7$). Stars show days significantly different from the same day with placebo (reproduced with permission from Bernardo-Faura et al. 2021) validation of co-druggable interactions upon fingolimod with the predicted kinase TAK1 in the animal model of MS.

7.4 Lessons from Network Analysis to Understand MS

By analyzing the signaling network in patients with MS, specific signaling reactions are found to be deregulated by the disease, showing higher activation of the NFkB pathway (TAK1 to IKKB), enhanced pro-survival effects of the trophic factor signaling pathway (SLP76 to AKT1), and modulation of the interferon pathway (JAK1 to STAT3). Furthermore, the NfkB pathway has been reported to be overactivated in PBMCs from patients with MS (Moreno et al. 2006; Chen et al. 2016) as well as to contribute to the genetic susceptibility of the disease (Housley et al. 2015; Hussman et al. 2016). These observations support the pro-inflammatory state of immune cells in MS. Furthermore, the trophic factor pathway involving SLP76 and AKT has been associated with MS (Kotelnikova et al. 2015). Additionally, the MS-susceptibility gene CD6 has been found to modulate SLP76 (Hassan et al. 2006; Johnson et al. 2010; International Multiple Sclerosis Genetics et al. 2013). Therefore, considering these pathways suggests that T and B cells are under a signaling state of pro-survival concerning inflammation of the microenvironment. Finally, there is an overactivation of the cytokine/interferon pathway (JAK1), previously reported in MS (Kotelnikova et al. 2015). Moreover, STAT3 has been confirmed as a susceptibility gene for the disease [International Multiple Sclerosis Genetics et al. 2013], and its activation is impaired in response to IL-10 in MS patients (Martinez-Forero et al. 2008), suggesting an inadequate response of regulatory Tr1 cells.

There are additional interactions that are downregulated, including the inhibition of LCK over STAT-1, suggesting impairment of the regulation of T/B cell signaling, and IL-2 trophic effects (Beyer et al. 2011) or cytotoxicity (Slavin-Chiorini et al. 2004), as well as the regulation of the ubiquitination system modulated by CBLB (Swaminathan and Tsygankov 2006). LCK is modulated by EVI5 and influences STAT-1 in our analysis, both of which are susceptibility genes for MS (International Multiple Sclerosis Genetics et al. 2013). Another inhibited interaction involved the MS-susceptibility gene CBL-B (Sanna et al. 2010), which regulates TCR, co-stimulatory signals, and immune tolerance through its ubiquitin E3-ligase activity. For MS patients, the expression of CBL-B is lowered, and the type I interferon pathway is modified in CD4 cells (Sellebjerg et al. 2012; Sturner et al. 2014). In addition, CBLB is activated by EGFR, leading to inhibition of several pathways by ubiquitination, including the EGFR itself (Galisteo et al. 1995; Tarcic et al. 2009). In summary, systems pharmacology analysis can verify the downregulation of the MS pathways mentioned in the previous studies above (Kotelnikova et al. 2015), suggesting both over-stimulation of pathways that trigger inflammatory responses and the suppression of pathways responsible for immune tolerance.

7.5 A System Pharmacology Approach for Designing Combination Therapies Beyond MS

The proposed algorithm identifies additional treatments to combine with ongoing therapies through analysis of signaling networks in PBMCs. The method searches for potential co-druggable interactions within the networks of the individual therapies, by considering those that were still not reverted to a healthy state after treatment (Fig. 7.1). In more concrete terms, the co-druggability score is used to identify these interactions, allowing the combination of therapies to be predicted from the topology of the signaling networks (Fig. 7.3). This pragmatic approach has the advantage of enabling the prediction of combinations between immunotherapies currently used in patients with cancer or autoimmune diseases and compounds that can stimulate signaling for those drugs that cannot fully revert the signaling activity to a healthy-like state.

Many strategies have been developed to predict drug combinations (Bulusu et al. 2016). Some have used phosphorylation data upon perturbation, using approaches such as applying principal component analysis and partial least square analysis (Lee et al. 2012), data-driven network inference (Korkut et al. 2015), or a combination of mechanistic and Bayesian network modeling (Halasz et al. 2016). For their simplicity, logic networks provide excellent topological descriptions of the signaling networks and simplify the dynamic analysis by assuming the interactions to be binary, providing a valuable framework to study and predict drug combinations (Flobak et al. 2015; Eduati et al. 2017; Silverbush et al. 2017).

References

Bernardo-Faura M, Massen S, Falk CS et al (2014) Data-derived modeling characterizes plasticity of MAPK signaling in melanoma. PLoS Comput Biol 10(9):e1003795

Bernardo-Faura M, Rinas M, Wirbel J et al (2021) Prediction of combination therapies based on topological modeling of the immune signaling network in multiple sclerosis. Genome Med 13(1):117

Beyer T, Busse M, Hristov K et al (2011) Integrating signals from the T-cell receptor and the interleukin-2 receptor. PLoS Comput Biol 7(8):e1002121

Bozic I, Reiter JG, Allen B et al (2013) Evolutionary dynamics of cancer in response to targeted combination therapy. elife 2:e00747

Bulusu KC, Guha R, Mason DJ et al (2016) Modelling of compound combination effects and applications to efficacy and toxicity: state-of-the-art, challenges and perspectives. Drug Discov Today 21(2):225–238

Butcher EC (2005) Can cell systems biology rescue drug discovery? Nat Rev Drug Discov 4(6): 461–467

Chen D, Ireland SJ, Remington G et al (2016) CD40-mediated NF-kappaB activation in B cells is increased in multiple sclerosis and modulated by therapeutics. J Immunol 197(11):4257–4265

Conway D, Cohen JA (2010) Combination therapy in multiple sclerosis. Lancet Neurol 9(3): 299–308

Cully M (2015) Combinations with checkpoint inhibitors at wavefront of cancer immunotherapy. Nat Rev Drug Discov 14(6):374–375

Davis MM, Tato CM, Furman D (2017) Systems immunology: just getting started. Nat Immunol 18(7):725–732
Eduati F, Doldan-Martelli V, Klinger B et al (2017) Drug resistance mechanisms in colorectal cancer dissected with cell type-specific dynamic logic models. Cancer Res 77(12):3364–3375
Flobak A, Baudot A, Remy E et al (2015) Discovery of drug synergies in gastric cancer cells predicted by logical modeling. PLoS Comput Biol 11(8):e1004426
Galisteo ML, Dikic I, Batzer AG et al (1995) Tyrosine phosphorylation of the c-cbl proto-oncogene protein product and association with epidermal growth factor (EGF) receptor upon EGF stimulation. J Biol Chem 270(35):20242–20245
Halasz M, Kholodenko BN, Kolch W, Santra T (2016) Integrating network reconstruction with mechanistic modeling to predict cancer therapies. Sci Signal 9(455):ra114
Hassan NJ, Simmonds SJ, Clarkson NG et al (2006) CD6 regulates T-cell responses through activation-dependent recruitment of the positive regulator SLP-76. Mol Cell Biol 26(17): 6727–6738
Housley WJ, Fernandez SD, Vera K et al (2015) Genetic variants associated with autoimmunity drive NFkappaB signaling and responses to inflammatory stimuli. Sci Transl Med 7(291): 291ra93
Hussman JP, Beecham AH, Schmidt M et al (2016) GWAS analysis implicates NF-kappaB-mediated induction of inflammatory T cells in multiple sclerosis. Genes Immun 17(5):305–312
International Multiple Sclerosis Genetics C, Beecham AH, Patsopoulos NA et al (2013) Analysis of immune-related loci identifies 48 new susceptibility variants for multiple sclerosis. Nat Genet 45(11):1353–1360
Jia J, Zhu F, Ma X et al (2009) Mechanisms of drug combinations: interaction and network perspectives. Nat Rev Drug Discov 8(2):111–128
Johnson BA, Wang J, Taylor EM et al (2010) Multiple sclerosis susceptibility alleles in African Americans. Genes Immun 11(4):343–350
Kholodenko BN, Hancock JF, Kolch W (2010) Signalling ballet in space and time. Nat Rev Mol Cell Biol 11(6):414–426
Kieseier BC, Stuve O (2011) Multiple sclerosis: combination therapy in MS–still a valid strategy. Nat Rev Neurol 7(12):659–660
Klinger B, Sieber A, Fritsche-Guenther R et al (2013) Network quantification of EGFR signaling unveils potential for targeted combination therapy. Mol Syst Biol 9:673
Kolch W, Halasz M, Granovskaya M, Kholodenko BN (2015) The dynamic control of signal transduction networks in cancer cells. Nat Rev Cancer 15(9):515–527
Korkut A, Wang W, Demir E et al (2015) Perturbation biology nominates upstream-downstream drug combinations in RAF inhibitor resistant melanoma cells. elife 4
Kotelnikova E, Bernardo-Faura M, Silberberg G et al (2015) Signaling networks in MS: a systems-based approach to developing new pharmacological therapies. Mult Scler 21(2):138–146
Lee MJ, Ye AS, Gardino AK et al (2012) Sequential application of anticancer drugs enhances cell death by rewiring apoptotic signaling networks. Cell 149(4):780–794
Martinez-Forero I, Garcia-Munoz R, Martinez-Pasamar S et al (2008) IL-10 suppressor activity and ex vivo Tr1 cell function are impaired in multiple sclerosis. Eur J Immunol 38(2):576–586
Milo R, Panitch H (2011) Combination therapy in multiple sclerosis. J Neuroimmunol 231(1–2): 23–31
Moreno B, Hevia H, Santamaria M et al (2006) Methylthioadenosine reverses brain autoimmune disease. Ann Neurol 60:323–334
Owens J (2007) Determining druggability. Nat Rev Drug Discov 6(3):187
Palacios R, Goni J, Martinez-Forero I et al (2007) A network analysis of the human T-cell activation gene network identifies JAGGED1 as a therapeutic target for autoimmune diseases. PLoS One 2(11):e1222
Poussin C, Mathis C, Alexopoulos LG et al (2014) The species translation challenge – a systems biology perspective on human and rat bronchial epithelial cells. Sci Data 1:140009

Pujol A, Mosca R, Farres J, Aloy P (2010) Unveiling the role of network and systems biology in drug discovery. Trends Pharmacol Sci 31(3):115–123
Ransohoff RM, Hafler DA, Lucchinetti CF (2015) Multiple sclerosis – a quiet revolution. Nat Rev Neurol 11(3):134–142
Saez-Rodriguez J, Alexopoulos LG, Epperlein J et al (2009) Discrete logic modelling as a means to link protein signalling networks with functional analysis of mammalian signal transduction. Mol Syst Biol 5:331
Sanna S, Pitzalis M, Zoledziewska M et al (2010) Variants within the immunoregulatory CBLB gene are associated with multiple sclerosis. Nat Genet 42(6):495–497
Sawcer S, Hellenthal G, Pirinen M et al (2011) Genetic risk and a primary role for cell-mediated immune mechanisms in multiple sclerosis. Nature 476(7359):214–219
Sellebjerg F, Krakauer M, Khademi M et al (2012) FOXP3, CBLB and ITCH gene expression and cytotoxic T lymphocyte antigen 4 expression on CD4(+) CD25(high) T cells in multiple sclerosis. Clin Exp Immunol 170(2):149–155
Silverbush D, Grosskurth S, Wang D et al (2017) Cell-specific computational modeling of the PIM pathway in acute myeloid leukemia. Cancer Res 77(4):827–838
Slavin-Chiorini DC, Catalfamo M, Kudo-Saito C et al (2004) Amplification of the lytic potential of effector/memory CD8+ cells by vector-based enhancement of ICAM-1 (CD54) in target cells: implications for intratumoral vaccine therapy. Cancer Gene Ther 11(10):665–680
Sturner KH, Borgmeyer U, Schulze C et al (2014) A multiple sclerosis-associated variant of CBLB links genetic risk with type I IFN function. J Immunol 193(9):4439–4447
Swaminathan G, Tsygankov AY (2006) The Cbl family proteins: ring leaders in regulation of cell signaling. J Cell Physiol 209(1):21–43
Tarcic G, Boguslavsky SK, Wakim J et al (2009) An unbiased screen identifies DEP-1 tumor suppressor as a phosphatase controlling EGFR endocytosis. Curr Biol 19(21):1788–1798
Terfve C, Cokelaer T, Henriques D et al (2012) CellNOptR: a flexible toolkit to train protein signaling networks to data using multiple logic formalisms. BMC Syst Biol 6:133
Trame MN, Biliouris K, Lesko LJ, Mettetal JT (2016) Systems pharmacology to predict drug safety in drug development. Eur J Pharm Sci 94:93–95
Villoslada P, Steinman L (2020) New targets and therapeutics for neuroprotection, remyelination, and repair in multiple sclerosis. Expert Opin Investig Drugs 29(5):443–459
Villoslada P, Steinman L, Baranzini SE (2009) Systems biology and its application to the understanding of neurological diseases. Ann Neurol 65(2):124–139

Approaches to Generating Virtual Patient Cohorts with Applications in Oncology

8

Anudeep Surendran, Justin Le Sauteur-Robitaille, Dana Kleimeier, Jana Gevertz, Kathleen Wilkie, Adrianne L. Jenner, and Morgan Craig

Abstract

Virtual clinical trials (VCTs) have gained popularity for their ability to rationalize the drug development process using mathematical and computational modelling, and to provide key insights into the mechanisms regulating patient responses to treatment. In this chapter, we cover approaches for generating virtual cohorts with applications in cancer biology and treatment. VCTs are an effective tool for predicting clinical responses to novel therapeutics and establishing effective treatment strategies. These VCTs allow us to capture interindividual variability

Anudeep Surendran, Justin Le Sauteur-Robitaille and Dana Kleimeier contributed equally with all other contributors.

A. Surendran
Department of Mathematics and Statistics, Université de Montréal, Montréal, Canada

Sainte-Justine University Hospital Research Centre, Montréal, Canada

Centre de recherches mathématiques, Montréal, Canada

J. Le Sauteur-Robitaille · M. Craig (✉)
Department of Mathematics and Statistics, Université de Montréal, Montréal, Canada

Sainte-Justine University Hospital Research Centre, Montréal, Canada
e-mail: morgan.craig@umontreal.ca

D. Kleimeier
Institute of Bioinformatics, University Medicine Greifswald, Greifswald, Germany

J. Gevertz
Department of Mathematics and Statistics, The College of New Jersey, Ewing, NJ, USA

K. Wilkie
Department of Mathematics, Toronto Metropolitan University, Toronto, Canada

A. L. Jenner
School of Mathematical Sciences, Queensland University of Technology, Brisbane, QLD, Australia

A. Cesario et al. (eds.), *Personalized Medicine Meets Artificial Intelligence*,
https://doi.org/10.1007/978-3-031-32614-1_8

(IIV) which can lead to diversity in patient drug responses. Here we discuss three main methodologies for capturing IIV with a VCT. First, we highlight the use of population pharmacokinetic (PopPK) models, which extrapolate from empirical data population PK parameters that best fits the individual variability seen in drug disposition using non-linear mixed effects models. Next, we show how virtual patients may be sampled from a normal distribution with mean and standard deviation informed from experimental data to estimate parameters in a mechanistic model that regulates drug PKs. Lastly, we show how optimization techniques can be used to calibrate virtual patient parameter values and generate the VCT. Throughout, we compare and contrast these methods to provide a broader view of the generation of virtual patients, and to aid the decision-making process for those looking to leverage virtual clinical trials in their research.

Keywords

Virtual clinical trials · Quantitative systems pharmacology · Virtual patient generation · Mathematical oncology · Drug development · Therapeutic regimen optimization

8.1 Introduction

Cancer is a heterogeneous disease with complex (sub)types, genetic compositions, and tumour spatial arrangements, all of which make designing and scheduling effective and minimally toxic cancer treatments more challenging. Despite the long-term concerted investment in highly intensive cancer research, the goal of precision and personalized medicine remains largely unrealized. The difficulty establishing new cancer treatment strategies is compounded by the complexity of the drug development pipeline, since getting a drug to market is a lengthy and expensive process. There have been many recent encouraging advances in cancer therapy, particularly the development of immunotherapies like T-VEC (an oncolytic virus to treat late-stage melanoma) (Andtbacka et al. 2015, 2016), and immune checkpoint inhibitors including nivolumab and pembrolizumab (Seidel et al. 2018). However, a number of disappointing trial results have highlighted the need for improved predictive and quantitative models to help guide clinical trials in oncology. Quantitative systems pharmacology (QSP) seeks to answer this call by developing systemic mathematical and computational models to explore dosing ranges and therapeutic regimens prior and concurrent to clinical trials.

Drug development in oncology and in general relies on the use of clinical trials. A randomized clinical trial evaluates new medical approaches by randomly dividing participants into separate groups, or arms. In these trials, either a new medical approach is compared to a placebo, or to an existing treatment (e.g. standard drug) in a non-inferiority trial. In a cross-over trial, multiple study arms may receive both treatments after a washout period which is calculated according to each drug's half-life (Brown 1980; Piantadosi 2017). In all scenarios, the trial population is randomly separated into study arms to satisfy the requirement of equally distributed cohorts for reproducibility and comparability (Bland and Altman 2011). This randomization is of course never perfect or identical, something which the virtual clinical trials we will describe later aim to address. In early phases, drug tolerability is tested and dose escalation is performed before a drug's efficacy in treating the target is measured (Lipsky and Sharp 2001). At each stage, between patient variability (e.g. genetic, physiological, etc.) leads to diversity in patient drug responses (Alfonso et al. 2020). Accounting for this inherent variability is often a significant obstacle for establishing effective and tolerable treatment schedules. Improperly estimating heterogeneity when planning trials, or observing a high degree of interindividual variability (IIV) contributes to decreased drug development success rates, which explains the high degree of attrition along the drug development pipeline (Kozłowska et al. 2019). In response, new approaches encompassing virtual clinical trials (VCT) have been increasingly integrated to (pre-)clinical drug development efforts as a means to quantify the effects of variable environmental, spatial, and genetic, etc., factors on therapeutic regimens and patients (Alfonso et al. 2020).

VCTs arose with the emergence of QSP (Polasek and Rostami-Hodjegan 2020; Ma et al. 2021) and enable finding potentially non-intuitive drug regimens that help to increase drug approval success rates and, in turn, reduce drug costs (Alfonso et al. 2020). To that end, VCTs have been shown to be an effective tool in mitigating various challenges arising at different stages of the drug development pipeline, particularly understanding heterogeneous responses to novel therapeutics, establishing efficacious treatment strategies and treatment schedules, and precision dosing in individual patients (Polasek and Rostami-Hodjegan 2020). Virtual clinical trials have been applied to a broad range of diseases including HIV (Stadeli and Richman 2013; Kirtane et al. 2018), tuberculosis (Pitcher et al. 2018), SARS-CoV-2 (Jenner et al. 2021a), sepsis (Clermont et al. 2004; An 2004; Vodovotz and Billiar 2013), diabetes (Visentin et al. 2014; Gyuk et al. 2019), cardiovascular diseases (Corral-Acero et al. 2020), and different tumours, including breast (Switchenko et al. 2019; Corral-Acero et al. 2020), brain (Agosti et al. 2018), melanoma (Barish et al. 2017; Cassidy and Craig 2019; Milberg et al. 2019), and lung (Jafarnejad et al. 2019; Sayama et al. 2021). The successful implementation of VCTs heavily depends on the ability to generate heterogeneous virtual patients that mimic a broad spectrum of patients with a multitude of disease presentations, as would be observed in real-world clinical studies.

As detailed in later sections, VCTs generally follow the same basic steps. First, a mathematical or computational model of a disease is constructed using prior information and domain expertise. Model parameters are estimated from existing biological studies, targeted experimentation, or ongoing or completed trials. Next, the sensitivity of model predictions to perturbations in parameter values is determined through local or global sensitivity analyses. To simulate a population-level response, with variations in individual patient responses (Cassidy and Craig 2019), the mathematical model is solved for each individual-based parameter set. These patient-mimicking parameter sets are informed by the sensitivity analyses and constructed either statistically, by imposing physiologically-appropriate ranges and distributions on the values, or by probabilistic data-fitting, to ensure that model predictions lie within ranges of experimental or clinical observations, using optimization schemes. Importantly, parameter value ranges may be pruned to ensure that model predictions recapitulate physiologically reasonable ranges and observations (Allen et al. 2016). If generated statistically, virtual patients are selected by randomly sampling each parameter value from the chosen distributions, and the resulting parameter set is accepted into the trial if predicted outcomes are within acceptable deviations from the known outcomes. If generated through probabilistic data-fitting, virtual patients are constructed as a set of parameter values by each successful fit of the optimization algorithm, and are thus automatically accepted into the trial. However generated, virtual patients accepted into the VCT are twinned/cloned and assigned to multiple identical cohorts. Therapeutic outcomes for each virtual patient in each cohort are then simulated and compared using a variety of statistical techniques. This virtual study design is evocative of a cross-over study, with the advantage that patients can be assigned to multiple cohorts at once, implying that differences observed between cohorts can be attributed to mechanistic causes (as cohorts are identical). VCTs using the strategies described above have been applied to non-chronic diseases (Alfonso et al. 2020; Jenner et al. 2021a) and to identify causal mechanisms controlling outcomes (Cassidy and Craig 2019; Alfonso et al. 2020; Jenner et al. 2021a).

Typically, for large models with many parameter values, not all parameters will be assigned to the patient-specific parameter sets, as leaving some parameters constant across the population can simplify the VCT construction and analysis. When choosing which model parameters to fix as constants for the whole population, and which to vary across virtual patients, there are three key aspects of the mathematical model to consider: identifiability, sensitivity, and the biological interpretation of parameters. Model identifiability refers to the ability of model parameter values to be uniquely determined by comparing to observations. For example, if data are limited or incomplete, certain parameter values may not be well constrained by data-fitting algorithms, resulting in a wide range of acceptable values that qualitatively fit the data. Performance may then be improved by identifying these parameters first and estimating their values from other sources or studies if possible. Or, these parameters may represent significant biological mechanisms that are a desirable addition to the patient-specific parameter set. Model sensitivity analysis compares the dependence of model predictions to small changes in each parameter value. A parameter with high sensitivity coefficient, for example, may be useful to capture population-level variance with minimal dimensionality in the patient-specific parameter sets. However, a parameter with low sensitivity coefficient may represent a significant biological mechanism that is desired in the variable parameter set. Thus, the decision of which parameters to fix, and which to allow to vary involves integration of all three above-mentioned key aspects. Once the composition of the variable parameter set is determined, these parameters may be used as potential biomarkers (Jafarnejad et al. 2019) to classify the virtual population.

For a number of reasons, VCTs are particularly attractive in cancer therapy development. Recent estimates put the probability that a candidate drug entering Phase I study today will obtain regulatory approval at around 10% (Hay et al. 2014), requiring around $3.953 billion USD for out-of-pocket and capitalized research and development costs (DiMasi et al. 2016) over 13 years (Mohs and Greig 2017). Yet, this probability is even lower in oncology (Hay et al. 2014), demonstrating the need for improved approaches in this area. Among the therapies that have met with disappointing real-world trial results are new cancer therapy strategies that aim to use the body's natural immune response against tumours. One reason for disappointing trial results with such immunotherapies is the complex, heterogeneous, and dichotomous nature of tumour–immune interactions (Wilkie and Hahnfeldt 2017; Jafarnejad et al. 2019; Wang et al. 2020). Another reason for the high failure rate in cancer clinical trials is the evolution of drug resistance (Bozic et al. 2010; Tirosh et al. 2016; Craig et al. 2019).

Drug resistance in particular can be addressed using VCTs, as in Emilia Kozłowska et al. 2018 where the authors constructed a model to find promising drug regimens to prevent platinum resistance of ovarian tumours typically treated with surgery and platinum-based chemotherapy. Unfortunately, relapses of ovarian tumours are highly frequent, but combination therapy with different platinum-taxanes (paclitaxel or docetaxel) can increase the amount of platinum sensitive cancer cells and the time to tumour relapse by administering drugs in six different combinations, optimally with three-to-four drugs (Emilia Kozłowska et al. 2018). A second study by Kozłowska et al. compared treatments using three different drugs. A combination of trientine, a copper chelating agent, and birinapant in biomarker-selected treatments, was compared with a biomarker-unselected treatment with Wee1 inhibitor resulting in an increased survival of Wee1 inhibitor-treated virtual patients. Other such VCTs have also been investigated in the context of preventing the evolution of resistance.

To further improve cancer treatment, VCTs have focused on the reduction of toxicity (and therefore increased tolerance) to improve therapeutic outcomes, and have studied new immunotherapies (Barish et al. 2017; Sové et al. 2020). For example, QSP-IO is a platform for modelling immuno-oncology (IO), which accommodates varying degrees of model complexity based on specific research questions (Sové et al. 2020). To build QSP-IO, the authors implemented several different modules, including aspects of T cell behaviour or the effects of immune checkpoint inhibitors (Jafarnejad et al. 2019; Ma et al. 2021). By applying those platforms on data from triple-negative breast cancer clinical trials with administration of placebo and nab-paclitaxel or atezolizumab and nab-paclitaxel concurrent therapies of atezolizumab and nab-paclitaxel were found to be the best therapy options (Wang et al. 2021). Similarly, VEPART (Barish et al. 2017) is a tool for identifying robust optimal treatment protocols integrating experimental data, mathematical modelling, and statistical analyses that was applied to a melanoma mouse model treated with immunostimulatory oncolytic viruses and dendritic cell vaccines to investigate optimal dose scheduling. Barish et al. 2017 found that, subject to a number of constraints on dose and treatment length, only one optimal combination (three days of oncolytic virus, followed by three days of dendritic cell therapy) led to total tumour eradication in the population average. Subsequent analyses have shown that this optimal strategy is located in a fragile region of the dosing space, suggesting that other treatment regimens would lead to more robust results in heterogenous cohorts (Jenner et al. 2021b).

Thus, VCTs have emerged as platforms with which to interrogate heterogeneous responses to drugs, delineate effective scheduling, and improve drug administration (Wang et al. 2008; Barish et al. 2017; Pérez-García et al. 2019; Cassidy and Craig 2019). In preclinical research, VCTs can also contribute to the decision-making process by distinguishing drug regimens leading to therapeutic successes and failures (Allen et al. 2016; Alfonso et al. 2020), and help to decipher individual patient risk classes and optimize drug-specific parameters (Viceconti et al. 2016; Boem et al. 2020). Across the multiple applications of virtual clinical trial strategies, the generation of virtual patient populations is paramount. Unfortunately, there is no solid consensus on the means of generating virtual patients. Here we address this specific issue by outlining popular approaches to virtual patient generation and highlighting their advantages and disadvantages using three case studies.

8.2 Using Population Pharmacokinetic Models to Generate Patients

In pharmacometric analysis, the standard method of evaluating and predicting the kinetics of plasma drug concentrations is through the assessment of drug pharmacokinetics ("what the body does to the drug"). Population pharmacokinetic (PopPK) models are built to discern population- and individual-level PKs using non-linear mixed effects (NLME) modelling. NLME models are statistical models that assume a fixed effect for the population and represent individual variation in the form of random effects. Let P be a set of parameters in the PK model. These parameters typically include factors like bioavailability (F), volume of distribution (V_d), transit rates (k_{ij}, where i and j denote model compartments), and clearance (CL). In its simplest form, a PopPK model for an individual k is given by the parameter vector P_k calculated as:

$$P_k = \overline{P}e^{\eta_k} \tag{8.1}$$

where $\overline{P}$ is the set of fixed (or population-level) parameters, η_k is a normally distributed random variable with mean of 0 and variance ω_p^2 ($\eta_k \sim \mathcal{N}\left(0, \omega_p^2\right)$), and e^{η_k} represents the resulting individual variability. Building a PopPK model using NLME models relies on population-level empirical data (usually from a clinical trial) from which the best pharmacokinetic model is defined. Here the "best" model is heuristic and determined by evaluating the calculated objective function after fitting several PK models. Multiple software packages, including R, NONMEM, and

Monolix, can be used to perform NLME estimates and establish population PK models from data.

As PopPK models are empirically established and not generally built from mechanistic principles, they are primarily relevant for the specific population upon which they were constructed. Nonetheless, during drug development, we may be interested in establishing dosing strategies, exploring the potential for toxicity due to reduced kidney or liver function, etc., within a given population that go beyond the scenarios explored in a clinical trial. Here, leveraging a PopPK model is particularly attractive because it inherently accounts for individual variation within the population studied. To demonstrate how a PopPK model can be used to generate a virtual cohort, we considered a simple model of Gompertzian tumour growth and its treatment by gemcitabine, a synthetic pyrimidine nucleoside prodrug used as a chemotherapeutic agent in a variety of solid tumours (Joerger et al. 2014).

Let $N(t)$ be the number of tumour cells at time t. The Gompertz model given by:

$$\frac{dN}{dt} = \beta \ln\left(\frac{K}{N_0}\right) e^{-\beta t} N, \tag{8.2}$$

models saturable sigmoidal tumour growth to a carrying capacity of K. Here, β denotes the tumour growth rate (that decreases exponentially in time), and N_0 the initial number of tumour cells. The PK model of gemcitabine is given by:

$$\frac{dG_1}{dt} = -k_{el}G_1(t) + k_{21}G_2(t) - k_{12}G_1(t), \tag{8.3}$$

$$\frac{dG_2}{dt} = -k_{21}G_2(t) + k_{12}G_1(t). \tag{8.4}$$

Each equation represents one of two compartments defined through a previous PopPK analysis (Jiang et al. 2007). These equations model the transfer of the drug between the central ($G_1(t)$) and peripheral ($G_2(t)$) compartments at rates k_{12} and k_{21}, respectively. Gemcitabine was modelled as being administered directly into the central compartment from which it is eliminated linearly at rate (k_{el}):

$$k_{12} = k_{21} = \frac{Q}{V} \tag{8.5}$$

$$k_{el} = \frac{CL}{V} \tag{8.6}$$

The between-subject variability (BSV%) for Q, CL, V and the range for body surface area (BSA) were previously estimated using NLME modelling (Jiang et al. 2007). Here, BSA is used to calibrate gemcitabine doses.

We modelled the cytotoxic effects of chemotherapy on tumour growth using an inhibitory Hill effect function given by:

Table 8.1 Parameter values for gemcitabine inhibition of tumour growth

Parameter	Units	Value	BSV%	Ref
β	day^{-1}	0.0251	–	Chosen
K	cells	5.157 × 10 ^ 8	–	Chosen
C_0	cells	2 × 10 ^ 6	–	Chosen
I_{max}	unitless	0.768	–	Chosen
h	unitless	0.518	–	Chosen
IC_{50}	μM	297	–	Chosen
Q	L/min	0.7	44	Jiang et al. (2007)
CL	L/min	2.7	31	Jiang et al. (2007)
V	L	15	39	Jiang et al. (2007)
BSA	m^2	Median: 1.8	Range: 1.2 – 2.5	Jiang et al. (2007)

Parameters with reported variation were used to generate the virtual cohort

$$E(D(t)) = 1 - I_{max}\frac{D(t)^h}{D(t)^h + IC_{50}^h}, \tag{8.7}$$

where $E(D(t))$ denotes the effect of a drug $D(t)$ at time t, I_{max} represents the drug's maximum inhibitory effect, IC_{50} the drug concentration at which inhibition is 50% its maximum, and h is the usual Hill coefficient controlling the slope of the curve. Integrating this into Eq. 8.2 we have the following model for the effects of gemcitabine on tumour growth:

$$\frac{dN}{dt} = \left(\beta \ln\left(\frac{K}{C_0}\right)e^{-\beta t}N\right)E(G_1(t)). \tag{8.8}$$

To generate virtual patients, we considered tumour cell parameters to be fixed at their previously estimated values and varied only the PK parameters according to the PopPK model for gemcitabine (Table 8.1). Assuming BSA to be normally distributed (Sacco et al. 2010), we calculated the mean and standard deviation required for the generation of BSA values (Hozo et al. 2005).

We then sampled a set of parameter values using between-subject variability specific to that parameter. For example, a vector of normally distributed values with mean (e.g. $\overline{Q}$) and standard deviation based on the BSV% of Q was used to generate a set of values $\{Q_k\}$, $k \in 1, 2, 3, \ldots, n$ (where n is the cohort size) using:

$$Q_k = \overline{Q}e^{\eta_k} \text{ where } \eta_k \sim \mathcal{N}\left(0, \left(\overline{Q} * BSV\%\right)^2\right). \tag{8.9}$$

The jth virtual patient is then given by the jth value from the PK vector such that each patient is described as a set $P_j = (Q_j, CL_j, V_j, BSA_j)$. Transit rates ($k_{12,\,j}$ and $k_{21,\,j}$) and elimination rates ($k_{el,\,j}$) for each patient j were calculated using Eqs. 8.5 and 8.6, and the dose for each patient j was calculated using the dosage 1000 mg/m^2, where BSA_j (m^2) is patient-specific body surface area. Using this process, we generated

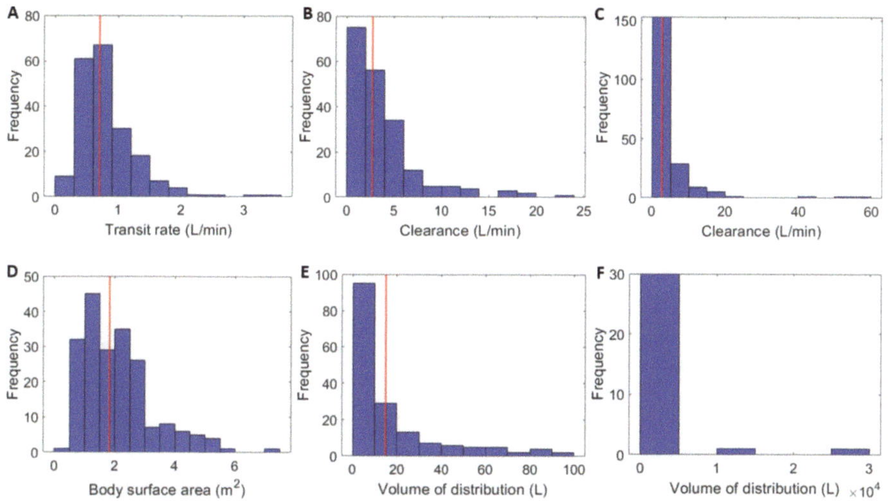

Fig. 8.1 Histograms of parameter values generated for 200 virtual patients. (**a**) Q, (**b**) CL under 30 L/min, (**c**) CL including outliers, (**d**) BSA, (**e**) V under 100 L, and (**f**) V outliers above 100 L. The red line on each plot represents the average or median value found in Table 8.1

200 virtual patients with their own dosage and model parameters tied to their assigned PK values (Fig. 8.1).

A schematic of the PopPK/PD model is provided in Fig. 8.2a. As expected, model simulations predicted that dynamics of cancer cell growth in the virtual cohort differed based on individual patients' PK parameters (Fig. 8.2). For the most part, we observed that drug concentrations fell within expected ranges in the central gemcitabine compartment (Fig. 8.2) and peripheral compartment (Fig. 8.2). The pharmacokinetics of gemcitabine has been shown to be linear up to 2500 mg/m^2 which coincides closely with maximum-tolerated dose in a dose escalation study (Fossella et al. 1997). The dynamic of such a high gemcitabine initial concentration can be seen in red in Fig. 8.2c–f. However, the virtual population contained obvious outliers (Fig. 8.2), with more than one virtual patient exhibiting markedly different drug concentrations from the rest of the cohort to the extent where toxic or lethal concentrations of gemcitabine were predicted. This highlights an obvious downfall of using PopPK models without integration of prior knowledge of inter-parameter relationships to generate virtual patients. Since the "top-down" approach (versus the "bottom-up" of mechanistic models) used here is not constrained by known mechanistic interactions, unrealistic (potentially dangerous) outcomes can be generated from what we may believe to be reasonable parameter ranges. This issue arises due to the method used to randomly sample parameter values from the previously established PopPK model with no built-in approach to verify that any specific combination of parameters (i.e. a virtual patient) is physiologically realistic. Though we may be able to clearly distinguish virtual patient outliers visually and remove

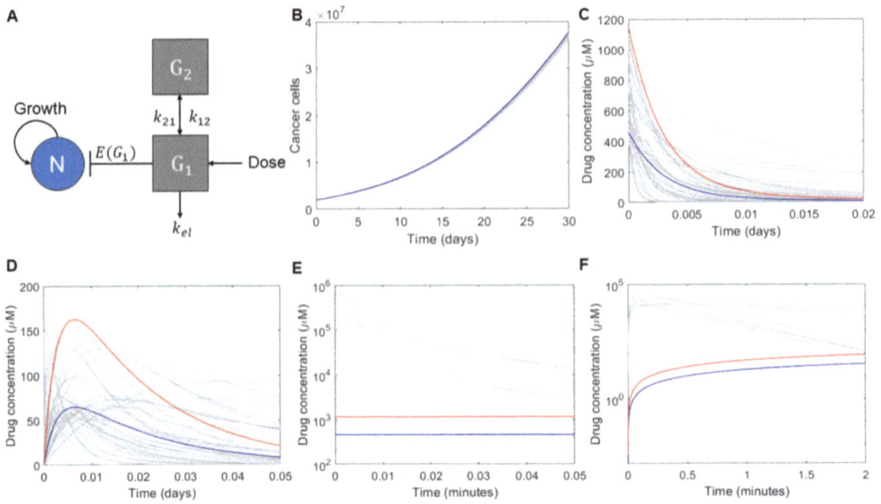

Fig. 8.2 Tumour growth and gemcitabine concentrations in a cohort of 200 virtual patients. (**a**) Schematic of the model and dynamics of (**b**) cancer cell growth ($N(t)$) over 30 days. (**c**) The central gemcitabine compartment ($G_1(t)$) and (**d**) the peripheral gemcitabine compartment ($G_2(t)$) for reasonable drug doses and concentrations below a near maximum dose of 2500 mg/m^2 (red). Outliers in (**e**) the central gemcitabine compartment ($G_1(t)$) and (**f**) the peripheral gemcitabine compartment ($G_2(t)$) compared to the average PopPK patient following different initial doses. Blue solid lines: average cohort response after 1000 mg/m^2 dose of gemcitabine; red solid lines: average cohort response after 2500 mg/m^2 dose of gemcitabine; grey solid lines: individual virtual patients

them from the cohort, a systematic approach to remove parameter sets that generate unrealistic individuals despite being drawn from realistic distributions is needed.

We can better understand how this random mismatch of parameters leads to potentially problematic virtual patients and outcomes by analyzing the PK parameters we generated and their implementation within the model considered here. As mentioned above, the elimination rate $k_{el,\,j}$ for each patient was calculated by dividing their clearance (CL_j) by their compartment volume (V_j) (Eq. 8.6). Theoretically, because all parameter values are randomly drawn from established ranges, it is possible that a high clearance and a small volume may be paired. However, at the extreme for each parameter, this combination is likely to be unrealistic. This problem is further compounded as dose sizes are calculated for each patient based on their BSA, and empirically-estimated correlates of BSA to the other model parameters were not reported. For example, it would be rare to have a patient with a lower BSA relative to the cohort mean, but with higher compartment volumes and low transit and thus elimination rates. The lack of explicit physiological constraints during the virtual patient generation phase provides no explicit guarantee that the random combinations of parameters will lead to realistic individuals.

Despite these shortcomings, the generation of virtual patients from PopPK models is a simple process that can be implemented rapidly and with ease. Therefore, to mitigate the risks of generating non-viable virtual individuals from PopPK models, we suggest a thorough investigation of the parameter combinations resulting from the sampling process. Furthermore, imposing limits on parameter ranges will help to ensure physiologically relevant sampling is achieved. This process should be guided by known physiology to not overly restrict sampling and virtual-patient generation, and thus reduce bias in the VCT.

8.3 Establishing Theoretical Bounds from Experimental Measurements

In the previous section, we explored the generation of virtual individuals using only PK parameters while physiological parameters remained fixed. The emergence of QSP approaches has increasingly integrated detailed mechanistic models of physiological systems and disease processes to PK/PD models to provide a more holistic understanding of the effects of xenobiotics. We have previously shown that QSP models account for interindividual PK variability by the nature of their construction (Craig et al. 2016; Le Sauteur-Robitaille et al. 2021). Therefore, it is reasonable to generate virtual patients by uniquely varying physiological parameters in the model, since it is primarily physiological heterogeneity driving variable responses to drugs.

To demonstrate the generation and use of a virtual clinical trial incorporating virtual patients generated by sampling physiological parameters from theoretically-defined parameter ranges, we considered a model for the interaction between a cytotoxic chemotherapy drug ($[drug]$ ng · ml^{-1}), a population of tumour cells ($S(t)$), and the immune system ($T(t)$). In this model, we investigated how the introduction of a chemotherapy drug may impact the antitumour immune response by reducing the pool of tumour cells and hence affecting the immune recruitment. We model cancer cells as growing logistically with proliferation rate r (day^{-1}) and carrying capacity K (cells). The effects of the immune system on the tumour are modelled by supposing that cancer cells undergo apoptosis through contact with immune cells at a rate κ ($\text{cells}^{-1}\ \text{day}^{-1}$). We assumed that immune cells are recruited at a rate proportional to the amount of tumour cells a (day^{-1}) and die at a rate d (day^{-1}), giving the model:

$$\frac{dS}{dt} = rS\left(1 - \frac{S}{K}\right) - \kappa ST - \frac{\delta[drug]}{[drug] + \eta}S, \tag{8.10}$$

$$\frac{dT}{dt} = aS - dT, \tag{8.11}$$

$$\frac{d[drug]}{dt} = -\rho[drug], \tag{8.12}$$

A schematic summary of the system can be found in Fig. 8.3a.

The model was parameterized using data of tumour growth in the absence of the immune system (Oh et al. 2017) (control tumour growth) and data of tumour growth

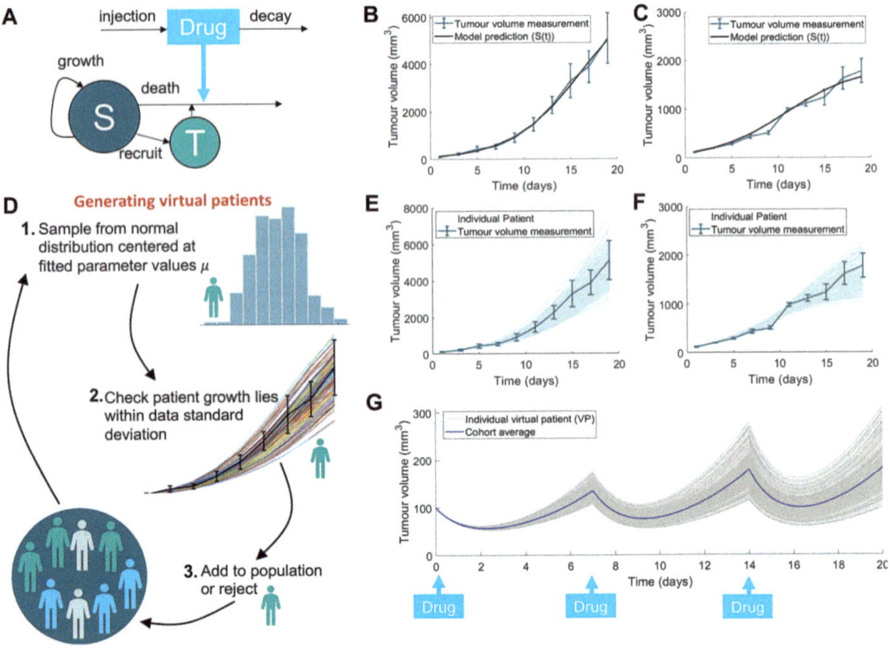

Fig. 8.3 Virtual cohort example in immune-tumour-anti-cancer drug model. (**a**) Model schematic. Cancer cells $S(t)$ grow logistically and are killed by a cytotoxic chemotherapy drug ($[drug]$) and through tumour–immune ($T(t)$) interactions. (**b**–**c**) Resulting model trajectories after parameter fits to untreated tumour growth without immune (**b**, control data; Oh et al. 2017) and untreated tumour growth in the presence of the immune system (**c**, immune-suppressed data; Kim et al. 2011). (**d**) Schematic overview of the generation of virtual patients informed by experimental data. (**e**–**f**) Predicted responses of tumour volume under control and treatment scenarios of the virtual patients accepted into the virtual patient cohort (light blue) with corresponding data measurements. (**g**) Evolution of virtual cohorts' tumour volume under chemotherapy treatment where the drug was administered every 7 days. Individual patient predictions (grey) and the cohort average evolution (blue) are illustrated.

in the presence of the immune system, but under suppressive virotherapy (Kim et al. 2011). We assumed the latter data to be representative of the immune-suppressed tumour growth model. Tumour volume was measured using callipers in mice and the average was used to fit parameters in the model. The data here were deployed solely for illustrative purposes, and we parameterized the model to the data using a simultaneous fitting approach (Gray and Coster 2016; Jenner et al. 2018).

Simultaneous fitting to the data was performed by setting the appropriate parts of the model to zero for each data set. In other words, for the control data, the drug and the immune population were set to zero and for the suppressed tumour growth data (immune-present data) the drug parameters and variables were set to zero. The remaining parameters in the drug-free model, r, K, κ, a and d were then fit simultaneously using non-linear least-squares fitting with the objective of minimizing the

Table 8.2 Parameter values for the tumour growth under immune suppression and chemotherapy treatment

Parameter	Units	Description	Value	Ref
r	day^{-1}	Tumour cell proliferation rate	0.2349	Fit Fig. 8.2
K	cells	Tumour carrying capacity	10190	Fit Fig. 8.2
κ	$\text{day}^{-1}\text{cells}^{-1}$	Immune cell induced tumour cell death rate	0.0021	Fit Fig. 8.2
a	day^{-1}	Immune cell recruitment rate	7.086	Fit Fig. 8.2
d	day^{-1}	Immune cell death rate	162.5	Fit Fig. 8.2
δ	day^{-1}	Drug-induced tumour cell death	2.64	Crosley et al. (2021)
η	$\text{ng} \cdot \text{ml}^{-1}$	Drug half-effect concentration	5	Crosley et al. (2021)
ρ	day^{-1}	Drug elimination rate	0.903	Crosley et al. (2021)

Parameters for the tumour growth and immune dynamics were obtained through fitting to data (Kim et al. 2011; Oh et al. 2017). The remaining parameters were estimated from PAC-1 (first procaspase activating compound) dynamics (Crosley et al. 2021)

residual of the model to both sets of data simultaneously. The fitted parameters can be found in Table 8.2, and the resulting model approximation to the data is in Fig. 8.3b and c. To parametrize the models for chemotherapy drug decay and drug-induced tumour cell death we used parameters for PAC-1 (first procaspase activating compound) fit previously from *in vitro* experiments (Crosley et al. 2021).

Next, to create n virtual patients, we sampled parameters from a normal distribution centred at μ with standard deviation σ, rejecting any negative parameters (Fig. 8.3d). We fixed μ to be the set of parameter values obtained from our fitting procedure for r, K, κ, a and d as we assumed individuals would vary in both the tumour growth and immune dynamics. The standard deviation σ was fixed to $\sigma = 0.2\mu$ to minimize the likelihood of sampling negative parameter values while maximizing the variation on the parameters chosen for our virtual population. Sampling with this σ resulted in 634 random samples being rejected to make 200 virtual patients. Increasing to $\sigma = 0.5\mu$ resulted in a higher number of negative parameter samples, which are set to zero, and also a higher number of samples rejected (1917). It is possible to also estimate σ from the standard deviation in the data (Jenner et al. 2021a).

To create a realistic representation of patients, we restricted the inclusion of virtual patients to those whose simulated control tumour growth and immune-suppressed tumour growth was within physiological reasonable regimes which we designated to be three standard deviations ($3\,\widehat{\sigma}$) of the mean control data and five standard deviations ($5\,\widehat{\sigma}$) of the immune-suppressed tumour growth data (Fig. 8.3b and c). Different thresholds were chosen for each case as it was not possible to generate patients within $3\widehat{\sigma}$ of tumour measurements at day 9 and 10 in the immune-suppressed data (Fig. 8.3f) with a fixed underlying parameter distribution for sampling. As such, we increased the tumour volume trajectories for the immune-suppressed data to lie within $5\hat{\sigma}$ of the data. Any virtual patient whose simulated tumour growth in the absence of drug lay outside these intervals was rejected from the virtual trial. Remaining patients were accepted into the trial and the distribution of all model-predicted tumour growths lay close to the data (Fig. 8.3e and f). This step mimics the selection process of clinical trials (for example, including only those patients with a specific grade of a cancer).

Following this generation approach, tumour growth in both the control and treatment scenarios was confirmed to fall within the prescribed heterogeneity bounds (Fig. 8.3e and f). Over time, we see the variation in virtual patient growths increasing with large variations of patient dynamics by day 19 in both the control and immune-suppressed case. Simulating drug administration every seven days starting at day 0, we observe a similar trend, where virtual patients respond similarly in the first cycle before more significant differences in tumour volume emerge (Fig. 8.3g).

As in the approach described in the previous section, the techniques described here are relatively straightforward to implement. However, in contrast to basing the virtual patient generation solely on PopPK parameters, this technique integrates constraints on the mechanistic (or physiological) model parameters through the integration of biological experimental data. Unfortunately, the procedure described above is still not able to capture parameter correlations (for example, a patient's tumour growth rate (r) and carrying capacity (K)), although it would be possible to address this shortcoming by performing statistical tests to guide the parameter sampling. The additional step of comparing model prediction to measured outputs responds to the limitation of creating unrealistic virtual patients. In practice, a large number of proposed virtual patients may need to be generated in order to ensure that the final trial cohort is sufficiently large In this section, virtual patients trajectories were restricted to lying between $3\hat{\sigma}$ and $5\hat{\sigma}$ for the control and immune-suppressed data, respectively. The difference in these ranges for the physiological regimes was driven by the inability to achieve trajectories within $3\hat{\sigma}$ of the immune-suppressed data. As described in the next section, this limitation can be overcome by adding a step within the virtual patient generation phase to ensure that all parameter samplings result in model trajectories that describe clinical observations.

8.4 Quantitative Systems Pharmacology Approaches

As more mechanisms are included, model complexity and the sparsity of relevant data complicate the implementation of VCTs. Accordingly, methods to generate virtual populations that reproduce the heterogeneity in patients as well as allow for the exploration of parametric uncertainty have been devised to overcome these challenges. One of the best-known approaches in this vein is that of Allen et al. 2016 who proposed a method to generate a large cohort of heterogeneous virtual patients by sampling a parameter set from a bounded interval informed by physiological constraints, and optimizing predicted trajectories to ensure model outcomes are within clinically-observed ranges. Therefore, this methodology expands upon the approach studied in the previous section, by explicitly integrating the constraint that model predictions for each virtual patient must fall within empirically-determined ranges, into the virtual patient generation process through an optimization step.

To demonstrate the generation of a virtual patient population using this approach, we will again use the cancer–immune interaction model introduced in Eqs. 8.10, 8.11, 8.12. Here we limit our focus solely on generating virtual patients using this alternative approach and thus ignore the effect of the drug, and reduce the model to a system of ODEs describing the time evolution of tumour ($S(t)$) and immune ($T(t)$) cells:

$$\frac{dS}{dt} = rS\left(1 - \frac{S}{K}\right) - \kappa ST,$$

$$\frac{dT}{dt} = aS - dT.$$

All parameters are as previously defined in Table 8.2. A schematic of various steps involved in virtual patient generation using the Allen et al. (2016) method is presented in Fig. 8.4.

A key component that determines the successful implementation of the Allen et al. (2016) method is the ability to define realistic bounds for the model parameters. In typical applications, these bounds can be inferred through empirical estimates from physiological experiments or through theoretical considerations. In our simulations, we consider parameter values within three standard deviations (3 σ) from mean (μ) values presented in Table 8.2 as plausible ranges for the model parameters, r, K, κ, a and d. As we described in the previous section, the mean parameter values are obtained by non-linear least-squares fitting. In the prior section, each parameter value was assumed to be normally distributed with standard deviation $\sigma = 0.2\mu$, so here we are considering a much broader range of plausible parameter values.

To start the patient generation process, we constructed an initial parameter set where each parameter is drawn from a uniform distribution bounded by the plausible ranges discussed above. Next, we optimized each set of parameters (r, K, κ, a and d) using simulated annealing (a probabilistic optimization algorithm implemented as the *simulannealbnd* function in MATLAB (Mathworks 2020) to ensure that predicted model trajectories fall within physiological ranges. For our purposes, these bounds on model outputs were assumed to be within three standard deviations of the control and immunosuppressed tumour growth data means. We defined the objective function of the simulated annealing scheme to be:

$$g(p) = \sum_i \max\left[\left(M(p,t) - \frac{l_i + u_i}{2}\right)^2 - \left(\frac{u_i}{2} - \frac{l_i}{2}\right)^2, 0\right],$$

and minimized $g(p)$. Here, $M(p, t)$ denotes the model output for parameter set p at time t and l_i and u_i denotes the ith plausible upper and lower bounds of the data. By defining the objective function in this fashion, we are guaranteed that the

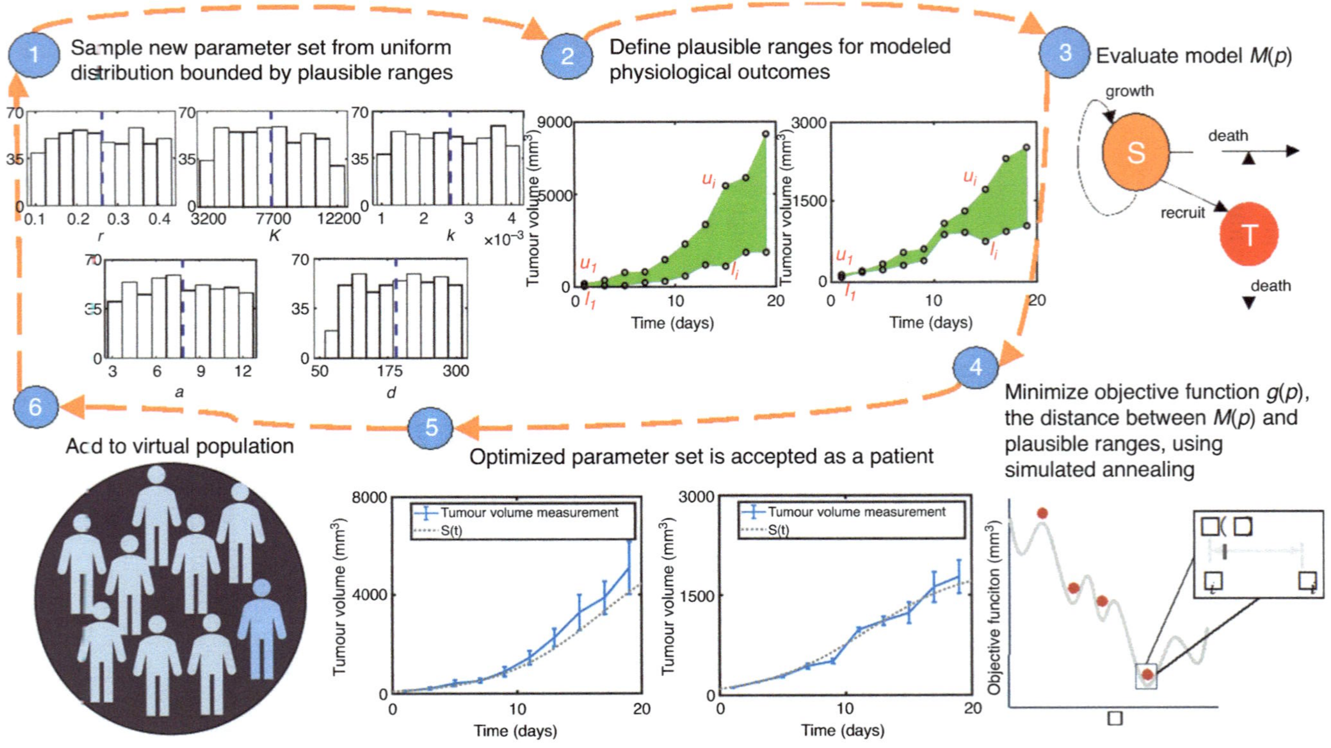

Fig. 8.4 Major steps in virtual patient generation using the Allen et al. method. (1) Initial model parameters are sampled from a uniform distribution bounded by plausible ranges. (2) Define an upper (u_i) and lower (l_i) bound for the observable model outcome (tumour volume) at each time point of the available data. (3) The model is evaluated with the initial parameter guesses sampled from the uniform distribution and obtain model output, $M(p, t)$. (4) An optimal parameter set that minimizes the objective function $g(p)$ is obtained using the simulated annealing algorithm. (5) The optimization of $g(p)$ using the simulated annealing algorithm generates a virtual patient whose disease dynamics lie within the physiologically plausible ranges. (6) The newly generated patent is added to the virtual patient population. The optimization routine is repeated until the required number of patients are generated

contribution of $M(p,t) - (l_i + u_i)/2$ is zero if the parameters lead to an outcome within the plausible range. If the optimization converges, the resulting parameter set was considered to belong to a physiologically-valid virtual patient and was added to the trial population.

Following these steps, we generated a virtual population of 500 patients using the Allen et al. (2016) method. The trajectories of all virtual patients were confirmed to lie within the physiological bounds determined for the model outputs (Fig. 8.5c and d). Thus, this method generates a large pool of patients with realistic disease dynamics without manual verification as was required in the methods discussed in Sects. 8.2 and 8.3. It should be noted that for specific applications, e.g. to consider patients with certain levels of tumour growth, we can subsample from the resulting virtual patient cohort. We may also exploit predetermined, empirical parameter distributions to create specific subpopulations of patients. For example, the posterior distribution of both r and κ was found to differ significantly from their prior (Fig. 8.5a and b), suggesting a mechanistic role of both parameters that may drive heterogeneity and segregate patient responses. Narrowing in on specific values of each (in isolation and in combination) could reveal specific subpopulations of virtual patients and allow for further tailoring of treatment strategies.

We have seen that the generation of VPs from PopPK models may lead to patients with disease dynamics beyond physiological limits, when using random parameter sampling without any built-in mechanism to incorporate physiological constraints and correlations. In contrast, the Allen et al. approach, which leverages physiologically-informed bounds on parameters and model outputs together with robust optimization using the simulated annealing algorithm, ensures the generation of realistic patient cohorts. While the second method we described based on obtaining bounds from experimental measurements is a reliable approach to get a realistic patient population with dynamics within the physiological bounds, the brute force way of generating a large number of candidate patients to eventually select the total size of the cohort is not efficient. On the other hand, the Allen et al. (2016) method is an alternative that produces a patient cohort without relying on generating a large candidate population. This comes, however, at a computational cost as every virtual patient must undergo optimization prior to being accepted into the virtual cohort. In response, recent extensions of the method involving augmenting optimization of the cost function (nested simulated annealing) or adopting alternative optimization routines (modified Metropolis–Hastings and genetic algorithms) have been shown to lead to more efficient generation of virtual patients (Rieger et al. 2018).

8.5 Discussion

Designing and developing new xenobiotics in immuno-oncology is complicated and costly, driven by our sometimes-limited knowledge of the mechanisms regulating therapeutic efficacy. Mathematical and computational modelling are increasingly integrated along the drug development pipeline and in fundamental studies to help

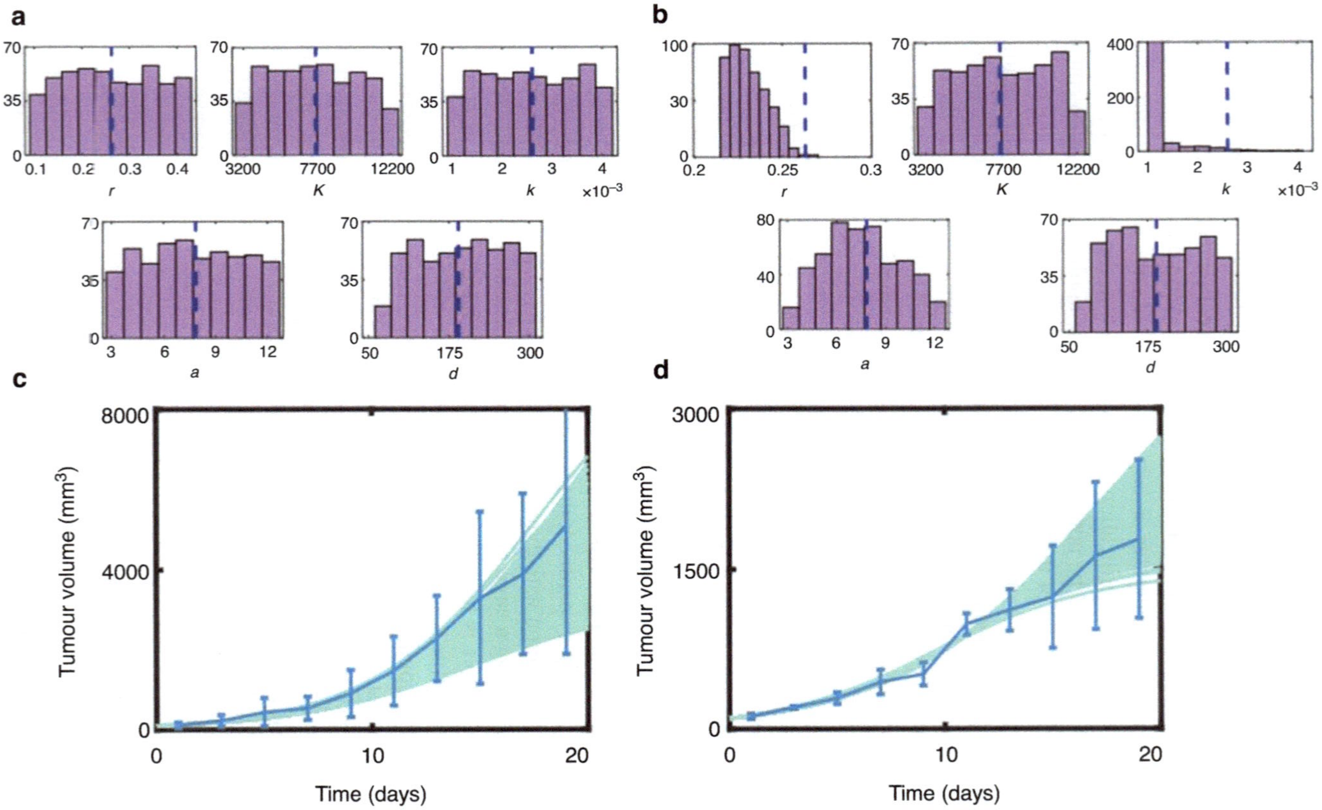

Fig. 8.5 A virtual population of 500 patients generated using the Allen et al. method. (**a**) shows the initial parameter value distributions (uniformly distributed) and (**b**) shows the final parameter value distributions after the optimization using the simulated annealing algorithm. The blue vertical line on each histogram represents the mean value of the distribution. (**c** and **d**) show the tumour growth in the control and in presence of the immune system, respectively. The virtual patient dynamics trajectories (cyan) are superimposed with the corresponding data measurements (blue). The error bars denote three standard deviations from the data mear

identify regulators of drug responses. Given the high degree of variability observed within patient populations, quantitative approaches that can also capture the heterogeneity in outcomes are now frequently deployed to assess the degree to which variability affects outcomes and to discern the sources of such heterogeneity.

In silico clinical trials are therefore well-situated to help guide the preclinical-to-clinical translation of candidate molecules, to assess the best candidate populations for a given treatment, to delineate successful combination strategies, and to establish optimal dosing schedules. Virtual patient populations are at the heart of these computational trials and must be reflective of the variability we observe in real patient populations. However, there is no universally accepted method of generating a virtual patient population, and each approach comes with its own set of advantages and limitations. Here we have described three widely-deployed methods to generate a virtual cohort, starting from the most simple implementation using empirically-defined population pharmacokinetic models to the more complex, and perhaps most robust, approach described by Allen et al. (2016) that verifies a virtual patient's trajectory within the generation step.

Whether the patient population is generated using statistical methods or probabilistic data-fitting methods, it is clear that the choice of which parameters are used to define the population, and thus which remain constant across the population, depends on the mathematical model and the application. For models with a large number of model parameters, a high-dimensional patient-specific parameter set may be challenging to both appropriately set-up and to optimize. Consideration of key aspects such as model identifiability, parameter sensitivity, and significant biological mechanisms can help inform which parameters to include in the patient-specific set. Additionally, the assumptions made during the set-up phase of VCT construction will have implications on the conclusions of the simulated study, and these implications should be explored. Further, as new methods for creating virtual patients continue to be proposed in the literature, it is important to establish standards for virtual clinical trials, in terms of the data needed to perform them, the dimensionality of the patient-specific parameter sets, how interpatient variability is represented by the data, and how the output of these trials is used and validated. Clearly there is still lots of work to be done in the theory of VCT design, implementation, and assessment. To that end, new approaches integrating machine learning and data dimensionality reduction techniques may also prove useful for selecting parameters of interest to generate the virtual cohort.

Ultimately, the choice of method for generating a virtual cohort depends on a number of factors, most predominately the level of complexity required, and the time allowed for model generation and implementation of the method. Once created, virtual patient populations and *in silico* clinical trials are powerful new tools that can provide biological insights that may be difficult or impossible to otherwise identify (Jenner et al. 2021a). When used in complement to experimental and clinical studies, virtual clinical trials have the potential to markedly decrease attrition rates along the drug development pipeline, helping to reduce disappointing trial results and improve patient outcomes.

Acknowledgements AS was funded by a CRM-ISM Postdoctoral Fellowship, JLR received funding from a CAMBAM fellowship and ISM scholarship, and an NSERC Discovery Grant to MC. KPW was supported by an NSERC Discovery Grant RGPIN-2018-04205. MC was funded by NSERC Discovery Grant RGPIN-2018-04546, and an FRQS J1 Research Scholar Grant. Some of the work in this chapter came from the tutorials that were run during the Centre de recherches mathématiques and Centre for Modelling in the Biosciences and Medicine funded workshop *Computational modelling of Cancer Biology and Treatments* supporting ALJ and MC.

References

Agosti A, Giverso C, Faggiano E et al (2018) A personalized mathematical tool for neuro-oncology: a clinical case study. Int J Non Linear Mech 107:170–181. https://doi.org/10.1016/j.ijnonlinmec.2018.06.004

Alfonso S, Jenner AL, Craig M (2020) Translational approaches to treating dynamical diseases through in silico clinical trials. Chaos An Interdiscip J Nonlinear Sci 30:123128. https://doi.org/10.1063/5.0019556

Allen RJ, Rieger TR, Musante CJ (2016) Efficient generation and selection of virtual populations in quantitative systems pharmacology models. CPT Pharmacometrics Syst Pharmacol 5:140–146. https://doi.org/10.1002/psp4.12063

An G (2004) In silico experiments of existing and hypothetical cytokine-directed clinical trials using agent-based modeling. Crit Care Med 32:2050–2060. https://doi.org/10.1097/01.CCM.0000139707.13729.7D

Andtbacka RHII, Kaufman HL, Collichio F et al (2015) Talimogene laherparepvec improves durable response rate in patients with advanced melanoma. J Clin Oncol 33:2780–2788. https://doi.org/10.1200/JCO.2014.58.3377

Andtbacka RHI, Ross M, Puzanov I et al (2016) Patterns of clinical response with talimogene laherparepvec (T-VEC) in patients with melanoma treated in the OPTiM phase III clinical trial. Ann Surg Oncol 23:4169–4177

Barish S, Ochs MF, Sontag ED, Gevertz JL (2017) Evaluating optimal therapy robustness by virtual expansion of a sample population, with a case study in cancer immunotherapy. Proc Natl Acad Sci 114:E6277–E6286

Bland JM, Altman DG (2011) Comparisons against baseline within randomised groups are often used and can be highly misleading. Trials 12:264. https://doi.org/10.1186/1745-6215-12-264

Boem F, Malagrinò I, Bertolaso M (2020) In silico clinical trials: a possible response to complexity in pharmacology. In: LaCaze A, Osimani B (eds) Uncertainty in pharmacology. Cham, Springer, pp 135–152

Bozic I, Antal T, Ohtsuki H et al (2010) Accumulation of driver and passenger mutations during tumor progression. Proc Natl Acad Sci 107:18545–18550. https://doi.org/10.1073/pnas.1010978107

Brown BW (1980) The crossover experiment for clinical trials. Biometrics 36:69. https://doi.org/10.2307/2530496

Cassidy T, Craig M (2019) Determinants of combination GM-CSF immunotherapy and oncolytic virotherapy success identified through in silico treatment personalization. PLoS Comput Biol 15:e1007495. https://doi.org/10.1371/journal.pcbi.1007495

Clermont G, Bartels J, Kumar R et al (2004) In silico design of clinical trials: a method coming of age. Crit Care Med. https://doi.org/10.1097/01.CCM.0000142394.28791.C3

Corral-Acero J, Margara F, Marciniak M et al (2020) The 'Digital Twin' to enable the vision of precision cardiology. Eur Heart J 41:4556–4564. https://doi.org/10.1093/eurheartj/ehaa159

Craig M, González-Sales M, Li J, Nekka F (2016) Impact of pharmacokinetic variability on a mechanistic physiological pharmacokinetic/pharmacodynamic model: a case study of neutrophil development, PM00104, and filgrastim. In Toni B (ed), Mathematical sciences with multidisciplinary applications, 91–112, Springer Science + Business Media: New York (NY)

Craig M, Kaveh K, Woosley A et al (2019) Cooperative adaptation to therapy (CAT) confers resistance in heterogeneous non-small cell lung cancer. PLoS Comput Biol 15:1–19. https://doi.org/10.1371/journal.pcbi.1007278

Crosley P, Farkkila A, Jenner AL et al (2021) Procaspase-Activating Compound-1 synergizes with TRAIL to induce apoptosis in established granulosa cell tumor cell line (KGN) and explanted patient granulosa cell tumor cells in vitro. Int J Mol Sci 22:4699. https://doi.org/10.3390/ijms22094699

DiMasi JA, Grabowski HG, Hansen RW (2016) Innovation in the pharmaceutical industry: new estimates of R&D costs. J Health Econ 47:20–33. https://doi.org/10.1016/j.jhealeco.2016.01.012

Emilia Kozłowska E, Färkkilä A, Vallius T et al (2018) Mathematical modeling predicts response to chemotherapy and drug combinations in ovarian cancer. Cancer Res 78:4036–4044

Fossella FV, Lippman SM, Shin DM et al (1997) Maximum-tolerated dose defined for single-agent gemcitabine: a phase I dose-escalation study in chemotherapy-naive patients with advanced non-small-cell lung cancer. J Clin Oncol 15:310–316. https://doi.org/10.1200/JCO.1997.15.1.310

Gray CW, Coster ACF (2016) The Akt switch model: is location sufficient? J Theor Biol 398:103–111. https://doi.org/10.1016/j.jtbi.2016.03.005

Gyuk P, Vassányi I, Kósa I (2019) Blood glucose level prediction for diabetics based on nutrition and insulin administration logs using personalized mathematical models. J Healthc Eng 2019:1–12. https://doi.org/10.1155/2019/8605206

Hay M, Thomas DW, Craighead JL et al (2014) Clinical development success rates for investigational drugs. Nat Biotechnol 32:40–51. https://doi.org/10.1038/nbt.2786

Hozo SP, Djulbegovic B, Hozo I (2005) Estimating the mean and variance from the median, range, and the size of a sample. BMC Med Res Methodol 5:1–10. https://doi.org/10.1186/1471-2288-5-13

Jafarnejad M, Gong C, Gabrielson E et al (2019) A computational model of neoadjuvant PD-1 inhibition in non-small cell lung cancer. AAPS J 21:79. https://doi.org/10.1208/s12248-019-0350-x

Jenner AL, Yun C-O, Kim PS, Coster ACF (2018) Mathematical modelling of the interaction between cancer cells and an oncolytic virus: insights into the effects of treatment protocols. Bull Math Biol 80. https://doi.org/10.1007/s11538-018-0424-4

Jenner AL, Aogo RA, Alfonso S et al (2021a) COVID-19 virtual patient cohort suggests immune mechanisms driving disease outcomes. PLoS Pathog 17:e1009753. https://doi.org/10.1371/journal.ppat.1009753

Jenner AL, Cassidy T, Belaid K et al (2021b) In silico trials predict that combination strategies for enhancing vesicular stomatitis oncolytic virus are determined by tumor aggressivity. J Immunother Cancer 9:e001387. https://doi.org/10.1136/jitc-2020-001387

Jiang X, Galettis P, Links M et al (2007) Population pharmacokinetics of gemcitabine and its metabolite in patients with cancer: effect of oxaliplatin and infusion rate. Br J Clin Pharmacol 65(3):326–333. https://doi.org/10.1111/j.1365-2125.2007.03040.x

Joerger M, Huitema ADR, Koeberle D et al (2014) Safety and pharmacology of gemcitabine and capecitabine in patients with advanced pancreatico-biliary cancer and hepatic dysfunction. Cancer Chemother Pharmacol 73:113–124. https://doi.org/10.1007/s00280-013-2327-2

Kim P-H, Sohn J-H, Choi J-W et al (2011) Active targeting and safety profile of PEG-modified adenovirus conjugated with herceptin. Biomaterials 32:2314–2326. https://doi.org/10.1016/j.biomaterials.2010.10.031

Kirtane AR, Abouzid O, Minahan D et al (2018) Development of an oral once-weekly drug delivery system for HIV antiretroviral therapy. Nat Commun 9:2. https://doi.org/10.1038/s41467-017-02294-6

Kozłowska E, Vallius T, Hynninen J et al (2019) Virtual clinical trials identify effective combination therapies in ovarian cancer. Sci Rep 9:18678. https://doi.org/10.1038/s41598-019-55068-z

Le Sauteur-Robitaille J, Yu ZS, Craig M (2021) Impact of estrogen population pharmacokinetics on a qsp model of mammary stem cell differentiation into myoepithelial cells. AIMS Math 6: 10861–10880. https://doi.org/10.3934/math.2021631

Lipsky MS, Sharp LK (2001) From idea to market: the drug approval process. J Am Board Fam Pract 14:362–367

Ma H, Pilvankar M, Wang J et al (2021) Quantitative systems pharmacology modeling of PBMC-humanized mouse to facilitate preclinical immuno-oncology drug development. ACS Pharmacol Transl Sci 4:213–225. https://doi.org/10.1021/acsptsci.0c00178

Mathworks 2020

Milberg O, Gong C, Jafarnejad M et al (2019) A QSP model for predicting clinical responses to monotherapy, combination and sequential therapy following CTLA-4, PD-1, and PD-L1 checkpoint blockade. Sci Rep 9:11286. https://doi.org/10.1038/s41598-019-47802-4

Mohs RC, Greig NH (2017) Drug discovery and development: role of basic biological research. Alzheimer's Dement Transl Res Clin Interv 3:651–657. https://doi.org/10.1016/j.trci.2017.10.005

Oh E, Oh J-E, Hong J et al (2017) Optimized biodegradable polymeric reservoir-mediated local and sustained co-delivery of dendritic cells and oncolytic adenovirus co-expressing IL-12 and GM-CSF for cancer immunotherapy. J Control Release 259:115–127. https://doi.org/10.1016/j.jconrel.2017.03.028

Pérez-García VM, Ayala-Hernández LE, Belmonte-Beitia J et al (2019) Computational design of improved standardized chemotherapy protocols for grade II oligodendrogliomas. PLoS Comput Biol 15:1–17. https://doi.org/10.1371/journal.pcbi.1006778

Piantadosi S (2017) Clinical trials: a methodologic perspective, 3rd edn. Wiley, New York (NY)

Pitcher MJ, Bowness R, Dobson S, Gillespie SH (2018) A spatially heterogeneous network-based metapopulation software model applied to the simulation of a pulmonary tuberculosis infection. Appl Netw Sci 3:33. https://doi.org/10.1007/s41109-018-0091-2

Polasek TM, Rostami-Hodjegan A (2020) Virtual twins: understanding the data required for model-informed precision dosing. Clin Pharmacol Ther 107:742–745. https://doi.org/10.1002/cpt.1778

Rieger TR, Allen RJ, Bystricky L et al (2018) Improving the generation and selection of virtual populations in quantitative systems pharmacology models. Prog Biophys Mol Biol 139:15–22. https://doi.org/10.1016/j.pbiomolbio.2018.06.002

Sacco JJ, Botten J, Macbeth F et al (2010) The average body surface area of adult cancer patients in the UK: a multicentre retrospective study. PLoS One 5:1.6. https://doi.org/10.1371/journal.pone.0008933

Sayama H, Marcantonio D, Nagashima T et al (2021) Virtual clinical trial simulations for a novel KRAS G12C inhibitor (ASP2453) in non-small cell lung cancer. CPT Pharmacometrics Syst Pharmacol 10:864–877. https://doi.org/10.1002/psp4.12661

Seidel JA, Otsuka A, Kabashima K (2018) Anti-PD-1 and anti-CTLA-4 therapies in cancer: mechanisms of action, efficacy, and limitations. Front Oncol 8:1–14. https://doi.org/10.3389/fonc.2018.00086

Sové RJ, Jafarnejad M, Zhao C et al (2020) QSP-IO: a quantitative systems pharmacology toolbox for mechanistic multiscale modeling for immuno-oncology applications. CPT Pharmacometrics Syst Pharmacol 9:484–497. https://doi.org/10.1002/psp4.12546

Stadeli KM, Richman DD (2013) Rates of emergence of HIV drug resistance in resource-limited settings: a systematic review. Antiretrovir Ther 18:115–123

Switchenko JM, Heeke AL, Pan TC, Read WL (2019) The use of a predictive statistical model to make a virtual control arm for a clinical trial. PLoS One 14:e0221336. https://doi.org/10.1371/journal.pone.0221336

Tirosh I, Izar B, Prakadan SM et al (2016) Dissecting the multicellular ecosystem of metastatic melanoma by single-cell RNA-seq. Science 352:189–196. https://doi.org/10.1126/science.aad0501

Viceconti M, Henney A, Morley-Fletcher E (2016) In silico clinical trials: how computer simulation will transform the biomedical industry. Int J Clin Trials 3:37. https://doi.org/10.18203/2349-3259.ijct20161408

Visentin R, Dalla Man C, Kovatchev B, Cobelli C (2014) The University of Virginia/Padova type 1 diabetes simulator matches the glucose traces of a clinical trial. Diabetes Technol Ther 16:428–434. https://doi.org/10.1089/dia.2013.0377

Vodovotz Y, Billiar TR (2013) In silico modeling. Crit Care Med 41:2008–2014. https://doi.org/10.1097/CCM.0b013e31829a6eb4

Wang Y, Bhattaram AV, Jadhav PR et al (2008) Leveraging prior quantitative knowledge to guide drug development decisions and regulatory science recommendations: impact of FDA pharmacometrics during 2004–2006. J Clin Pharmacol 48:146–156. https://doi.org/10.1177/0091270007311111

Wang H, Milberg O, Bartelink IH et al (2020) In silico simulation of a clinical trial with anti-CTLA-4 and anti-PD-L1 immunotherapies in metastatic breast cancer using a systems pharmacology model. R Soc Open Sci 6:190366. https://doi.org/10.1098/rsos.190366

Wang H, Ma H, Sové RJ et al (2021) Quantitative systems pharmacology model predictions for efficacy of atezolizumab and nab-paclitaxel in triple-negative breast cancer. J Immunother Cancer 9:e002100. https://doi.org/10.1136/jitc-2020-002100

Wilkie KP, Hahnfeldt P (2017) Modeling the dichotomy of the immune response to cancer: cytotoxic effects and tumor-promoting inflammation. Bull Math Biol 79:1426–1448. https://doi.org/10.1007/s11538-017-0291-4

Artificial Intelligence and Deep Phenotyping in COVID-19

9

Luciano Giacó, Bertrand De Meulder, Vincenzo Valentini, Giovanni Scambia, Alfredo Cesario, and Charles Auffray

Abstract

In this chapter, we will provide an overview of the research and applications of artificial intelligence and deep phenotyping on and around the COVID-19 pandemic. Although the COVID-19 pandemic has found us ill prepared for an immediate reaction, it came to a time when the data, tools, rationales, and collaboration schemas were ripe for a worldwide application of machine learning, artificial intelligence, big data analytics, and deep phenotyping. It is our hope that those examples will be repeated in the future, not motivated by the urgency to manage and extinguish this pandemic but as standards of collaborative science.

L. Giacó (✉)
Bioinformatics Research Core Facility, Gemelli Science and Technology Park (G-STeP), Fondazione Policlinico Universitario A. Gemelli IRCCS, Rome, Italy
e-mail: luciano.giaco@policlinicogemelli.it

B. De Meulder · C. Auffray
Association European Institute for Systems Biology and Medicine, Vourles, France
e-mail: cauffray@eisbm.org

V. Valentini
Gemelli Generator, Fondazione Policlinico Universitario A. Gemelli IRCCS, Department of Diagnostic Imaging, Radiation Oncology and Hematology, Fondazione Policlinico Universitario A. Gemelli IRCCS, Rome, Italy

G. Scambia
Scientific Directorate, Fondazione Policlinico Universitario A. Gemelli IRCCS, Rome, Italy

Dipartimento di Scienze della Salute della Donna e del Bambino e di Sanità Pubblica, Fondazione Policlinico Universitario A. Gemelli IRCCS, Rome, Italy
e-mail: giovanni.scambia@policlinicogemelli.it

A. Cesario
Open Innovation Unit, Scientific Directorate, Fondazione Policlinico Universitario A. Gemelli IRCCS, Rome, Italy
e-mail: alfredo.cesario@policlinicogemelli.it

A. Cesario et al. (eds.), *Personalized Medicine Meets Artificial Intelligence*,
https://doi.org/10.1007/978-3-031-32614-1_9

Keywords

COVID-19 · Pandemic · Deep phenotyping · Artificial intelligence · Variants · Epidemiology

9.1 SARS-CoV-2 and COVID-19

The past two decades were marked with the outbreaks of many viral diseases such as Chikungunya, Ebola, Zika, Nipah, H7N9 Bird Flu, H1N1, SARS, and MERS. In late 2019, a new severe acute respiratory disease appeared in the Hubei province of China, caused by a coronavirus identified as SARS-CoV-2. This virus is a beta coronavirus of zoonotic origin belonging to the subgenus *Sarbecovirus* in the *Orthocoronavirinae* subfamily of the family *Coronaviridae* transmitted to humans in a spillover event. Bats are thought to be the animal reservoir of SARS-CoV-2 but the other likely intermediate animal host is yet to be formally identified, although authors have suggested the Kolonok (a bat-hunting mustelid) (Maurin et al. 2021), snakes or pangolin (Mahdy et al. 2020) as possible intermediates.

The World Health Organization (WHO) named the disease caused by the SARS-CoV-2 virus "COVID-19", an acronym of Coronavirus Disease 2019, on the 11th of February 2020, then declared a pandemic status on 11th of March 2020 (WHO 2020a). As of November second, 2021, 246,594,191 confirmed cases have been reported, resulting in 4,998,784 deaths (WHO 2020b), numbers likely to be vastly underestimated.

The COVID-19 pandemic has found us ill prepared for an immediate reaction to a situation that had not been anticipated properly. However, new technologies such as machine learning, deep phenotyping, and artificial intelligence (AI) have been used fruitfully to help monitor and tackle this pandemic. Here, we present an overview of deep phenotyping and AI models, with a special focus on the very timely and relevant research around COVID-19 variants.

9.2 Overview of Deep Phenotyping in COVID-19

While the medical community has gained insight into the epidemiology of COVID-19 (using AI tools, see further), important questions remain about the clinical complexities and underlying mechanisms of disease phenotypes. Deep phenotyping—or the precise and comprehensive characterization of observable traits representing unique morphological, biochemical, physiological, or behavioural properties of the identified patient populations (Weng et al. 2020; Robinson 2012)—has been used extensively to try and elucidate the mechanisms of the COVID-19 infection.

Deep phenotyping efforts have been conducted in the past two years using various sources of data. First and mainly, the clinical records of COVID-19 patients can and are being used to characterize the clinical presentations of patients in the hospital and identify risk factors for more severe forms of the disease (Burn et al. 2020; Planchuelo-Gómez et al. 2020; Deer et al., 2021), but also to differentiate the COVID-19 disease from previous viral respiratory diseases (SARS, MERS).

Deep phenotyping based on clinical data of hospitalized patients has facilitated tailoring therapies and improved outcomes. One study in the USA has used EHR data for 1022 patients admitted between March and August 2020 across 14 hospitals (Lusczek et al. 2021). They performed ensemble clustering using 33 clinically-relevant variables and identified 3 phenotypes, with different levels of severity and mortality. They then associated comorbidities, clinical complications and outcomes with the clusters through regression models and suggested this model could be used as a guide for clinical decision support.

Another study gathered records for 34,128 patients across the USA, South Korea, and Spain, comparing their clinical characteristics between centres and comparing COVID-19 characteristics with those of Influenza using data for more than 85,000 patients hospitalized between 2014 and 2019 (Burn et al. 2020).

9.3 Deep Phenotyping of COVID-19 Variants

9.3.1 Epitope: Antibody Background

Antigens recognition by the antibodies plays a key role in the humoral immune system response against pathogens. The antigen part recognized by the antibody Fragment Antigen Binding (FAB) domain, is known as B-cell epitope and the structural properties of these epitopes, or antigenic determinants, lead to the specificity of the antibody/epitope interaction. The knowledge of these interactions is the first fundamental step to design and develop therapeutics, peptide-based vaccines and diagnostics tools. On this basis, the computational approach becomes pivotal to investigate the molecular basis of epitope recognition from antibodies. Several computational tools were developed in the last years, using different methods, to study and predict epitope–antibody interactions.

The SARS-CoV-2 pandemic triggered an effort in the scientific community to develop therapies to fight the COVID-19 disease. Some of these, like RNA-based vaccines and monoclonal antibody therapies, are based on the interaction between antibody and the SARS-CoV-2 Spike protein. Monoclonal antibody therapy could be a good strategy to mitigate COVID-19 respiratory illness and the RNA-vaccines are designed to stimulate the humoral immune system to produce antibodies against the Spike protein. This antibody-based approach revealed a decrease of infection (Menni et al. 2021) and illness (Sterlin et al. 2021) related to SARS-CoV-2 and indicates this strategy as a good approach to exit from the global pandemic.

9.3.2 Spike Protein and Variants

Coronavirus 2 Spike protein (S protein) is a trimeric surface glycoprotein that, interacting with Angiotensin-converting enzyme 2 (ACE2) of host cells, allows to infect human cells as: ciliate cells in lung, proximal tubular cells in kidneys, cardiomyocytes in heart tissue. The Spike structure is composed of S1 and S2 principal domains and can be found in open and closed states. The S1 domain is involved in protein–host interaction, while the S2 domain contains several sub-domains involved in the membrane host fusion process. The receptor binding domain (RBD) is located in the S1 subunit and its up-conformation is related to the protein open state. In this structural conformation, the protein is able to bind ACE2 and infect the host cells. The S2 subunit contains HR1 and HR2 sub-domains, required for the fusion to the host cellular membrane; these two sub-domains constitute a six-helix bounded structure (6-HB) high conserved between SARS-CoV and SARS-CoV-2.

The RBD-ACE2 interaction is the critical step during virus-host infection; in the "down" state, RBD is not able to interact with the ACE2 protein. The conformational transition from "down" to "up" state makes the Spike protein able to engage the host, forming a Spike-ACE2 complex. Interfering into this flow could be a strategy to avoid the host infection, like monoclonal antibodies able to work as competitors, binding the RBD and breaking the infection mechanism. Understanding the Spike-ACE2 complex, as the sub-domains interaction, is useful to know which RBD domains can be an antibody target.

WHO provides to the scientific community the "state of the art" of SARS-CoV-2 variants circulating over the countries updated on a regular basis (CDC 2020a). WHO classified the variants in three categories: Variant of Interest (VOI), Variants of Concern (VOC), and Variants Under Monitoring (VUM). Spike protein substitutions and clinical attributes are provided for each variant.

The VOI includes all variants related to a change of RBD Spike domain which is observed as a neutralization reduction of antibodies produced by vaccination or against infection. A VOI is associated with an increased potential transmissibility, disease severity, and diagnostic impact. In VOC are included all variants where there exists evidence of increased transmissibility, reduced neutralizing antibodies and vaccine treatments effectiveness and diagnostic detection failure. In the VUM are included all those variants for which the epidemiological impact is unclear and requires more evidence to understand their impact.

9.3.3 Spike–Antibody Interaction

The therapeutic option of monoclonal antibodies, in parallel with vaccines, is justified by the important technical progress in this field that allows the development of targeted therapies in a short period of time, with high confidence in their safety. It is possible to evaluate the interaction of antibodies against the RBD domain of Protein S and study its stability through molecular dynamics experiments.

The RBD domain is in fact immunodominant and represents the target of 90% of the neutralizing activity present in SARS-CoV-2 immune sera (Menni et al. 2021).

The number of structure complex RBD/Antibody stored in the Protein Data Bank (PDB) is increasing and some complexes involve monoclonal antibodies used in the therapy against COVID-19. On the other hand, a great number of RBD/Antibody isolated from patients are available and the number of structures could be the starting point to understand the antibodies interaction mechanisms against the RBD Spike domain.

In this context, the synergy of two computational methodologies, Molecular Docking and Molecular Dynamics, plays a role to anticipate the behaviour of variants with the antibodies in disease care using monoclonal therapy.

9.4 Overview of AI for COVID-19

Many AI and advanced modelling approaches have been and are being used to date in the context of the fight against COVID-19. AI is being successfully used in the identification of disease clusters, monitoring of cases, prediction of future outbreaks, mortality risk, diagnosis for COVID-19, disease management by resource allocation, facilitating training, record maintenance, and pattern recognition for studying the disease trend.

9.4.1 AI in Prediction and Tracking

Using social media platforms, online discussion, and news sites, AI can forecast the spread of viruses and help develop early warning systems (Hussain and Sheikh 2021). It can then provide useful information about vulnerable regions and for prediction of morbidity and mortality. For example, Andreadis et al. (2021) have developed an analytics platform to collect and analyse tweets about the pandemic in Italy, using deep learning to geotag the tweets and predict places at risk and cases of overcrowding (Chen et al. 2020). Another group (Cresswell et al. 2021) used AI to perform a sentiment-based analysis to understand public views and concerns based on Facebook and Twitter content. In general, access to the appropriate data and using the proper models is crucial in building a useful AI model (Santosh 2020).

9.4.2 AI in Contact Tracing and Population Control

Contact tracing has proven to be key to slowing the spread of COVID-19 (CDC 2020b). The example set by Singapore in its control of epidemic risks, with the support of technology, is certainly unique and difficult to extend to other countries because of the social acceptance of restrictive measures. In China, AI has been extensively used in support of such mass surveillance policies, such as using devices to measure temperature remotely or facial recognition (COE 2020). In Israel, South

Korea, Taiwan, or Italy, smartphone apps have been developed to track people's location using the phone's geolocalization and AI to alert when people are either not complying with lockdown or isolation periods, or visiting crowded places (Laurent 2020).

9.4.3 AI in Monitoring of COVID-19 Cases

AI techniques are applied for monitoring patients in clinical settings and prediction of course of treatment. Based on the data derived from vital statistics and clinical parameters, AI provides support for clinical decision-making and resource allocation in Intensive Care Units (ICU) (Arora et al. 2020), or help doctors perform remote monitoring of vital signs and symptoms of COVID-19 (Rohmetra et al. 2021), thereby avoiding contacts with infected patients, decreasing contaminations and freeing time and resources in overburdened hospitals.

9.4.4 AI in Early Diagnosis

AI can help harnessing data from mobile health applications, for example to distinguish cough from patients with COVID-19 from other illnesses (Imran et al. 2020). Data from wearable devices (temperature, oxygen saturation levels, breathing rate, cough and lung sound monitoring, ECG, blood pressure...) combined with AI has been used to speed up the screening process of the spread of the virus or distinguishing mild from severe infections (Channa et al. 2021). Additionally, AI can be used to speed up the diagnosis process, for example in reading Computed Tomography (CT) scans of patients presenting at the hospital (Li et al. 2020).

9.4.5 AI in Reducing the Burden for Healthcare Staff and Medical Practitioners

AI-based triage systems can help in reducing the work burden of medical staff and healthcare professionals by automating several processes such as helping in training practitioners, treatment decision support, digitization of patients reports and minimizing contacts with infected patients, distant monitoring through chatbots or remote monitoring, and classification of patients based on medical records (Iwendi et al. 2020; Miner et al. 2020; Rasheed et al. 2020; Wu et al. 2020).

9.4.6 AI in Curbing the Spread of Misinformation

There is an overwhelming amount of sometimes conflicting information and various doubts and conspiracies spreading online and having an impact on the fight against the pandemic. AI has been and is being used to identify and monitor false information and help curtail rumours and misinformation (Khan et al. 2020). AI techniques can be further used to clearly and usefully present updated information to fight misinformation and identify gaps (Samuel et al. 2020). AI can then further be used to track new evidence in diagnosis, treatment, and outcomes, helping clinicians make the best possible choice to treat their patients and help the public overcome fear and doubts.

9.4.7 AI in COVID-19 Variants

Molecular dynamic simulations have employed to compare the relative dynamics, stabilities, and energy differences between the SARS-CoV and the SARS-CoV-2/ ACE2 complexes (Pavlova et al. 2021). Recent advances in hardware and software made molecular simulations possible, by harnessing the large amount of data available and machine learning or AI algorithms to perform those simulations.

9.5 Future Directions

One of the few positive aspects of the COVID-19 pandemic has been the accrued collaboration in the scientific world, including in data sharing and in building models. As shown above, AI has benefitted from this trend, and demonstrated it can yield significant and useful results in many aspects around the pandemic and its impact on the health systems around the world. We hope this effort can be applied with benefits to a broader set of clinical and research settings, and that the general population will become better informed about AI and the positive contribution it can have on our lives. In research, we hope that the trend of data sharing will continue to be adopted and strengthened, as has been the case, for example, at our hospital by the "Gemelli against COVID-19" group (Murri et al. 2021) and within the COVID-19 Disease Map Community (Ostaszewski et al. 2020, 2021) so that more big data, machine learning, and AI analyses and modelling can take place, for the benefit of all.

References

Andreadis S, Antzoulatos G, Mavropoulos T et al (2021) A social media analytics platform visualising the spread of COVID-19 in Italy via exploitation of automatically geotagged tweets. Online Soc Netw Media 23:100134

Arora N, Banerjee AK, Narasu ML (2020) The role of artificial intelligence in tackling COVID-19. Future Virol 15:717–724

Burn E, You SC, Sena AG et al (2020) Deep phenotyping of 34,128 adult patients hospitalised with COVID-19 in an international network study. Nat Commun 11:5009

Centers for Disease Control and Prevention (CDC) (2020a) Coronavirus Disease 2019 (COVID-19). Source: https://www.cdc.gov/coronavirus/2019-ncov/variants/variant-info.html. Last consultation: 02/07/2021

Centers for Disease Control and Prevention (CDC) (2020b) COVID-19 and Your Health. Source: https://www.cdc.gov/coronavirus/2019-ncov/daily-life-coping/contact-tracing.html Last consultation: 02/07/2021
Channa A, Popescu N, Skibinska J, Burget R (2021) The rise of wearable devices during the COVID-19 pandemic: a systematic review. Sensors 21:5787
Chen E, Lerman K, Ferrara E (2020) Tracking social media discourse about the COVID-19 pandemic: development of a public coronavirus Twitter Data Set. JMIR Public Health Surveill 6:e19273
Council of Europe (COE) (2020) AI and control of Covid-19 coronavirus. Artificial Intelligence. Source: https://www.coe.int/en/web/artificial-intelligence/ai-and-control-of-covid-19-coronavirus Last consultation: 02/07/2021
Cresswell K, Tahir A, Sheikh Z et al (2021) Understanding public perceptions of COVID-19 contact tracing apps: artificial intelligence–enabled social media analysis. J Med Internet Res 23: e26618
Deer RR, Rock MA, Vasilevsky N et al (2021) Characterizing long COVID: deep phenotype of a complex condition. medRxiv 2021.06.23.21259416
Hussain A, Sheikh A (2021) Opportunities for artificial intelligence–enabled social media analysis of public attitudes toward covid-19 vaccines. NEJM Catal Innov Care Deliv 23(4):e26627
Imran A, Posokhova I, Qureshi HN et al (2020) AI4COVID-19: AI enabled preliminary diagnosis for COVID-19 from cough samples via an app. Inform Med Unlocked 20:100378
Iwendi C, Bashir AK, Peshkar A et al (2020) COVID-19 patient health prediction using boosted random forest algorithm. Front Public Health 8:357
Khan R, Shrivastava P, Kapoor A et al (2020) Social media analysis with AI: sentiment analysis techniques for the analysis of Twitter COVID-19 data. J Crit Rev 7:2020
Laurent A (2020) COVID-19: des États utilisent la géolocalisation pour savoir qui respecte le confinement. Source: https://usbeketrica.com/fr/article/covid-19-la-geolocalisation-pour-savoir-qui-respecte-confinement. Last consultation: 02/07/2021
Li L, Qin L, Xu Z et al (2020) Using artificial intelligence to detect COVID-19 and community-acquired pneumonia based on pulmonary CT: evaluation of the diagnostic accuracy. Radiology. https://pubs.rsna.org/doi/10.1148/radiol.2020200905
Lusczek ER, Ingraham NE, Karam BS et al (2021) Characterizing COVID-19 clinical phenotypes and associated comorbidities and complication profiles. PLoS One 16:e0248956
Mahdy MAA, Younis W, Ewaida Z (2020) An overview of SARS-CoV-2 and animal infection. Front Vet Sci 7
Maurin M, Fenollar F, Mediannikov O et al (2021) Current status of putative animal sources of SARS-CoV-2 infection in humans: wildlife, domestic animals and pets. Microorganisms 9:868
Menni C, Klaser K, May A et al (2021) Vaccine side-effects and SARS-CoV-2 infection after vaccination in users of the COVID Symptom Study app in the UK: a prospective observational study. Lancet Infect Dis 21:939–949
Miner AS, Laranjo L, Kocaballi AB (2020) Chatbots in the fight against the COVID-19 pandemic. NPJ Digit Med 3:1–4
Murri R, Lenkowicz J, Masciocchi C et al (2021) A machine-learning parsimonious multivariable predictive model of mortality risk in patients with Covid-19. Sci Rep 11:21136
Ostaszewski M, Mazein A, Gillespie M et al (2020) COVID-19 Disease Map, building a computational repository of SARS-CoV-2 virus-host interaction mechanisms. Scientific Data 7(1):136
Ostaszewski M, Niarakis A, Mazein A et al (2021) COVID19 Disease Map, a computational knowledge repository of virus–host interaction mechanisms. Mol Syst Biol 17:e10387
Pavlova A, Zhang Z, Acharya A et al (2021) Machine learning reveals the critical interactions for SARS-CoV-2 spike protein binding to ACE2. J Phys Chem Lett 12:5494–5502
Planchuelo-Gómez Á, Trigo J, de Luis-García R et al (2020) Deep phenotyping of headache in hospitalized COVID-19 patients via principal component analysis. Front Neurol 11:583870

Rasheed J, Jamil A, Hameed AA et al (2020) A survey on artificial intelligence approaches in supporting frontline workers and decision makers for the COVID-19 pandemic. Chaos Solitons Fractals 141:110337

Robinson PN (2012) Deep phenotyping for precision medicine. Hum Mutat 33:777–780

Rohmetra H, Raghunath N, Narang P et al (2021) AI-enabled remote monitoring of vital signs for COVID-19: methods, prospects and challenges. Computing. https://doi.org/10.1007/s00607-021-00937-7

Samuel J, Ali GGMN, Rahman MM et al (2020) COVID-19 public sentiment insights and machine learning for tweets classification. Information 11:314

Santosh KC (2020) AI-driven tools for coronavirus outbreak: need of active learning and cross-population train/test models on multitudinal/multimodal data. J Med Syst 44:93

Sterlin D, Mathian A, Miyara M et al (2021) IgA dominates the early neutralizing antibody response to SARS-CoV-2. Sci Transl Med 13:eabd2223

Weng C, Shah NH, Hripcsak G (2020) Deep phenotyping: embracing complexity and temporality – towards scalability, portability, and interoperability. J Biomed Inform 105:103433

WHO (2020a) WHO Director-General's opening remarks at the media briefing on COVID-19–11 March 2020. Source: https://www.who.int/director-general/speeches/detail/who-director-general-s-opening-remarks-at-the-media-briefing-on-covid-19%2D%2D-11-march-2020 Last consultation: 02/07/2021

WHO (2020b) WHO Coronavirus (COVID-19) Dashboard. Source: https://covid19.who.int Last consultation: 02/07/2021

Wu J, Zhang P, Zhang L et al (2020) Rapid and accurate identification of COVID-19 infection through machine learning based on clinical available blood test results. medRxiv. https://doi.org/10.1101/2020.04.02.20051136

Multilevel Modelling with AI: The Synergy-COPD Endeavour

10

Rubèn González-Colom, Isaac Cano, Jordi Piera-Jiménez, and Josep Roca

Abstract

Key lessons learnt in the Synergy-COPD project, that aimed at a systems medicine approach to patients with chronic obstructive pulmonary disease (COPD), contributed to formulate the concept of multisource predictive modelling for enhanced clinical risk assessment described in the chapter. Further research and innovation developments in the field, as well as practicalities learnt during the process of digitalization of the regional health system in Catalonia have been main sources for the current report that aims to provide a summary description of the steps needed for implementation and adoption of a Learning Healthcare System.

Keywords

Digitalization · Healthcare · Health sciences · Predictive modelling · Preventive medicine · Systems medicine

R. González-Colom · I. Cano · J. Roca (✉)
Hospital Clínic de Barcelona, Institut d'Investigacions Biomèdiques August Pi i Sunyer (IDIBAPS), Universitat de Barcelona, Barcelona, Spain
e-mail: jroca@clinic.cat

J. Piera-Jiménez
Catalan Health Service, Barcelona, Spain

Digitalization for the Sustainability of the Healthcare (DS3) - IDIBELL, Barcelona, Spain

Faculty of Informatics, Telecommunications and Multimedia, Universitat Oberta de Catalunya, Barcelona, Spain

A. Cesario et al. (eds.), *Personalized Medicine Meets Artificial Intelligence*,
https://doi.org/10.1007/978-3-031-32614-1_10

10.1 Multisource Predictive Modelling for Enhanced Clinical Risk Assessment

Physician's best judgement, combined with rules-based management, are the two components of conventional clinical decision-making across all healthcare tiers. While health professionals' judgement relies on knowledge, training, and experience; rules-based management consists of thresholds of target variables articulating evidence-based decision criteria, often emerging from randomized controlled trials (RCTs).

Current clinical practice is facing major challenges due to the enormous evolution of computational sciences, as well as quick progress of digital medicine towards large-scale clinical application. Computational modelling is already a powerful tool extensively used for highly standardized medical procedures, like analysis of imaging techniques. However, it is expected that multilevel predictive modelling will offer valuable support to enhance clinical decision-making, complementing but not substituting clinical judgement (Rajpurkar et al. 2022).

The use of dynamic multisource predictive modelling approaches for clinical decision support that establishes relationships between multilevel and multiscale sets of predictors, targeting specific health outcomes by the use of statistical techniques and/or Artificial Intelligence/Machine Learning (AI/ML), is still in its infancy (Doos et al. 2016). However, it is a natural step towards customization of care to individual patient's needs.

Several interconnected factors, such as: current changes in healthcare paradigm, well-identified complexities in the healthcare scenario, as well as complexities of multilevel data integration and data security and privacy, explain the barriers encountered to define, deploy, and adopt operational strategies to facilitate preventive, value-based (Porter 2008) healthcare for acute and chronic patients using computational modelling in real-world clinical settings, with a twofold aim: (i) to slow-down chronic patients' progression towards the tip of the population-based risk stratification pyramid through cost-effective preventive strategies; and (ii) to enhance reliable decision-making.

It has become widely accepted that health risk assessment for patient stratification is a relevant component in the strategies for regional adoption of integrated care because of its contribution in the design of healthcare policies and services using a population-based approach, as well as for enhanced clinical management for individual patients.

The foundations of health risk assessment proposed in the chapter rely on two concepts generated within the Synergy-COPD project (Synergy-COPD Consortium 2010). Briefly, Cano et al. (2014a) reported on the concept and operational aspects of a Digital Health Framework (DHF), defined as the articulation of open and modular digital components supporting the interplay among four types of data sources: (i) patients' self-tracking data including lifestyles, environmental, behavioural aspects, and sensors; (ii) healthcare data from electronic health records (HER); (iii) biomedical research data; and (iv) population-based registry data. The basic idea is that an operational DHF could overcome current health-related silos of information,

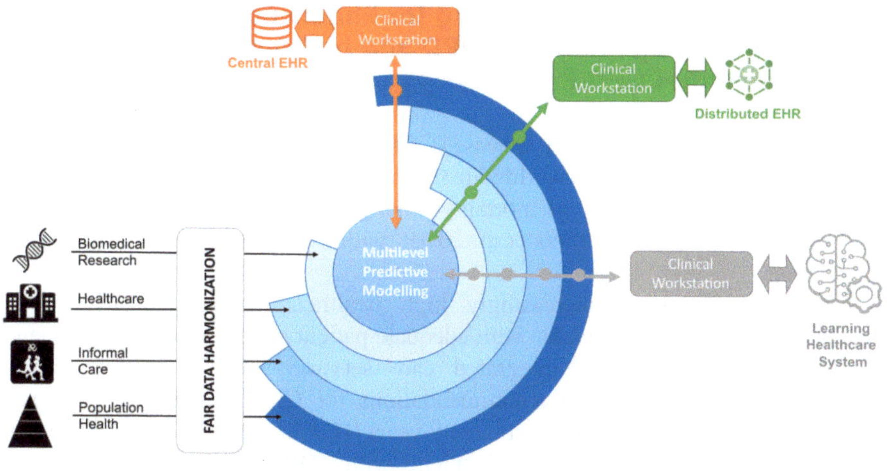

Fig. 10.1 Dynamic enhancement of multilevel clinical predictive modelling feeding clinical decision support systems (CDSS). Development of enhanced clinical predictive modelling requires consideration, and eventual integration, of computational modelling of four different dimensions: (i) Underlying biological mechanisms (biomedical research); (ii) Current evidence-based clinical knowledge (healthcare); (iii) Patients' self-tracked data, including sensors, behavioural, environmental, and social information (informal care); and (iv) Population-based health risk assessment data (population health). Figure taken from Roca et al. (2020)

being the core component of a Learning Healthcare System (LHS). The concept of LHS was first formulated in 2012 by the Institute of Medicine as a strategy to improve the quality and efficiency of healthcare (Ferguson 2012). Thereafter, the American Heart Association (AHA) (Maddox et al. 2017) further developed the practicalities of the LHS concept and proposed specific steps to make it operational and evaluate its implementation.

The second major pillar (Dueñas-Espín et al. 2016) refers to the huge potential for enhancing clinical risk predictive modelling by incorporating the four categories of variables alluded to above as covariates using a multilevel approach, as described in Fig. 10.1.

The analyses of facilitators and barriers expected in the deployment of multilevel clinical predictive modelling within a DHF clearly indicate that achievement of personalized management of patients into real-world scenarios will be a stepwise, medium-term, process requiring proper adoption strategies that must necessarily consider the different dimensions described in (Dueñas-Espín et al. 2016).

10.2 Computational Modelling for Enhanced Understanding and Management of COPD and Its Co-morbidities: The Synergy-COPD Project

Chronic Obstructive Pulmonary Disease (COPD) is a disorder that generates a high burden on healthcare systems worldwide (Murray and Lopez 2013; Blumenthal et al. 2016) being the third cause of mortality among chronic conditions, causing 3.23 million deaths in 2019 (Mathers et al. 2019), with a prevalence of 9–10% in adult population.

COPD generates an increasingly high healthcare impact mostly due to hospitalizations, partly avoidable with adequate patient stratification strategies leading to better selection of integrated care services. Despite highly relevant contributions of international recommendations for COPD management, mostly pulmonary-oriented, during the last twenty years (Halpin et al. 2021), it is nowadays widely accepted that optimal care of patients with COPD requires a systems medicine approach (as proposed in Roca et al. 2020; Roca et al. 2014). This is due to a combination of factors, such as important heterogeneities of patients' phenotypes, high rate of co-morbidities, and overlapping of diagnosis with other obstructive pulmonary diseases, with under- and overdiagnosis of the disorder (Diab et al. 2018).

The EU project Synergy-COPD (Synergy-COPD Consortium 2010), running from 2011 to 2014 (FP7-ICT-270086), was a systems research programme on multimorbidity taking COPD as a use case. The project focused on non-pulmonary phenomena often seen in patients with COPD addressing unknown aspects of skeletal muscle dysfunction/wasting (Maltais et al. 2014) and the phenomenon of co-morbidity clustering (Barnes 2015). The research was designed as an iterative process wherein data from several sources, encompassing animal experimentation (Davidsen et al. 2014), human studies (Rodríguez et al. 2011, 2012), epidemiological research and registry information (Vela et al. 2018; Gomez-Cabrero et al. 2016), were articulated and analysed using different, and in some cases complementary, computational modelling techniques. The details of the research design and methodological issues were reported in a dedicated monograph (Gomez-Cabrero et al. 2014) and the project outcomes addressing three biomedical areas: (i) Skeletal muscle dysfunction; (ii) COPD co-morbidities; and (iii) Proposals for enhanced transfer of knowledge into clinical practice, have been described in different scientific publications (Marín De Mas et al. 2017; Tényi et al. 2018a, b; Cano et al., 2014b).

Briefly, the project findings contributed to better understand the interplay of factors modulating non-pulmonary manifestations in patients with COPD. Abnormalities in co-regulation of core biological pathways (i.e. bioenergetics, inflammation, and tissue remodelling) at systemic level seem to play a central role on both skeletal muscle dysfunction and co-morbidity clustering (Fig. 10.2), with evidence of the relevant role of oxidative stress as a characteristic mechanism in these patients (Barnes 2015).

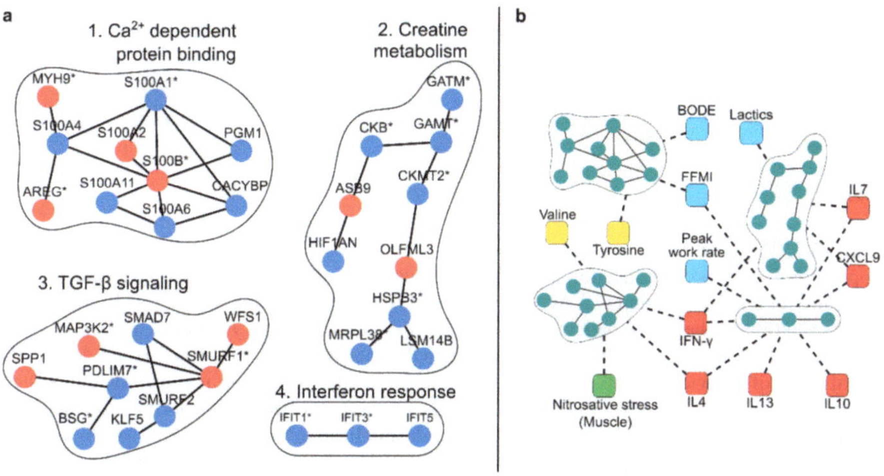

Fig. 10.2 Disease effects network modules. Panel (**a**) depicts the four network modules associated to COPD disease effects on skeletal muscle and their composing genes. Genes are coloured according to their differential regulation, namely: up regulation—red nodes; and down regulation—blue nodes. Significantly, differentially expressed genes are indicated by * (FDR ≤ 0.05). Panel (**b**) shows the significant correlations of independent measurements with any of the network modules' first three principal components. Blue squares depict exercise related independent variables; red squares show blood cytokines levels; yellow squares correspond to serum amino acids levels; and green squares represents redox biomarkers. It is noted that abnormal skeletal muscle findings, associated to poor patients' prognosis, showed significant correlations with aerobic capacity, but not with lung function measurements at rest (Tényi et al. 2018a). Figure taken from Roca et al. (2020)

Synergy-COPD generated experience on integration of records from approximately 13 million patients from the Medicare database with disease-gene maps that were derived from several resources including a semantic-derived knowledgebase (Gomez-Cabrero et al. 2016). The results demonstrated higher prevalence of most of the diseases, as comorbid conditions, seen in elderly patients with COPD compared with non-COPD subjects, an effect confirmed latter (Tényi et al. 2018b) in a regional EU dataset (1.4 million patients). Moreover, the analysis of temporal order of disease diagnosis showed that comorbid conditions in elderly patients with COPD tend to appear after the diagnosis of the obstructive disease, rather than before it. Overall, the results (Vela et al. 2018) demonstrated high impact of COPD co-morbidities on health risk stratification with major negative impact on mortality, hospitalizations, and use of healthcare resources (Fig. 10.3) and highly encourage developments of AI/ML tools using health registries and data from EHR to build robust health risk stratification strategies.

Figure 10.4 displays the distribution of individual costs per year in patients with COPD based on their multimorbidity level: from low (left column) to very high risk associated to co-morbidities (right column), indicating huge heterogeneities among patients' healthcare expenditure per year, explained by multimorbidity. The analysis of distribution of costs clearly indicates the high impact of hospitalizations and

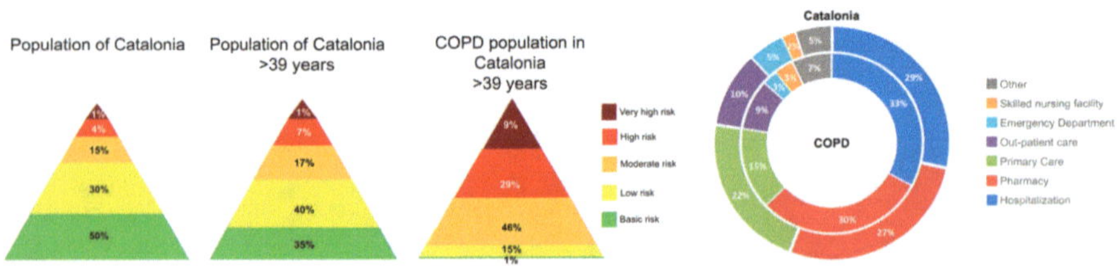

Fig. 10.3 Regional population-based study of patients with COPD. Left panel depicts three population-based risk stratification pyramids build using AMG as multimorbidity index: (i) Left, the entire regional population (7,7 M); (ii) Centre, citizens above 39 years; and, (iii) Right, display the distribution of the 264 k patients with COPD in the region across AMG risk grades: baseline (1%), low (15%), moderate (46%), high (29%), and very high risk (9%). Right panel depicts the distribution of individual costs per year comparing overall cost for the regional Health System expressed as percentages (outer circle) and the relative costs ascribed to patients with COPD (inner circle) in the left-hand side figure indicates that Hospitalization costs (€ 2291.8 M, 29%, and € 356.6 M, 33%, respectively) and, Pharmacy costs (€ 2193.4 M, 27%, and € 325.8 M, 33%) are relatively higher in COPD patients than in the overall health system; whereas, Primary Care costs (€ 1745.0 M, 22%, and € 158.9 M, 15%) are relatively lower in COPD than in the overall health system. The item: Others, includes home-based respiratory therapies, dialysis, outpatient rehabilitation, and non-urgent healthcare transportation (Vela et al. 2018)

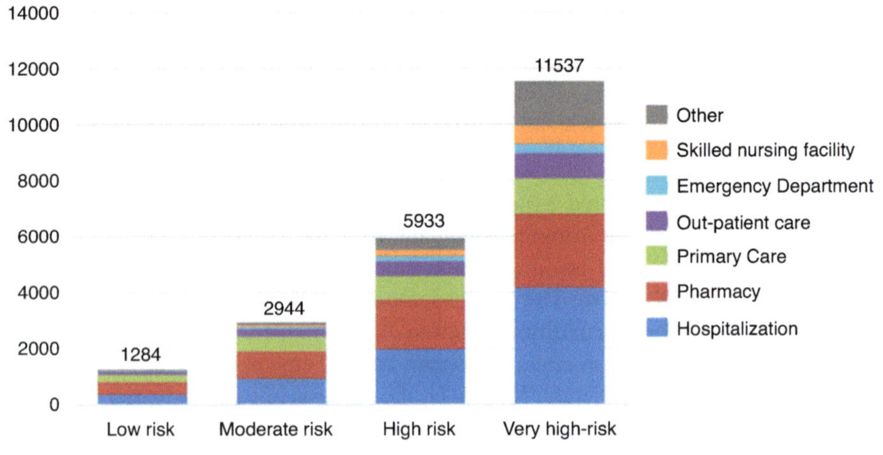

Fig. 10.4 Global yearly expenditure, expressed in €, of COPD patients by morbidity scoring at regional level (Vela et al. 2018) (264 k patients with COPD)

pharmacy on overall costs, as well as the relatively reduced impact of primary care on overall patients' cost.

All in all, the research strongly pointed out the need for a broader vision in the care and management of COPD by adopting a patient-oriented approach that addresses much more than just the pulmonary manifestations of the disease.

One of the major strengths of the Synergy-COPD project was the combination of well-defined biomedical goals with parallel technological developments beyond the state of the art in terms of novel modelling approaches, knowledge generation tools, and digital technologies supporting care coordination.

10.3 Multilevel Data Integration and Advanced AI/ML: Beyond Synergy-COPD

Beyond the project lifespan, developments on multilevel data integration and advanced AI/ML are allowing to further explore factors modulating multimorbidity aiming at transferring novel knowledge into the clinical arena. Two specific areas raising high expectations are:

The use of sparse Bayesian Direct Morbidity Maps (BDMM) to improve construction of comorbidity patterns (Marx et al. 2017). The method shows clear advantages compared to conventional hypothesis-free exploration of comorbid conditions using pairwise methods often leading to confounders due to large number of pairwise associations arising indirectly through other comorbidity associations (Fig. 10.5).

Recent studies are benefiting from the experience of using BDMM for the analysis of multilevel datasets (Trajectome 2020). Consolidated achievements in the analysis of temporal disease map-based stratification of depression-related multimorbidity can be transferred to other chronic conditions, such as COPD to enhance our understanding of the use case, but also to improve management of co-morbidities in general.

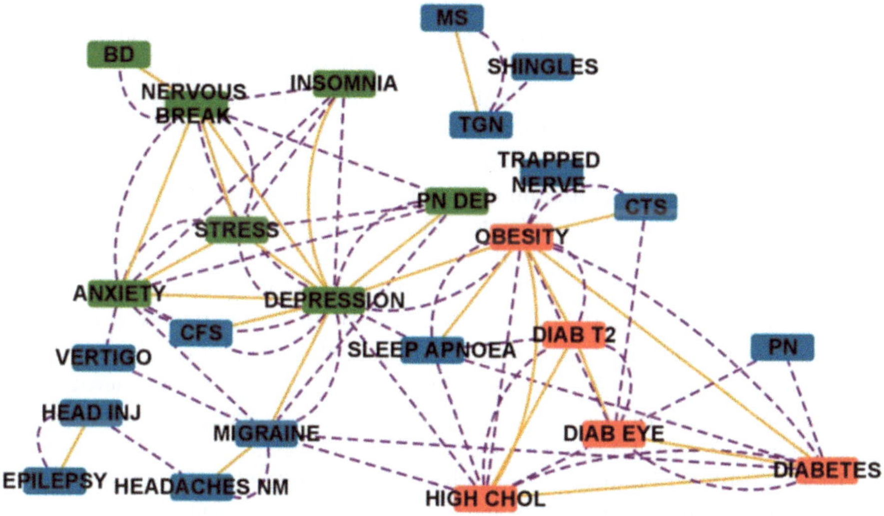

Fig. 10.5 Network representation of disease-disease comorbid relations assessed with pairwise χ^2 statistical associations (purple) and Bayesian Direct Multimorbidity Maps (BDMM – gold) in the UK Biobank dataset (Marx et al. 2017). In red metabolic syndromes, in blue diseases of the nervous system, and in green mental and behavioural disorders

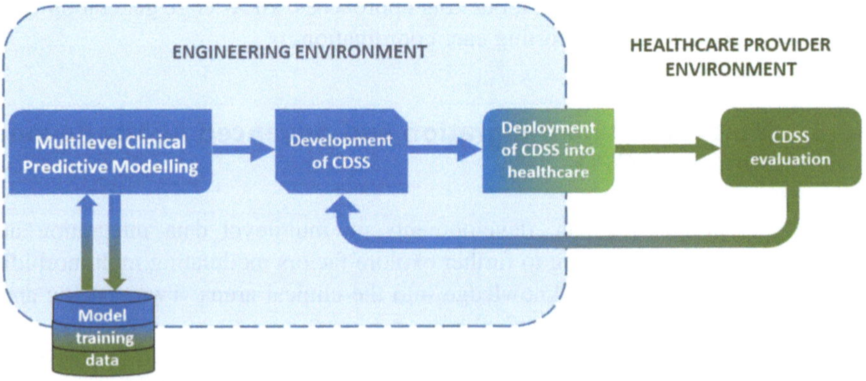

Fig. 10.6 Federated Learning approach for Multilevel data integration and advanced AI/ML. Development of enhanced dynamic clinical predictive modelling will require consideration, and eventual integration, of computational modelling of different dimensions, as described in Fig. 10.1. The current figure illustrates the steps from model elaboration, model training, evaluation and feeding clinical decision support systems (CDSS) to be applied in a real-world healthcare setting

A second area of interest is the use of the Adjusted Morbidity Groups (AMG) morbidity index (Monterde et al. 2016, 2018, 2020; Vela et al. 2021) as covariate in multilevel computational modelling (Calvo et al. 2020). It is of note that AMG is an open, publicly owned algorithm, weighted by the real impact of morbidities in each healthcare system. AMG offers clear advantages against all other morbidity indices. The algorithm is already extensively used for both policy makers and clinicians. Its site transferability has been proven and is currently being successfully tested at EU level within the ongoing Joint Action on implementation of digitally enabled integrated person-centred care (JADECARE 2020). Moreover, knowledge generated from BDMM, and disease trajectories could be used to enrich the current AMG tool to improve management of multimorbidity in general, beyond COPD.

Progress in this field needs to take advantage of Federated Learning (FL) (Rajpurkar et al. 2022) to decentralize AI/ML across data controllers to collaboratively learn a shared prediction model that ultimately could feed a clinical decision support system (CDSS) (Fig. 10.6). To this end, Bayesian multilevel systems-based analysis from consolidated methodological developments (Marx et al. 2017; Trajectome 2020) can be used to address fusion of heterogeneous information sources (registry data, clinical information, genetic information, and other biological markers) and outcomes from data owners.

However, further biomedical research is still needed to identify causal factors of co-morbidities clustering in COPD and to gain insight on the heterogeneities seen in these patients. Specific examples of target aspects requiring research are: (i) in-born genetic susceptibility; (ii) epigenetic changes associated with unhealthy lifestyles; and (iii) unknown interactions with gut microbiome, among others. Likewise, identification of plasma metabolomics patterns facilitating early identification of subsets of patients with COPD that are candidates for secondary prevention of co-morbidities would also be a major achievement to significantly reduce the burden of multimorbidity.

With the perspective of some years after completion of the Synergy-COPD project, we can conclude that targeting COPD as a use case and adopting a systems approach to address the analysis of non-pulmonary phenomena in these patients was a right choice because it facilitated to the researchers involved to think "outside of the box". On the other hand, the concurrence of three intertwined phenomena: (i) relatively poor knowledge of underlying mechanisms; (ii) marked heterogeneities of these patients; and (iii) taxonomy problems (Celli and Augustì 2018) perfectly justified the choice for a systems approach.

Moreover, the analysis of co-morbidities in these patients has provided relevant new knowledge on multimorbidity in general, with high positive impact on management strategies for chronic patients aiming at effectively reduce the burden of non-communicable conditions on health systems. A major lesson learnt was the huge potential of multilevel integrative analyses of registry data, biomedical research information, EHR and patients' self-tracking data for enhanced clinical decision support, as displayed in Fig. 10.1. It clearly constitutes a high priority to pave the way towards enhanced clinical management and personalized medicine for patients with chronic disorders (Dueñas-Espín et al. 2016).

10.4 From Systems Medicine to Integrated Care

All the above results indicate that convergence between a systems medicine approach to chronic disorders and care coordination, integrated care (JADECARE 2020), may conform an optimal scenario to foster cross-fertilization between biomedical research and clinical practice (Ferguson 2012; Maddox et al. 2017). The two approaches have several common aspects: (i) holistic and multidisciplinary approach; (ii) use of computational modelling; and (iii) digitalization as enabler. It is clear, however, that optimal efficiencies can only be obtained by incorporating a new healthcare setting represented by LHS, as alluded to above.

In a LHS, digitalization of healthcare is a fundament pillar to foster quick transfer and application of scientific knowledge into the clinical scenario. It simultaneously facilitates data collection and gain novel insights from real-world settings towards academia promoting both healthcare discovery and scientific innovation. Such that a LHS generates a virtuous cycle stimulating value-based healthcare as well as scientific excellence in an iterative manner. It is of note that the model, LHS, implies strong complementarities, and synergies, between classical study designs to generate evidence on efficacy, such as randomized clinical trials, and novel methodological approaches targeting generation of evidence in real-life scenarios. The process ultimately results in a necessary reduction of the efficacy-effectiveness gap seen in clinical interventions, which is often limiting healthcare value generation.

The LHS relies on the existence of an operational DHF including two main components. One of them is accessibility of interoperable health-related data covering different domains: (i) clinical data across healthcare layers, (ii) population-health registries; (iii) patient's self-tracking data encompassing citizens reported outcomes and experience of care, sensor monitoring and environmental information; and (iv) biological research data relevant for clinical purposes. The second key component of the DHF are tools that process data, such as predictive modelling, defined care paths and clinical decision support embedded into care paths. Such tools should contribute to gain on accessibility, personalization, as well as predictive and preventive approach to value-based healthcare. A building blocks strategy for implementation and sustainable adoption of such a system is needed to ensure interoperability of reliable data, technological maturity, compliance with the regulatory frame and a prepared workforce ensuring professionals engagement and active participation of citizens. Computed patient risk then can be used to stratify patients to intervention groups that help in the optimal service selection for the patient with a preventive approach. The real challenge is to define and implement appropriate strategies fostering evolution towards the new health scenario.

10.5 Deployment and Adoption Strategies

While the conceptual frame of multilevel clinical predictive modelling, as well as the final desirable healthcare scenario, is well-defined; deployment and adoption of the novel approach are exceedingly challenging with similar barriers and facilitators already mentioned for deployment of LHS. However, there are specific steps for any given computational modelling that should be considered for a successful deployment and adoption of enhanced multilevel clinical risk assessment, as briefly described below:

1. *Digitalization and standardization to a common data model.* There is a need for resolving how enhanced risk prediction can be implemented within country and organizational boundaries in a manner that supports federated AI/ML learning and that has a standard base, international user-base, and data volume content base large enough to warrant investment. A choice is to build on existing standards promoted by the European Health Data & Evidence Network (EHDEN) for data harmonization to a common data model that can scale.
2. *Data acquisition using federated learning.* Healthcare organizations should adopt a new framework to facilitate a shift from a "break-fix" to a "predict-prevent" model of healthcare to deliver better patient outcomes, while preserving data security and privacy to ensure citizen's trust. It should be achieved providing healthcare organizations a decentralized federated learning model, as the underlying GDPR-compliant framework for harmonizing existing and newly acquired datasets. Such a federated learning model allows the creation of a suite of tools and the testing of data AI/ML readiness by supplying existing risk prediction and patient stratification algorithms to local teams without exchanging the data itself.
3. *Co-design and development of a collaborative learning framework for accelerating the use of multilevel assessment in clinical care.* Use of currently

available tools from the Observational Health Data Sciences and Informatics (OHDSI) multi-stakeholder, interdisciplinary and collaborative programme to drive implementation of multilevel predictive models. By achieving this specific objective, clinicians and medical professionals will be involved throughout an AI/ML development process that will conclude with validated tools for health risk assessment and patient stratification.

4. *To drive inclusive and equitable utilization of data and risk prediction models.* The development of Best Practices to act as beacons of excellence internationally to ensure risk prediction models can be relied upon for fair outputs that aid decision-making and translate to daily life in support of clinical care. To this end, it will foster a transparent data and AI/ML eco-system that can be open to ethical and technical challenge to build trust and utilization. Health care professionals should be able to utilize robust, trustworthy, and privacy-preserving computational modelling that provide quantitative indicators valuable to identify and prioritize individuals with higher risk.
5. *To perform proof of concept, as well as clinical validation, studies in real-world settings.* There is a need for organizing evaluation studies in real-world settings to in silico assess the technical robustness of the developed AI/ML tools for risk assessment. Based on the evaluation studies, specific personalized preventive interventions should be piloted to assess healthcare value generation in comparison to the standard-of-care. Maturity of AI/ML tools in terms of Technology Readiness Level (TRL = 9) and health value generation will ultimately determine adoption provided that regulatory acceptability prior to deployment is demonstrated. Case-related reimbursement models to incentivize adoption could be envisaged.

10.6 Conclusions

Current evidence fosters multilevel integrative analyses including registry data, biomedical research information, EHR and patients' self-tracking data to elaborate, assess, and deploy predictive modelling, using AI/ML tools, for patients' stratification to properly pave the way towards enhanced clinical management and truly predictive, preventive, personalized, and participatory medicine, or "P4 medicine" (Auffray et al. 2010).

Acknowledgments We acknowledge the support of the following projects: TRAJECTOME (ERAPERMED2019-108); JADECARE (UE/19/3HP/JA/951442) and FIS-Smart PITeS (PI18/00841); as well as AGAUR research groups (2009SGR911 and 2014SGR661), and CERCA Programme/Generalitat de Catalunya.

References

Auffray C, Charron D, Hood L (2010) Predictive, preventive, personalized and participatory medicine: back to the future. Genome Med 2(8):57

Barnes PJ (2015) Mechanisms of development of multimorbidity in the elderly. Eur Respir J 45(3): 790–806. https://doi.org/10.1183/09031936.00229714

Blumenthal D, Chernof B, Fulmer T et al (2016) Caring for high-need, high-cost patients-an urgent priority caring for high-need, high-cost patients. NEJM 375(10):909–911. https://doi.org/10.1056/NEJMp1608511

Calvo M, Gonzales R, Seijas N et al (2020) Health outcomes from home hospitalization: multi-source predictive modeling. J Med Internet Res 22(10):e21367. https://doi.org/10.2196/21367

Cano I, Lluch-Ariet M, Gomez-Cabrero D et al (2014a) Biomedical research in a digital health framework. J Transl Med 12(Suppl 2):S10. https://doi.org/10.1186/1479-5876-12-S2-S10

Cano I, Selivanov V, Gomez-Cabrero D et al (2014b) Oxygen pathway modeling estimates high reactive oxygen species production above the highest permanent human habitation. PLoS One 9(11):e111068. https://doi.org/10.1371/JOURNAL.PONE.0111068

Celli BR, Augustì A (2018) COPD: time to improve its taxonomy? ERJ Open Res 4(1):117–2017. https://doi.org/10.1183/23120541.00132-2017

Davidsen PK, Herbert JM, Antczak P et al (2014) A systems biology approach reveals a link between systemic cytokines and skeletal muscle energy metabolism in a rodent smoking model and human COPD. Genome Med 6(8):1–16. https://doi.org/10.1186/S13073-014-0059-5/FIGURES/6

Diab N, Gershon AS, Sin DD et al (2018) Underdiagnosis and overdiagnosis of chronic obstructive pulmonary disease. Am J Respir Crit Care Med 198(9):1130–1139. https://doi.org/10.1164/RCCM.201804-0621CI

Doos L, Packer C, Ward D et al (2016) Past speculations of the future: a review of the methods used for forecasting emerging health technologies. BMJ Open 6(3):e010479. https://doi.org/10.1136/bmjopen-2015

Dueñas-Espín I, Vela E, Pauws S et al (2016) Proposals for enhanced health risk assessment and stratification in an integrated care scenario. BMJ Open 6(4):e010301. https://doi.org/10.1136/BMJOPEN-2015-010301

Ferguson TB (2012) The institute of medicine committee report Best care at lower cost: The path to continuously learning health Care. Circ Cardiovasc Qual Outcomes 5(6). https://doi.org/10.1161/CIRCOUTCOMES.112.968768

Gomez-Cabrero D, Lluch-Ariet M, Tegner J et al (2014) Synergy-COPD: a systems approach for understanding and managing chronic diseases. J Transl Med 12(Suppl 2):S2. https://doi.org/10.1186/1479-5876-12-S2-S2

Gomez-Cabrero D, Menche J, Vargas C et al (2016) From comorbidities of chronic obstructive pulmonary disease to identification of shared molecular mechanisms by data integration. BMC Bioinformatics 17. https://doi.org/10.1186/S12859-016-1291-3

Halpin DMG, Criner GJ, Papi A et al (2021) Global initiative for the diagnosis, management, and prevention of chronic obstructive lung disease. Am J Respir Crit Care Med 203(1):24–36. https://doi.org/10.1164/RCCM.202009-3533SO/SUPPL_FILE/DISCLOSURES.PDF

JADECARE (2020) Joint Action on implementation of digitally enabled integrated person-centred care. Source: https://www.jadecare.eu/ Last consultation: 24/08/2022

Maddox TM, Albert NM, Borden WB et al (2017) The learning healthcare system and cardiovascular care: a scientific statement from the American heart association. Circulation 135(14):e826–e857. https://doi.org/10.1161/CIR.0000000000000480

Maltais F, Decramer M, Casaburi R et al (2014) An official American thoracic society/European respiratory society statement: update on limb muscle dysfunction in chronic obstructive pulmonary disease. Am J Respir Crit Care Med 189(9). https://doi.org/10.1164/RCCM.201402-0373ST/SUPPL_FILE/EXECUTIVE_SUMMARY.PDF

Marín De Mas I, Fanchon E, Papp B et al (2017) Molecular mechanisms underlying COPD-muscle dysfunction unveiled through a systems medicine approach. Bioinformatics 33(1):95–103. https://doi.org/10.1093/BIOINFORMATICS/BTW566

Marx P, Antal P, Bolgar B et al (2017) Comorbidities in the diseasome are more apparent than real: what Bayesian filtering reveals about the comorbidities of depression. PLoS Comput Biol 13(6). https://doi.org/10.1371/journal.pcbi.1005487

Mathers C, Hogan D, Stevens G (2019) Global health estimates: modelling and predicting health outcomes. In: Macfarlane S, AbouZahr C (eds) The Palgrave handbook of global health data methods for policy and practice. Palgrave Macmillan, London, pp 469–486

Monterde D, Vela E, Clèries M (2016) Adjusted morbidity groups: a new multiple morbidity measurement of use in Primary Care. Aten Primaria 48(10):674–682. https://doi.org/10.1016/J.APRIM.2016.06.003

Monterde D, Vela E, Clèries M et al (2018) Validity of adjusted morbidity groups with respect to clinical risk groups in the field of primary care. Aten Primaria 51(3):153–161. https://doi.org/10.1016/J.APRIM.2017.09.012

Monterde D, Vela E, Clèries M et al (2020) Multimorbidity as a predictor of health service utilization in primary care: a registry-based study of the Catalan population. BMC Fam Pract 21(1):39. https://doi.org/10.1186/s12875-020-01104-1

Murray CJL, Lopez AD (2013) Measuring the global burden of disease. NEJM 369(5):448–457. https://doi.org/10.1056/NEJMRA1201534/SUPPL_FILE/NEJMRA1201534_DISCLOSURES.PDF

Porter ME (2008) Value-based health care delivery. Ann Surg 248(4):503–509. https://doi.org/10.1097/SLA.0B013E31818A43AF

Rajpurkar P, Chen E, Banerjee O, Topol EJ (2022) AI in health and medicine. Nat Med 28(1): 31–38. https://doi.org/10.1038/s41591-021-01614-0

Roca J, Vargas C, Cano I et al (2014) Chronic Obstructive Pulmonary Disease heterogeneity: challenges for health risk assessment, stratification and management. J Transl Med 12(Suppl 2): S3. https://doi.org/10.1186/1479-5876-12-S2-S3

Roca J, Tenyi A, Cano I (2020) Digital health for enhanced understanding and management of chronic conditions: COPD as a use case. In: Wolkenhauer O (ed) Systems medicine: integrative, qualitative and computational approaches. Academic Press, London, pp 264–274

Rodríguez DA, Alcarraz-Visan G, Diaz-Morali S et al (2011) Plasma metabolic profile in COPD patients: effects of exercise and endurance training. Metabolomics 8(3):508–516. https://doi.org/10.1007/S11306-011-0336-X

Rodriguez DA, Kalko S, Puig-Vilanova E et al (2012) Muscle and blood redox status after exercise training in severe COPD patients. Free Radic Biol Med 52(1):88–94. https://doi.org/10.1016/J.FREERADBIOMED.2011.09.022

Synergy-COPD Consortium (2010) Modelling and simulation environment for systems medicine: Chronic obstructive pulmonary disease (COPD) as a use case. FP7-ICT-270086. Source: http://www.synergy-copd.eu Last consultation: 24/08/2022

Tényi Á, Cano I, Marabita F et al (2018a) Network modules uncover mechanisms of skeletal muscle dysfunction in COPD patients. J Transl Med 16(1):34. https://doi.org/10.1186/s12967-018-1405-y

Tényi Á, Vela E, Cano I et al (2018b) Risk and temporal order of disease diagnosis of comorbidities in patients with COPD: a population health perspective. BMJ Open Respir Res 5(1):e000302. https://doi.org/10.1136/BMJRESP-2018-000302

Trajectome (2020) Temporal disease map-based stratification of depression-related multimorbidities: towards quantitative investigations of patient trajectories and predictions of multi-target drug candidates. Source: https://semmelweis.hu/trajectome/en/ Last consultation: 24/08/2022

Vela E, Tenji A, Cano I et al (2018) Population-based analysis of patients with COPD in Catalonia: a cohort study with implications for clinical management. BMJ Open 8(3):e017283. https://doi.org/10.1136/BMJOPEN-2017-017283

Vela E, Cleries M, Monterde D et al (2021) Performance of quantitative measures of multimorbidity: a population-based retrospective analysis. BMC Public Health 21(1):1–9. https://doi.org/10.1186/S12889-021-11922-2

Artificial Intelligence and Radiotherapy: Impact on Radiotherapy Workflow and Clinical Example

11

Luca Tagliaferri, Bruno Fionda, Valeria Masiello, Frank-Andre Siebert, Rafael Martínez-Monge, and Andrea Damiani

Abstract

Radiation oncology is a discipline in which a treatment is delivered to the tumor using three main modalities: external beam radiotherapy (ERT), which is based on the use of highly technological linear accelerators producing ionizing radiation (electrons or X-rays); interventional radiotherapy (IRT, brachytherapy), which is characterized by placing radioactive sources nearby or directly inside the tumor volume to treat, thus allowing us to obtain the optimal therapeutic ratio; and metabolic radiotherapy, which is characterized by the administration of radiopharmaceuticals (orally or intravenously) with a high affinity and selectivity for binding to tumor cells. The role of artificial intelligence (AI) in ERT has been described in several publications, while in the field of IRT and metabolic radiotherapy it has gained great attention only recently among researchers. There are at least four distinct areas where AI could be useful to the ERT and IRT's workflow, including providing clinical decision support, mining "omics" and analyzing data, facilitating repetitive tasks (thus optimizing time), and modeling behaviors in heterogeneous contexts). Large databases are very useful tools for generating evidence, especially in the field of AI-based software which needs to be further

L. Tagliaferri · B. Fionda (✉) · V. Masiello
Department of Diagnostic Imaging, Radiation Oncology and Hematology, Fondazione Policlinico Universitario A. Gemelli IRCCS, Rome, Italy
e-mail: bruno.fionda@policlinicogemelli.it

F.-A. Siebert
Department of Radiotherapy, Christian-Albrechts-Universität zu Kiel, Kiel, Germany

R. Martínez-Monge
Department of Radiation Oncology, Clínica Universidad de Navarra, Navarra, Spain

A. Damiani
Gemelli Generator "Real-World Data", Gemelli Science and Technology Park, Fondazione Policlinico Universitario A. Gemelli IRCCS, Rome, Italy

A. Cesario et al. (eds.), *Personalized Medicine Meets Artificial Intelligence*,
https://doi.org/10.1007/978-3-031-32614-1_11

implemented and integrated into clinical practice. Applications are different in ERT and IRT, but AI will certainly play an increasingly important role in the future of both.

Keywords

Interventional radiotherapy · Brachytherapy · Artificial intelligence

11.1 Introduction

Radiation oncology (RO) is a discipline in which a treatment is delivered to the tumor using three main modalities:

1. External beam radiotherapy (ERT), which is based on the use of highly technological linear accelerators producing ionizing radiation (electrons or X-rays)
2. Interventional radiotherapy (IRT, brachytherapy), which is characterized by placing radioactive sources nearby or directly inside the tumor volume to treat, thus allowing us to obtain the optimal therapeutic ratio
3. Metabolic radiotherapy, which is characterized by the administration of radiopharmaceuticals (orally or intravenously) with a high affinity and selectivity for binding to tumor cells

The role of artificial intelligence (AI) in ERT is described in several publications, while in IRT it has recently gained great attention among the radiation oncology community (Fionda et al. 2020). In fact, differently from ERT—which is facing a phase of renovation by incorporating and exploiting the precious contributions provided by AI, IRT has remained mainly based on the operator's knowledge and experience until nowadays. Such a technological gap, however, allows for a huge potential for embodying AI into IRT, thus making it possible to save time especially in repetitive tasks (Banerjee et al. 2021). In recent years, the role of AI is also explored in the frame of metabolic radiotherapy (Currie and Iqbal 2021).

There are at least four distinct areas where AI could be useful to the IRT's workflow, in detail:

1. In providing clinical decision support
2. In mining "omics" and analyzing data
3. In facilitating repetitive tasks, thus optimizing time
4. In modeling behaviors in heterogeneous contexts

11.2 Impact of AI and Automation in External Beam Radiotherapy Workflow

Artificial intelligence can support the entire process of external beam radiotherapy. According to the ERT workflow, we can identify six phases: first patient consultation, delineation, planning, setup, treatment session delivery, and end of treatment. A summary of AI contributions in each phase of the process is shown in Fig. 11.1.

In the following paragraphs, we present the most recent evidence in the literature for each phase, supporting the use of AI in ERT. Many diseases with multidisciplinary approach and different variables to take into account for clinical approach can benefit from AI supporting. In particular, breast and prostate cancer are relevant fields for AI application, due to the complexity of these pathologies.

11.2.1 First Patient Consultation

Defining the clinical indication for a radiation treatment based on different parameters can be very complex. AI can be used for guiding the decision process, also in a multidisciplinary team. An example is the application of AI during the decision process. Decision tools based on AI algorithms use evidence from big data to personalize treatments according to patient and tumor's individual characteristics (Kazmierska et al. 2020). In general, AI tools for first consultation are models for outcome prediction to support decision calculating benefit from such an intervention (Luo et al. 2019). Outcome prediction models can be output from traditional classifiers, such as logistic regressions, up to more complex models such as random forests, Bayesian networks, and neural networks. An example of a logistic regression model is the *Salvage Radiation Therapy Nomogram,* predicting whether a recurrence of prostate cancer after radical prostatectomy can be treated successfully with

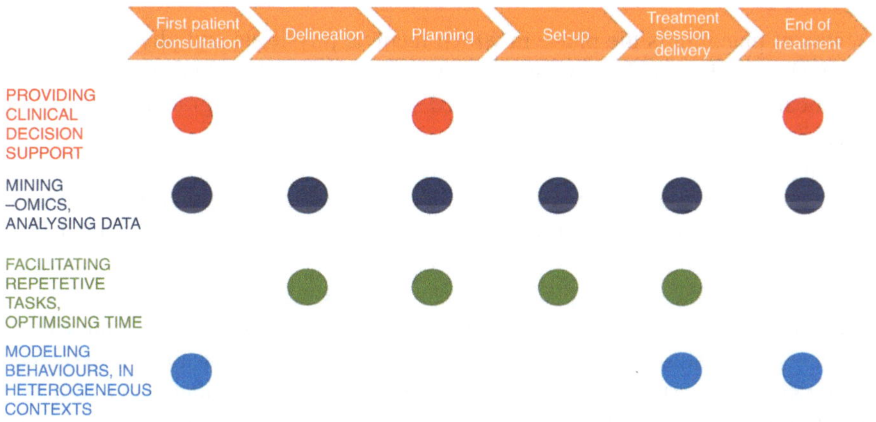

Fig. 11.1 Potential contribution of artificial intelligence in external beam radiotherapy. Modified from Fionda et al. (2020)

salvage radiation therapy (external beam radiation given after the prostate cancer returns) (Stephenson et al. 2007).

Another example of outcome prediction model was developed for two-year survival prediction in lung cancer patients treated with radiotherapy (Jayasurya et al. 2010). This tool uses a Bayesian network that also includes clinical data that otherwise could have been overlooked.

11.2.2 Delineation

In this phase of ERT, physicians identify target volumes and organs at risk. Delineation is usually based on a radiological atlas but may vary between operators. AI tools developed to aid in this phase have the purpose of facilitating a repetitive task and optimize time while collecting retrospective data for prediction tools. A consortium of French centers has recently developed a deep learning (DL) auto-contouring system for breast cancer radiotherapy, based on the ESTRO recommendations. A DL algorithm is an artificial neural network (ANN) with more than one hidden layer. This AI tool has limited human interactions and the workflow is quite fluid (Robert et al. 2021). Another extensive experience is found in the field of automatic segmentation using DL for radiotherapy of lung cancer. A review of these experiences confirms that Dice similarity indexes close to 0.9 are obtained for large OARs, 0.8 for GTVs, and between 0.7 and 0.8 for small OARs like esophagus (Liu et al. 2021). These findings support the efficacy of a time optimization AI application, but there are still some issues to be solved, in particular low contrast zones, dataset extension according to stage, and possible modification in consensus guidelines.

11.2.3 Treatment Planning

Treatment planning is the phase in which medical physicists and physicians elaborate treatment plans with treatment plan systems (TPS) and evaluate them according to clinical feasibility and quality. AI-based software can be integrated to TPS within specific tools.

AI applications in the treatment planning phase have two different purposes: time optimization and quality improvement of treatment plans. Time optimization can be obtained through software that reduce time and human intervention in plan elaboration. In breast cancer, there are several experiences of the RapidPlan (Varian, a Siemens Healthineers company in Palo Alto, CA) software application for optimizing treatment plan. RapidPlan was tested to predict if left-sided supine breast cancer patients would benefit from the deep inspiration breath-hold (DIBH) technique (Rice et al. 2019). This application addresses treatment plan elaboration with DIBH only of patients who benefit from it, resulting in time optimization. Another application of RapidPlan for breast cancer was addressed in volumetric modulated arch therapy (VMAT). Rago et al. (2021) created a knowledge-based model of treatment plan (TP) optimization on RapidPlan, with and without optimization structures. Results showed that all TP elaborated by these models were better than the original ones.

Another potentially useful contribution of AI to the TP phase is the construction of mixed models, both from clinical and dosimetric data to predict toxicity and create prospective optimization systems. An interesting experience conducted by Johns Hopkins University led to learning a classification and regression tree (CART) prediction model for weight loss (WL) in head and neck cancer (HNC) patients treated with radiation therapy (RT) (Cheng et al. 2017). The CART prediction model, based on a decision tree, includes anatomic tumor site, dosimetric parameters, and age.

11.2.4 Setup

In the setup phase, a radiation therapist positions the patients for ERT. In this phase, the physical positioning and the radiographic control of the positioning take place with corrections when needed. Given the high rate of human intervention dependence of this important phase, no deep implementations of AI have been published. It would be desirable to have systems automatically identify patient's features and implement the setup phase, although this remains as of today highly operator dependent. In particular, researchers have shown great interest toward 3D surface positioning systems, a field in which AI seems to be extremely promising (Zhao et al. 2021).

11.2.5 Treatment Delivery

Treatment delivery is the phase in which RT is administered and intrafraction changes are tracked. This is the phase in which AI offers very interesting solutions to improve the accuracy and personalization of treatments. Even more, images acquired during therapy are used for models to predict the response in order to modulate the treatments themselves.

In an experience by de Jong et al., cone beam CTs (CBCT) were used during therapy to adapt the TP on daily anatomy variations. Thanks to AI, applications based on structure-guided deformation and the synthetic CT scan contours were adapted by the system to match the anatomy on the CBCT. Specifically in this case the algorithm generated delineations of the rectum and bladder, called "influencer structures" as they influence the deformation of the target volumes using structure-guided deformable registration (de Jong et al. 2021). Another similar experience was conducted by Romaguera et al. (2021). The authors developed a model trained to simultaneously create a low-dimensional manifold representation of 3D non-rigid deformations and to predict the motion of the treatment target. The purpose of these models is to deliver a more accurate ERT.

Another field of interest in this phase is the response prediction during radiotherapy. In particular, fractal-based radiomic features studies applied to imaging acquired during magnetic resonance-guided radiotherapy (MRgRT) could predict pathological response in neoadjuvant treatments (Boldrini et al. 2019).

11.2.6 End of Treatment

After ERT treatment, beyond site and purpose of treatments, machine learning (ML) can be used to predict patients' characteristics associated with both toxicity and local or symptoms control. The CART model also included a second model based on a prediction tree for weight loss prediction after ending of head and neck cancer radiation therapy (Cheng et al. 2017). Variables included in this model were patient reported quality of life, dosimetry parameters, RT toxicity during therapy, and shape relationship (i.e., distance between PTV and larynx). Finally, complex models such as multi-objective Bayesian networks (MO-BNs) approach can correlate toxicity and outcomes variables not only to RT parameters and patients' characteristics but also to microenvironmental features. An example of MO-BNs was developed by Luo et al. (2018) with the purpose of prediction for response-adapted personalized treatment planning, also evaluating single nucleotide polymorphisms (SNPs), microRNAs, pretreatment cytokines, and pretreatment PET radiomics together with lung and tumor characteristics.

A complete list of the studies discussed in this chapter about AI and ERT is reported in Table 11.1.

Table 11.1 Complete list of the studies discussed about AI and ERT

Phase	Author	Year of Publication
First patient consultation	Stephenson et al. Jayasurya et al.	2007 2010
	Luo et al. Kazmierska et al.	2019 2020
Delineation	Robert et al.	2021
	Liu et al.	2021
Planning	Cheng et al.	2017
	Rice et al.	2019
	Rago et al.	2021
Treatment session delivery	Boldrini et al. de Jong et al. Romaguera et al.	2018 2021 2021
End of treatment	Cheng et al. Luo et al.	2017 2018

11.3 Impact of AI and Automation in Interventional Radiotherapy (IRT) Workflow

All the aforementioned contributions may be considered also along the entire IRT workflow. In comparison with ERT, a possible contribution of AI for IRT can be considered during implantation, while the setup phase is negligible. AI in IRT can be considered starting from the first patient consultation, during the implant, the delineation, planning, and treatment session delivery until the end of treatment. A summary of the contributions in each phase of the process is shown in Fig. 11.2.

We will now go into deeper details for each phase summarizing the latest evidence available in the literature supporting the use of AI in IRT. The most clinically relevant sites are by far prostate, breast, and cervix cancer; therefore, we will mainly focus on them, also including other sites in the following discussion.

11.3.1 First Patient Consultation

At the moment, there is limited evidence about the role of AI in guiding the decision process either within the multidisciplinary setting or in the usual workflow. A good example is the case of prostate cancer where artificial neural networks may be useful in aiding the decision process to evaluate whether IRT may be used as the sole treatment strategy or if there is a high risk of lymph node involvement, therefore requiring the addition of ERT (Gamito et al. 2000). In fact, ERT is required in those cases where there is a high risk of lymph node involvement in order to provide a better disease-free survival.

Another interesting application is about the chance to help identify early stages of disease for breast cancer that, after undergoing a surgical approach, may benefit from receiving a complementary accelerated partial breast irradiation by inserting plastic

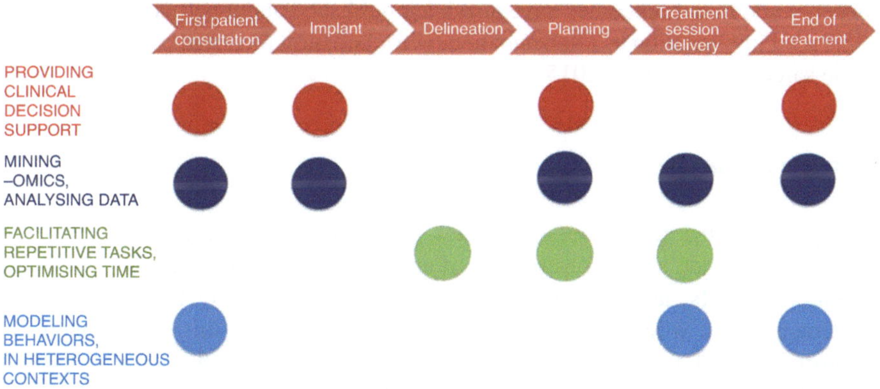

Fig. 11.2 Potential contribution of artificial intelligence in interventional radiation oncology. Modified from Fionda et al. (2020)

tubes within the tumor bed and, then, by using a dedicated remote after-loader irradiating the tumor bed (Polgar et al. 2010).

11.3.2 Implant

Regarding this phase of the IRT workflow, there is extensive evidence in the literature about the valuable addition that AI may provide specifically about the optimization of the applicators' location. The first reports about this topic go back to the late 1990s and considered the location of seeds implantation for low-dose-rate (LDR) IRT in prostate cancer (Yu et al. 1999); the first experiences focused on simulated annealing and artificial neural networks (Miller et al. 2001).

Nicolae et al. (2017) used ML to extract preplans which were optimized from a dosimetrical point of view in terms of seeds location; when comparing the results to those obtained by expert physicians, they found that the quality of the plan was comparable in terms of target coverage, normal tissue avoidance, implant confidence, and the need for plan modifications; moreover, a consistent reduction in terms of planning time had been achieved: the average time required by the ML approach was 0.84 ± 0.57 minutes, compared to 17.88 ± 8.76 minutes for the expert planner.

Great attention has also been paid to cervix cancer where there are preliminary experiences with DL used to guide the type of applicator for high-dose-rate (HDR) IRT; more specifically Stenhouse et al. have validated the process to choose between interstitial and intracavitary applicators considering the shape and geometry of high-risk clinical target volumes (HR-CTV) (Stenhous et al. 2021).

Also within the gynecological field there are interesting papers on the use of ML in preventing the formation of rectovaginal fistulas for patients receiving interstitial IRT (Tian et al. 2019). Another interesting experience, again with the aim to predict and reduce toxicities, has been released about side effects of CT-guided implantation of 125-I seeds for recurrent malignant tumors of the head and neck assisted by 3D printing of a non-co-planar template (Jiang et al. 2018). Focusing on other types of cancer, some initial outcomes have also been published about permanent interstitial 125-I seed implantation as a salvage therapy for pediatric recurrent or metastatic soft tissue sarcoma (Yao et al. 2015).

11.3.3 Delineation

This phase of the IRT process includes the identification of the organs at risk (OARs), the target, and reconstruction of applicator/catheters. Regarding the delineation of OARs and target volumes, so far there have been no clinical implementations, whereas the greater amount of evidence in this case is about the reconstruction of catheters and applicators. For example, DL may be used to adequately reconstruct tandem and ovoid applicators for cervix cancer in less than 25 s (Defuel et al. 2020); similarly DL may also be used to reconstruct interstitial needles (Jung et al. 2019). Interestingly, DL methods have shown promising results with MRI-based catheter reconstruction in procedures up to 35 catheters (Zaffino et al. 2019).

11.3.4 Treatment Planning

Regarding prostate cancer, several investigations have been conducted according to the different implantation techniques used by the centers (ultrasound or CT based); in particular, ultrasound-based preplanning protocols and techniques turned out to be similar, whereas computerized tomography-based post-implant dosimetry varied because of differing estimations of the prostate volume (Al-Qaisieh 2003). ML has been used in prediction of prostate IRT rectal toxicity (Leydon et al. 2015) but also to prove the non-inferiority of an ML-based planning workflow for LDR IRT in terms of time savings and operational efficiencies compared to conventional treatment planners (Nicolae et al. 2020).

Another potentially revolutionary contribution in this setting has been given in cervix cancer where Shen et al. developed an inverse treatment planning for HDR-IRT based on a weight-tuning policy network (WTPN) that observes dose-volume histograms of a plan and outputs an action to adjust organ-weighting factors, like the behaviors of an expert physician. The WTPN was trained by end-to-end reinforcement learning neural network (Shen et al. 2019).

11.3.5 Treatment Session Delivery

About this phase of IRT workflow, there are some very interesting reports regarding cervix cancer. In fact, there is data about the use of artificial neural networks (ANN)-based models to predict intra-fractional OAR dose-volume histogram parameters variations during intracavitary IRT. The models were trained to propose an adapted treatment plan to compensate for dosimetrical changes between applicators and OARs keeping the prescribed dose at the target volume (Jaberi et al. 2017).

11.3.6 End of Treatment

After IRT treatment, ML may be useful in identifying patients' characteristics associated with recurrence in prostate cancer. Valdes et al. have developed decision tree-based algorithms including classification and regression trees, MediBoost, and random forests which allowed us to reach the conclusion that patients with a fraction of positive cores ≥ 0.35 and a disease-free interval <4.12 years after their initial treatment are at higher risk of biochemical failure (Valdes et al. 2018).

A complete list of the studies discussed in this chapter about AI and IRT is reported in Table 11.2.

Table 11.2 Complete list of the studies discussed about AI and IRT

Phase	Author	Year of publication
First patient consultation	Gamito et al.	2000
	Polgár et al.	2010
Implant	Yu et al.	1999
	Miller et al.	2001
	Yao et al.	2015
	Nicolae et al.	2017
	Jiang et al.	2018
	Tian et al.	2019
	Stenhouse et al.	2021
Delineation	Jung et al.	2019
	Zaffino et al.	2019
	Deufel et al.	2020
Planning	Al-Qaisieh et al.	2003
	Leydon et al.	2015
	Shen et al.	2019
	Nicolae et al.	2020
Treatment session delivery	Jaberi et al.	2017
End of treatment	Valdes et al.	2018

11.4 Large Databases

Large databases are very useful tools for generating evidence, especially in the field of artificial intelligence. Large databases can be monocentric or multicentric. Direct connection with hospital systems and data repositories, such as that achieved by Gemelli Generator Real-World Data (Damiani et al. 2021) is fundamental. "GENERATOR Breast DataMart" is an example of a Breast Cancer Data Discovery System for Research and Monitoring, a computerized system to manage information and encourage the generation of evidence. The GENERATOR Breast DataMart was created for supporting breast cancer pathways of care. An AI-based process automatically extracts data from different sources and uses them for generating trend studies and clinical evidence. Two PoCs were performed, by which waiting time interval for radiotherapy and performance index of breast unit were tested and made available inside the hospital system (Marazzi et al. 2021).

The AI approach to knowledge extraction from data in radiotherapy is strongly encouraged by the European Commission, with the approval of Horizon projects such as iHelp: Personalised Health Monitoring and Decision Support Based on Artificial Intelligence and Holistic Health Records, approved in 2020 under the *Digital transformation in Health and Care* track (Pagliara et al. 2020). The iHelp project aims at building full-spectrum AI-enabled models of pancreatic cancer from the active prevention phase to the therapeutic or palliative path for diagnosed patients. In particular, the project will study the integration of clinical data with real life data obtained via smartphones and Internet-of-Things (IoT) devices to follow patients during the external radiotherapy period to improve prediction of adverse effects, toxicities, and overall decrease of quality of life based on both objective and subjective elements, thus adding totally new dimensions to the already available traditional hospital data (Manias et al. 2021).

Regarding the field of interventional and metabolic radiotherapy, “COBRA” is an example of multicenter large database and avatar implementation in clinical practice. Large databases are a natural extension of traditional statistical approaches, a valuable and increasingly necessary tool for modern healthcare system. The COBRA Ontology is a good solution to the multidimensional criticalities of data collection, retrieval, and usability. It allows us to create a software for large multicentric databases with implementation of specific remapping functions wherever necessary. Future analysis of the collected multinational and multicenter data will show whether the use of the system can produce high-quality evidence to support multidisciplinary management and utilizing this information for personalized treatment decisions (Tagliaferri et al. 2018; Lancellotta et al. 2020; Pagliara et al. 2020).

11.5 Conclusion

First applications for usage of AI in radiotherapy, either in ERT or IRT, are already published for various phases in the treatment process. Applications are different both in ERT and IRT. AI-based software needs to be further implemented and integrated into clinical practice. AI will play an increasingly important role in the future of radiotherapy.

References

Al-Qaisieh B (2003) UK Prostrate brachytherapy group. Pre- and post-implant dosimetry: an intercomparison between UK prostate brachytherapy centres. Radiother Oncol 66(2):181–183

Banerjee S, Goyal S, Mishra S et al (2021) Artificial intelligence in brachytherapy: a summary of recent developments. Br J Radiol 94(1122):20200842

Boldrini L, Cusumano D, Chiloiro G et al (2019) Delta radiomics for rectal cancer response prediction with hybrid 0.35 T magnetic resonance-guided radiotherapy (MRgRT): a hypothesis-generating study for an innovative personalized medicine approach. Radiol Med 124(2):145–153

Cheng Z, Nakatsugawa M, Hu C et al (2017) Evaluation of classification and regression tree (CART) model in weight loss prediction following head and neck cancer radiation therapy. Adv Radiat Oncol 3(3):346–355

Currie GM, Iqbal BM (2021) Re-modelling 99m-technetium pertechnetate thyroid uptake; statistical, machine learning and deep learning approaches. J Nucl Med Technol, jnmt.121.263081. https://doi.org/10.2967/jnmt.121.263081

Damiani A, Masciocchi C, Lenkowicz J, et al (2021) Building an artificial intelligence laboratory based on real world data: the experience of Gemelli generator. Front Comput Sci https://doi.org/10.3389/fcomp.2021.768266

de Jong R, Visser J, van Wieringen N et al (2021) Feasibility of Conebeam CT-based online adaptive radiotherapy for neoadjuvant treatment of rectal cancer. Radiat Oncol 16(1):136

Deufel CL, Tian S, Yan BB et al (2020) Automated applicator digitization for high-dose-rate cervix brachytherapy using image thresholding and density-based clustering. Brachytherapy 19(1): 111–118

Fionda B, Boldrini L, D'Aviero A, Lancellotta V et al (2020) Artificial intelligence (AI) and interventional radiotherapy (brachytherapy): state of art and future perspectives. J Contemp Brachytherapy 12(5):497–500

Gamito EJ, Stone NN, Batuello JT, Crawford ED (2000) Use of artificial neural networks in the clinical staging of prostate cancer: implications for prostate brachytherapy. Tech Urol 6(2): 60–63

Jaberi R, Siavashpour Z, Aghamiri MR et al (2017) Artificial neural network based gynaecological image-guided adaptive brachytherapy treatment planning correction of intra-fractional organs at risk dose variation. J Contemp Brachytherapy 9(6):508–518

Jayasurya K, Fung G, Yu S et al (2010) Comparison of Bayesian network and support vector machine models for two-year survival prediction in lung cancer patients treated with radiotherapy. Med Phys 37(4):1401–1407

Jiang Y, Ji Z, Guo F et al (2018) Side effects of CT-guided implantation of 125I seeds for recurrent malignant tumors of the head and neck assisted by 3D printing non co-planar template. Radiat Oncol 13(1):18

Jung H, Shen C, Gonzalez Y et al (2019) Deep-learning assisted automatic digitization of interstitial needles in 3D CT image based high dose-rate brachytherapy of gynecological cancer. Phys Med Biol 64(21):215003

Kazmierska J, Hope A, Spezi E et al (2020) From multisource data to clinical decision aids in radiation oncology: the need for a clinical data science community. Radiother Oncol 153:43–54

Lancellotta V, Guinot JL, Fionda B et al (2020) SKIN-COBRA (consortium for brachytherapy data analysis) ontology: the first step towards interdisciplinary standardized data collection for personalized oncology in skin cancer. J Contemp Brachytherapy 12(2):105–110

Leydon P, Sullivan F, Jamaluddin F (2015), "Machine learning in prediction of prostate brachytherapy rectal dose classes at day 30", in Dahyot R, Lacey G, Dawson-Howe K, Pitie F, Moloney D (Eds.), Proceedings of the 17th Irish machine vision and image processing conference. Irish Pattern Recognition and Classification Society, pp 105–109

Liu X, Li KW, Yang R, Geng LS (2021) Review of deep learning based automatic segmentation for lung cancer radiotherapy. Front Oncol 11:717039

Luo Y, McShan DL, Matuszak MM et al (2018) A multiobjective Bayesian networks approach for joint prediction of tumor local control and radiation pneumonitis in nonsmall-cell lung cancer (NSCLC) for response-adapted radiotherapy. Med Phys 45:3980. https://doi.org/10.1002/mp.13029

Luo Y, Tseng HH, Cui S et al (2019) Balancing accuracy and interpretability of machine learning approaches for radiation treatment outcomes modeling. BJR Open 1(1):20190021

Manias G, op der Akker H, Azqueta A, et al. (2021) iHELP: Personalised Health Monitoring and Decision Support Based on Artificial Intelligence and Holistic Health Records. In: 2021 IEEE Symposium on Computers and Communications (ISCC). https://doi.org/10.1109/ISCC53001.2021.9631475

Marazzi F, Tagliaferri L, Masiello V et al (2021) GENERATOR breast DataMart-the novel breast cancer data discovery system for research and monitoring: preliminary results and future perspectives. J Pers Med 11(2):65

Miller S, Bews J, Kinsner W (2001) Brachytherapy cancer treatment optimization using simulated annealing and artificial neural networks. Can Conf Electr Comput Eng 1:649–654

Nicolae A, Morton G, Chung H et al (2017) Evaluation of a machine-learning algorithm for treatment planning in prostate low-dose-rate brachytherapy. Int J Radiat Oncol Biol Phys 97(4):822–829

Nicolae A, Semple M, Lu L et al (2020) Conventional vs machine learning-based treatment planning in prostate brachytherapy: results of a phase I randomized controlled trial. Brachytherapy 19(4):470–476

Pagliara MM, Tagliaferri L, Lenkowicz J et al (2020) AVATAR: analysis for visual acuity prediction after eye interventional radiotherapy. In Vivo 34(1):381–387

Polgár C, Van Limbergen E, Pötter R et al (2010) Patient selection for accelerated partial-breast irradiation (APBI) after breast-conserving surgery: recommendations of the Groupe Européen de Curiethérapie-European Society for Therapeutic Radiology and Oncology (GEC-ESTRO) breast cancer working group based on clinical evidence (2009). Radiother Oncol 94(3):264–273

Rago M, Placidi L, Polsoni M et al (2021) Evaluation of a generalized knowledge-based planning performance for VMAT irradiation of breast and locoregional lymph nodes-internal mammary and/or supraclavicular regions. PLoS One 16(1):e0245305

Rice A, Zoller I, Kocos K et al (2019) The implementation of RapidPlan in predicting deep inspiration breath-hold candidates with left-sided breast cancer. Med Dosim 44(3):210–218

Robert C, Munoz A, Moreau D, et al. (2021), "Clinical implementation of deep-learning based auto-contouring tools-experience of three French radiotherapy centers", Cancer Radiother, S1278-3218(21)00122–0

Romaguera LV, Mezheritsky T, Mansour R et al (2021) Predictive online 3D target tracking with population-based generative networks for image-guided radiotherapy. Int J Comput Assist Radiol Surg 16(7):1213–1225

Shen C, Gonzalez Y, Klages P et al (2019) Intelligent inverse treatment planning via deep reinforcement learning, a proof-of-principle study in high dose-rate brachytherapy for cervical cancer. Phys Med Biol 64(11):115013

Stenhouse K, Roumeliotis M, Banerjee R et al (2021) Development of a machine learning model for optimal applicator selection in high-dose-rate cervical brachytherapy. Front Oncol 11:611437

Stephenson AJ, Scardino PT, Kattan MW et al (2007) Predicting the outcome of salvage radiation therapy for recurrent prostate cancer after radical prostatectomy. J Clin Oncol 25(15): 2035–2041

Tagliaferri L, Budrukkar A, Lenkowicz J et al (2018) ENT COBRA ONTOLOGY: the covariates classification system proposed by the Head & Neck and Skin GEC-ESTRO Working Group for interdisciplinary standardized data collection in head and neck patient cohorts treated with interventional radiotherapy (brachytherapy). J Contemp Brachytherapy 10(3):260–266

Tian Z, Yen A, Zhou Z et al (2019) A machine-learning-based prediction model of fistula formation after interstitial brachytherapy for locally advanced gynecological malignancies. Brachytherapy 18(4):530–538

Valdes G, Chang AJ, Interian Y et al (2018) Salvage HDR brachytherapy: multiple hypothesis testing versus machine learning analysis. Int J Radiat Oncol Biol Phys 101(3):694–703

Yao L, Wang J, Jiang Y et al (2015) Permanent interstitial 125I seed implantation as a salvage therapy for pediatric recurrent or metastatic soft tissue sarcoma after multidisciplinary treatment. World J Surg Oncol 13:335

Yu Y, Anderson LL, Li Z et al (1999) Permanent prostate seed implant brachytherapy: report of the American Association of Physicists in Medicine Task Group No. 64. Med Phys 26(10): 2054–2076

Zaffino P, Pernelle G, Mastmeyer A et al (2019) Fully automatic catheter segmentation in MRI with 3D convolutional neural networks: application to MRI-guided gynecologic brachytherapy. Phys Med Biol 64(16):165008

Zhao W, Shen L, Islam MT et al (2021) Artificial intelligence in image-guided radiotherapy: a review of treatment target localization. Quant Imaging Med Surg 11(12):4881–4894

Artificial Intelligence in Surgery

12

Marika D'Oria, Pietro Mascagni, Ivo Boškoski, Nicolas Padoy, Sergio Alfieri, Guido Costamagna, and Giovanni Scambia

Abstract

Modern surgery is a highly effective sociotechnical process susceptible to errors. Surgical data science is a novel transdisciplinary research field aiming at improving surgical care using advanced data analytics such as artificial intelligence (AI). This chapter presents several applications of AI tools in perioperative and postoperative surgery that can help to achieve personalized precision surgery while considering current limitations and ethical aspects.

Keywords

Artificial intelligence · Surgical data science · Deep learning · Computer vision · Robotic surgery

M. D'Oria · G. Scambia
Fondazione Policlinico Universitario A. Gemelli IRCCS, Rome, Italy
e-mail: giovanni.scambia@policlinicogemelli.it

P. Mascagni (✉)
Department of Medical and Surgical Sciences, Fondazione Policlinico Universitario A. Gemelli IRCCS, Rome, Italy

Institute of Image-Guided Surgery, IHU-Strasbourg, Strasbourg, France
e-mail: pietro.mascagni@ihu-strasbourg.eu

I. Boškoski · S. Alfieri · G. Costamagna
Department of Medical and Surgical Sciences, Fondazione Policlinico Universitario A. Gemelli IRCCS, Rome, Italy

N. Padoy
Institute of Image-Guided Surgery, IHU-Strasbourg, Strasbourg, France

ICube, University of Strasbourg, CNRS, Strasbourg, France

A. Cesario et al. (eds.), *Personalized Medicine Meets Artificial Intelligence*,
https://doi.org/10.1007/978-3-031-32614-1_12

12.1 Introduction

In the last decade, the effectiveness and volume of surgery has significantly increased with more than 300 million procedures performed in 2012 (Weiser et al. 2015). Still, a large part of medical mistakes, the third leading cause of death in the USA (Makary and Daniel 2016), happens in surgery (Zegers et al. 2011). Fortunately, more than half of these surgical adverse events are considered preventable (Zegers et al. 2009). Given the prevalence of modern surgery and the great opportunity for quality improvement, surgical safety is today considered a global health priority.

In the era of personalized and precision medicine, great attention is given to assuring "the right cure to the right patient, at the right time" (Nimmesgern et al. 2017). Similarly, in the context of modern surgery, it is important to give the surgeon "the right assistance, at the right time" (Maier-Hein et al. 2017). In fact, when performing a surgical procedure, surgeons interact with a highly specialized team of collaborators, while recalling and projecting surgical principles into the present scenario, capturing, and interpreting several signals from high-tech devices, anticipating consequences of decisions, and acting timely for the benefit of patients. The proper coordination of this complex socio-technical process requires physical, relational, and cognitive efforts: any imperfection can be potentially harmful for patients (Schreuder et al. 2020; Mascagni 2021). Moreover, errors have a tremendous cost on patients, surgeons, and healthcare systems (Berci et al. 2013).

Surgical adverse events derive from technical errors as well as from perceptual illusions which are, for example, the leading cause of major bile duct injuries (BDIs) (Way et al. 2003).

The growing availability of digital data from surgical procedures using endoscopy, laparoscopy, surgical microscopy, robotic minimally invasive surgery, and hybrid platforms which also combine advanced intraoperative imaging technologies, together with the increased accessibility to advanced analytics, such as artificial intelligence (AI) and machine learning (ML), could converge to improve surgical safety and efficiency. In addition, knowledge generation in surgery can be enriched with extra data generated in the operating room (OR) (e.g., anesthesia monitors, OR devices usage, environmental cameras, microphones) and during perioperative care (e.g., patients' clinical history and outcomes stored in electronic medical records, imaging studies in *Picture Archiving and Communication System* [PACS]) (Mascagni 2021).

This has led computer scientists and surgeons to partner in surgical data science (SDS), a new transdisciplinary field that aims at improving the quality of surgery and other interventional disciplines using data and advanced analytics (Maier-Hein et al. 2022). Recent developments in this field have transformed the way experts envision the future of surgery. In fact, while *offline* analysis of surgical data allows accumulating detailed knowledge on the surgical process, intraoperative *online* inference could give surgeons actionable insights to ameliorate surgical care (Mascagni 2021). However, while an increasing number of data-driven approaches and clinical applications have been studied in radiological and clinical data science, translational success stories are still lacking in surgery (Maier-Hein et al. 2022).

In this chapter, we introduce basic concepts of AI and overview its possible applications in the perioperative and intraoperative settings, envisioning the way forward for achieving personalized precision surgery.

12.2 Surgical Data and Fundamentals AI Concepts

Imaging is one of the pillars for the ongoing evolution of surgical precision (Mascagni et al. 2018). In fact, digital endoscopic videos guide surgical actions and decisions and are a natural source of unbiased information on intraoperative events. If properly analyzed, these videos are more reliable than operator dictated postoperative reports (Eryigit et al. 2020; van de Graaf et al. 2019). Moreover, the growing number of intraoperative imaging techniques constantly introduced within the surgical practice could be interpreted with algorithms to inform surgeons about the best strategies for improving safety, becoming in turn an important input for AI algorithms generation (Mascagni et al. 2018).

As described in Alapatt et al. (2020), AI and related methods are rapidly evolving and set to become an important tool to enable the vision of SDS and precision surgery. If a traditional computer software is explicitly programmed with certain functions to analyze data and generate outputs for a desired information, in ML those functions are learned by the algorithm itself, without the need for explicitly knowing the function. The ability of ML to learn how to perform complex tasks by analyzing high-quality data is what makes AI revolutionary: it would be unpractical to program a priori every function needed to perform complex tasks such as interpreting complex unstructured data like surgical images. ML can serve several purposes ranging from data acquisition to information generation and from visual and motor assistance to case-specific update and follow-up. Powerful ML tools are artificial *neural networks*, algorithmic architectures inspired by the biological one and conceived as a collection of interconnected "neurons," each of them taking a set of inputs, processing them, and sending a signal (a numerical one in this case) that in turn serves as inputs to other neurons. Each network consists of an input layer, one or more hidden layers used to extract meaningful information from the input, and finally an output layer that aggregates information into the desired form. For example, if the input layer is a CT scan slice of a suspicious lesion, the output could represent its probability of malignancy.

When multiple hidden layers occur between the input and output layers, the network is called *deep neural network* and its training process is known as deep learning (DL). In DL, lower layers can extract simple information (e.g., edges of images) and higher layers build on that information to develop an understanding of more complex shapes and concepts. Most DL tasks on images fit into one of following three categories based on the type of output expected from the model (Fig. 12.1):

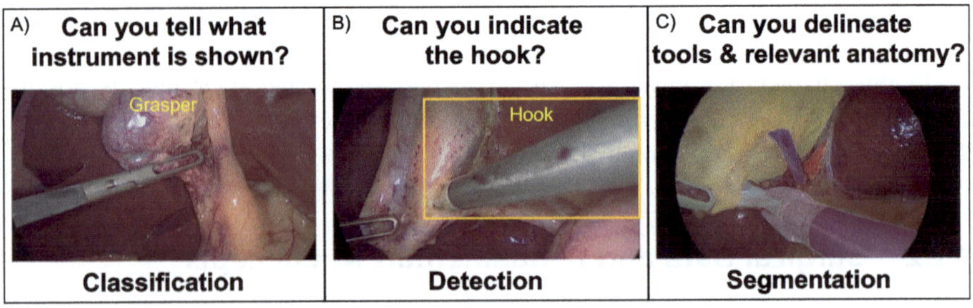

Fig. 12.1 Examples of common computer vision tasks in minimally invasive surgery. (**a**) Classification of surgical tools; (**b**) Detections of surgical tools; (**c**) Anatomy and tools segmentation

- *Classification* refers to the task of categorizing a given input into two or more possible classes (e.g., discrimination between benign or malignant lesions).
- *Detection* refers to the task of identifying the category of an object of interest (e.g., a malignant lesion on a tissue) in the input and localizing its position spatially (e.g., contouring the malignant lesion with a bounding box).
- *Semantic segmentation* refers to the task of classifying every pixel of an image into a particular category (e.g., highlighting healthy tissue versus malignant tissue).

12.3 Potential Application of AI in Surgery

12.3.1 Perioperative Applications

In perioperative settings, AI could be used for several applications such as patient stratification, operative planning, resource allocation, documentation, and training.

For example, an ensemble of specifically trained algorithms could identify a laparoscopic cholecystectomy (LC)[1] case in the surgical waiting list, use text mining techniques to analyze digital medical record and computer vision (CV) to extract features from preoperative images such as ultrasound or CT/MRI scans to estimate the case complexity, and present it to a surgeon recognized as having the right expertise and technical skills (Vannucci et al. 2022). Similarly, algorithms could also stratify postoperative risks and prognosis to better tailor patients' treatment and follow-up. Moreover, AI/ML solutions could automatically reserve an appropriate

[1]The most common abdominal surgical procedure (Pucher et al., 2018) is usually performed laparoscopically. This minimally invasive procedure is associated with less pain, less scarring, and a faster return to normal activities; despite research confirming its safety (ibidem), surgeries can have various complications that may impact the life of patients, like iatrogenic bile duct injuries (BDIs) that still complicate 0.32–1.5% of LCs (Törnqvist et al.; 2012; Pucher et al., 2018), a rate that is higher than the incidence commonly reported in open surgery (Southern Surgeons Club, 1991).

OR slot for the patient so as to optimize OR scheduling and resource allocation. Once in the OR, a virtual "assistant" can inform the OR staff about the case peculiarities and expected events and guides the WHO *Surgical Safety Checklist* time-out before skin incision (Conley et al. 2011; Mascagni 2021). This scenario envisions a personalization of the whole surgical pathway, from training to patient's presentation and discharge.

Furthermore, DL models can be translated into surgery for the automatic generation of documentation (e.g., surgical reports) and the production of short videos clips which selectively document critical moments of procedures, such as the critical view of safety (CVS) in LC where it is possible to reliably locate the time of the cystic duct division and efficiently video document CVS despite the highly variable workflows (Mascagni et al. 2021a, 2022).

AI could serve surgical training by automatizing and expediting performance assessment, so to provide unbiased formative feedback to foster and personalize the learning experience. A recent systematic review on the topic (Pedrett et al. 2022) has analyzed 37 original papers using AI for technical skills assessment in minimally invasive surgery, finding that most approaches leverage either endoscopic videos or kinematics data from robots and sensors to assess tasks execution, to this day mostly in simulated settings.

12.3.2 Intraoperative Applications

In intraoperative settings, AI can be used for performance assessment, improving OR staff communication, providing guidance and better visualization, and robotic assistance.

OR black boxes and surgical control towers have been proposed to systematically capture broad data on surgical procedures and to oversee and assist during OR activities, respectively (Mascagni and Padoy 2021). In this regard, AI could be used to make sense of these big surgical data to assess surgical performances in terms of technical judgments, team routine, and communication patterns while offering the opportunity to conduct prospective intraoperative studies of human performance and allows postoperative discussion, review, and teaching (Guerlain et al. 2005).

OR data can be streamed continuously to an external surgical control room from which senior surgeons could remotely proctor multiple cases and administrators can monitor OR status (Mascagni and Padoy 2021). Thanks to the current availability of powerful computational resources and advances in algorithms efficiency, surgical data analysis could be performed in real time and online, to give the surgeons feedbacks during procedures. This would allow us to translate all the data available into actionable information with case-specific feedbacks during procedures as well as recognizing phases and tools to efficiently coordinate OR staff and inform on ORs status (Mascagni and Padoy 2021).

AI-driven devices could provide guidance and better visualization to surgeons while operating. For example, CV could support intraoperative navigation within complex and deformable environments (such as the abdominal cavity), especially when equipped with ML/DL solutions to identify critical events such as bleeding (Hashimoto et al. 2018; Mascagni et al. 2022). In addition, CV could be used to provide timely intraoperative cognitive aids, to make sure that the right information is recalled at the appropriate moment and delivered to the right team member (Treadwell et al. 2014; Mascagni et al. 2021b).

The ML/DL powered analysis presented could be integrated in robotic platforms. In this scenario, surgical robots could serve as data capture platforms and, in turn, be augmented by AI tools. These robotic surgical systems powered by AI could deliver a paradigm shift in which AI-assisted robotic surgery first automates repetitive tasks to ease surgeons' workload and finally performs complex maneuvers to deliver more precise and better care to patients. However, while AI may help with several assistance functions, the surgeon will always remain in the loop to oversee and control decisions and gestures (Sapre et al. 2022).

12.4 The Future Ahead

The use of AI to analyze and improve surgical activities together with the integration of precise robotic platforms will provide surgeons with smart assistance in perceptual, cognitive, and physical tasks. To date, AI in surgery has mostly focused on understanding the context and workflow of procedures, team behaviors, and factors that affect surgical safety (Padoy 2019; Mascagni and Padoy 2021).

Some key aspects should be addressed before translating and exploiting the full potential of AI within surgical practice. In the white paper "Surgical Data Science: A consensus perspective" drafted following the *1st International Workshop on Surgical Data Science*, the authors suggest that several matters still hinder the application of AI in surgery. For example, data privacy, security, and ownership, especially when treating highly sensitive health information, are still open issues, requiring solutions for advanced patient digital data acquisition, anonymization, storage, and handling (Maier-Hein 2018). Information and communication technology (ICT) infrastructures should be safe by design for data collection, storage, and access, and fully compliant with national and international regulations and procedures, including data protection.

Another relevant issue is that annotating raw data with surgical knowledge requires rare and expensive physician time, an important bottleneck slowing the development of DL models in surgery is the scarcity of well-annotated datasets (Mascagni 2021). There is the need for methods to facilitate large-scale data annotation that could be based on concepts of crowdsourcing, expert data augmentation, or self-supervised learning.

Validation and ethical issue will also need to be solved. Indeed, when a machine is more precise than humans, but it is validated by labels provided by them, the intrinsic nature of the learned process (and its practical application) is called into ethical questions (Rudzicz and Saqur 2021). Before AI will be fully integrated within surgery practices, algorithms must undergo a strong scientific validation to promote to guarantee generalization of performance across practices and surgical scenarios. Altogether, algorithms should be monitored for their fairness and biases.

To achieve personalization and precision in medicine, further technological advances should be made, such as the improvement of methods to analyze multimodal and heterogeneous datasets (e.g., with genetics, biomarkers, imaging, peri- and intraoperative information). Such holistic modeling of sparse clinical information is key for advancing knowledge-based medicine as well as applying these tools in clinical practice.

Finally, surgical–technical partnerships should be encouraged to develop and evaluate ML/DL models to improve surgical safety. We believe that education plays a crucial role in bringing awareness on the potential and the pitfalls of the available AI and SDS solutions, and these outlooks should be promoted toward academic courses and hands-on immersive training (Edu4SDS 2022).

References

Alapatt N, Mascagni P, Author N et al (2020) Neural networks and deep learning. In: Hashimoto DA (ed) Artificial intelligence in surgery: a primer for surgical practice. McGraw Hill, New York

Berci G, Hunter J, Morgenstern L et al (2013) Laparoscopic cholecystectomy: first, do no harm; second, take care of bile duct stones. Surg Endosc 27(4):1051–1054. https://doi.org/10.1007/s00464-012-2767-5

Conley DM, Singer SJ, Edmondson L, Berry WR, Gawande AA (2011) Effective surgical safety checklist implementation. J Am Coll Surg 212(5):873–879

Edu4SDS (2022) *Homepage*, Source: www.edu4sds.org Last consultation 03/01/2023

Eryigit O, van de Graaf FW, Nieuwenhuijs VB et al (2020) Association of Video Completed by audio in laparoscopic cholecystectomy with improvements in operative reporting. JAMA Surg 155(7):617–623. https://doi.org/10.1001/jamasurg.2020.0741

Guerlain S, Adams RB, Turrentine FB et al (2005) Assessing team performance in the operating room: development and use of a 'black-box' recorder and other tools for the intraoperative environment. J Am Coll Surg 200(1):29–37. https://doi.org/10.1016/j.jamcollsurg.2004.08.029

Hashimoto DA, Rosman G, Rus D, Meireles OR (2018) Artificial intelligence in surgery: promises and perils. Ann Surg 268(1):70–76. https://doi.org/10.1097/SLA.0000000000002693

Madani A, Namazi B, Altieri MS et al (2021) Artificial intelligence for intraoperative guidance: using semantic segmentation to identify surgical anatomy during laparoscopic cholecystectomy. Ann Surg 276(2):363–369. https://doi.org/10.1097/SLA.0000000000004594

Maier-Hein L (2018) Surgical Data Science: A Consensus Perspective. source: https://arxiv.org/pdf/1806.03184.pdf. Last consultation: 09/11/2022

Maier-Hein L, Vedula SS, Speidel S et al (2017) Surgical data science for next-generation interventions. Nat Biomed Eng 1:691–696. https://doi.org/10.1038/s41551-017-0132-7

Maier-Hein L, Eisenmann M, Sarikaya D et al (2022) Surgical data science - from concepts toward clinical translation. Med Image Anal 76:102306. https://doi.org/10.1016/j.media.2021.102306

Makary MA, Daniel M (2016) Medical error - the third leading cause of death in the US. BMJ 353. https://doi.org/10.1136/bmj.i2139

Martin JA, Regehr G, Reznick R et al (1997) Objective structured assessment of technical skill (OSATS) for surgical residents. Br J Surg 84(2):273–278. https://doi.org/10.1046/j.1365-2168.1997.02502.x

Mascagni P (2021) Surgical data science for safe laparoscopic cholecystectomy. Doctoral Dissertation in Clinical, Cellular and Molecular Research, AA 2020/2021. Università Cattolica del Sacro Cuore, Rome

Mascagni P, Padoy N (2021) OR black box and surgical control tower: recording and streaming data and analytics to improve surgical care. J Visc Surg 158(3, Suppl.):S18–S25. https://doi.org/10.1016/j.jviscsurg.2021.01.004

Mascagni P, Longo F, Barberio M et al (2018) New intraoperative imaging technologies: innovating the surgeon's eye toward surgical precision. J Surg Oncol 118(2):265–282. https://doi.org/10.1002/jso.25148

Mascagni P, Vardazaryan A, Alapatt D et al (2020) Artificial intelligence for surgical safety: automatic assessment of the critical view of safety in laparoscopic cholecystectomy using deep learning. Ann Surg 275(5):955–961. https://doi.org/10.1097/SLA.0000000000004351

Mascagni P, Alapatt D, Urade T et al (2021a) A computer vision platform to automatically locate critical events in surgical videos: documenting safety in laparoscopic cholecystectomy. Ann Surg 274(1):e93. https://doi.org/10.1097/SLA.0000000000004736

Mascagni P, Rodriguez-Luna MR, Urade T, et al. (2021b) Intraoperative time-out to promote the implementation of the critical view of safety in laparoscopic cholecystectomy: a video-based assessment of 343 procedures. J Am Coll Surg, S1072-7515(21)00509-3. https://doi.org/10.1016/j.jamcollsurg.2021.06.018

Mascagni P, Alapatt D, Laracca GG et al (2022) Multicentric validation of EndoDigest: a computer vision platform for video documentation of the critical view of safety in laparoscopic cholecystectomy. Surg Endosc 36(11):8379–8386. https://doi.org/10.1007/s00464-022-09112-1

Nimmesgern E, Benediktsson I, Norstedt I (2017) Personalized medicine in Europe. Clin Transl Sci 10:61–63. https://doi.org/10.1111/cts.12446

Padoy N (2019) Machine and deep learning for workflow recognition during surgery. Minim Invasive Ther Allied Technol 2:82–90. https://doi.org/10.1080/13645706.2019.1584116

Pedrett R, Mascagni P, Beldi G, et al (2022) Technical skill assessment in minimally invasive surgery using artificial intelligence: a systematic review. medRxiv. https://doi.org/10.1101/2022.11.08.22282058

Pucher PH et al (2018) Outcome trends and safety measures after 30 years of laparoscopic cholecystectomy: a systematic review and pooled data analysis. Surg Endosc 32:2175–2183

Rudzicz F, Saqur R (2021) Ethics of artificial intelligence in surgery. In Hashimoto DA, Meireles OR, Rosman G (Eds.), Artificial intelligence in surgery: understanding the role of AI in surgical practice. McGraw Hill

Sapre N, Shah TT, Dasgupta P (2022) Current and upcoming robotic surgery platforms and adjunctive technologies. In: Wilkund P, Mottrie A, Gundeti MS, Patel V (eds) Robotic urology surgery, 3rd edn. Springer, pp 11–20

Schreuder AM, Busch OR, Besselink MG et al (2020) Long-term impact of iatrogenic bile duct injury. Dig Surg 37(1):10–21. https://doi.org/10.1159/000496432

Southern Surgeons Club (1991) A prospective analysis of 1518 laparoscopic cholecystectomies. N Engl J Med. 324(16):1073–1078. https://doi.org/10.1056/NEJM199104183241601

Törnqvist B, Strömberg C, Persson G, Nilsson M (2012) Effect of intended intraoperative cholangiography and early detection of bile duct injury on survival after cholecystectomy: population based cohort study. BMJ 345:e6457

Treadwell JR, Lucas S, Tsou AY (2014) Surgical checklists: a systematic review of impacts and implementation. BMJ Qual Saf 23(4):299–318

Twinanda AP, Yengera G, Mutter D et al (2018) RSDNet: learning to predict remaining surgery duration from laparoscopic videos without manual annotations. IEEE Trans Med Imaging 3(4): 1069–1078

van de Graaf FW, Lange MM, Spakman JI et al (2019) Comparison of systematic video documentation with narrative operative report in colorectal cancer surgery. JAMA Surg 154(5):381–389

Vannucci M, Laracca GG, Mercantini P, Perretta S, Padoy N, Dallemagne B, Mascagni P (2022) Statistical models to preoperatively predict operative difficulty in laparoscopic

cholecystectomy: a systematic review. Surgery 171(5):1158–1167. https://doi.org/10.1016/j.surg.2021.10.001
Way LW, Stewart L, Gantert W et al (2003) Causes and prevention of laparoscopic bile duct injuries: analysis of 252 cases from a human factors and cognitive psychology perspective. Ann Surg 237(4):460
Weiser TG, Haynes AB, Molina G et al (2015) Estimate of the global volume of surgery in 2012: an assessment supporting improved health outcomes. Lancet 385:S11
Zegers M, de Bruijne MC, Wagner C et al (2009) Adverse events and potentially preventable deaths in Dutch hospitals: results of a retrospective patient record review study. BMJ Qual Saf 18(4): 297–302. https://doi.org/10.1136/qshc.2007.025924
Zegers M, de Bruijne MC, de Keizer B et al (2011) The incidence, root-causes, and outcomes of adverse events in surgical units: implication for potential prevention strategies. Patient Saf Surg 5(1):13

Part III

Emerging and Future Technologies

How Start-ups and Established Organisations Together Can Drive Meaningful Healthcare Innovation in Personalised Medicine and AI

13

Fabrizio Bellina and Sven Jungmann

Abstract

While there is no cookbook recipe to developing an artificial intelligence solution from conception to market launch in healthcare, some key principles must be observed just like in other areas of healthcare. The preliminary phases of design and setting the cross-functional team with the right capabilities, assets, and financial resources are crucial for making a project solid and speeding up the subsequent phases. Examples include the need of deal flow with methodological rigour and incremental scientific research grounded in robust evidence to which we can apply new technologies. It is also critical to continuously review the ever-evolving needs of doctors and patients throughout the project lifecycle and the dynamic business model and go to the market model with an entrepreneurial mindset to make the AI solution scalable and sustainable overtime. The end users should be an integral part of the co-design process as they will ultimately drive the adoption of any new solution. Start-ups, patients, facilities, and investment funds today have everything to achieve the expected impact of AI in the health world. Will we be able to overcome the barriers that, fortunately less and less often, do not lead to collaborations in the medium to long term? If we look back, it is better to have remorse for a failure (and learn) than regret for not doing it. There is a tendency to seek early proof that an "s-curve" type of scalability can be achieved. We strongly recommend resisting this temptation and instead focus on experiments that start small—even if the big, scalable idea should be kept in mind at every design stage. This means making agreements revisable and flexible and combining them with a well-defined exit strategy to avoid wasting resources

F. Bellina (✉)
Digital Health Solutions and Open Innovation in HC&LS Expert, Milano, Italy
e-mail: fabrizio.bellina@me.com

S. Jungmann
Halitus GmbH, Berlin, Germany

A. Cesario et al. (eds.), *Personalized Medicine Meets Artificial Intelligence*,
https://doi.org/10.1007/978-3-031-32614-1_13

on satisfying needs that have proven throughout the project to no longer be the key priority. Especially in more complex collaborations, there is a tendency to "stick to past" promises even if they appear less sensible in hindsight. Setting the right foundation means creating milestones based on clinical evidence and efficiency to speed up stop-and-go decisions. It also reduces efforts and capital needed to scale the solution further down the road.

Keywords

Start-up · Artificial intelligence · Healthcare · Open innovation · MedTech

13.1 Landscape: Why Are Partnerships Between Start-ups and Established Organisations Relevant for Personalised Medicine and AI?

Artificial intelligence (AI) has the potential to revolutionise healthcare and help its systems address existing challenges in increasing care delivery efficiency and in achieving superior health outcomes for patients. There are several definitions of AI, but for our purposes, we draw on the concise and helpful wording proposed by the European Parliament (2016):

> AI is the capability of a computer program to perform tasks or reasoning processes that we usually associate with intelligence in a human being.

Because of its potential, AI is now at the forefront for healthcare decision-makers, governments, investors, innovators, and the European Union itself.

13.1.1 Problem

Digital, big data, and AI are currently impacting the healthcare pathway in connection with remote care, connectivity, real-time analytics, and automation. Even if these tools come to full fruition, healthcare remains "one of the industries *with the lowest automation potential*, with only 35 per cent of time spent estimated as potentially automatable and only 15 percent predicted to be automated" (EIT Health 2020). Given this, the role of AI has become over-hyped, so to speak, in regard to the expectation of a peak in the next 5 to 10 years (as proposed in 2018).

If we look at the current reality with an eye towards collaboration and partnership around these solutions, example AI use cases are present across the entire healthcare value chain (chronic care management, self-care/prevention/wellness, care delivery, triage and diagnosis, diagnostics, and clinical decision support) and in improving the following three main areas:

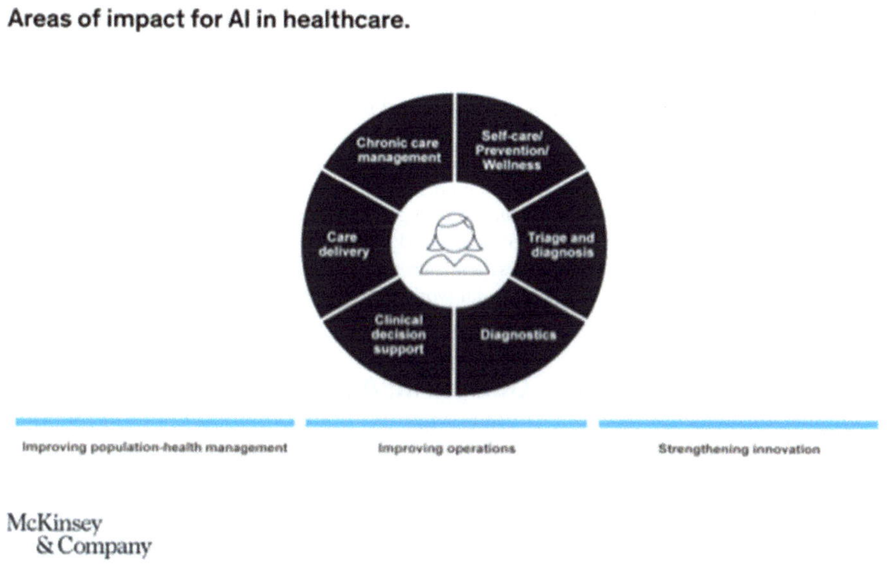

Fig. 13.1 AI areas of impact in healthcare (EIT Health 2020)

1. Population-health management
2. Operations
3. Innovation

Worries exist about AI replacing jobs, but healthcare is a complex ecosystem that requires human attention regardless of innovation. Further, the field of medicine currently faces a significant workforce gap (and likely will in the future) that AI has the potential to help bridge. Indeed, the impact of AI on healthcare workers extends far beyond losing or gaining jobs. The greatest changes will be felt in the nature of their work, and this presents opportunities for substantially improved patient care, as expressed in Fig. 13.1.

While AI is considered a tool to free doctors and healthcare staff from burdensome administrative task (estimated to take up to 70% of practitioner time), AI also has the potential to enhance several areas of clinical activities that will lead to improved care quality and superior patient outcomes:

- Assisting healthcare workers access patient information
- Enabling remote monitoring
- Faster and easier access to knowledge
- Faster and more accurate diagnostics
- Facilitates self-care options for patients

Empowering AI for these solutions is no small task and requires not only training and new skills but changing healthcare education as we know it. As stated in a joint report from EIT Health and McKinsey & Company (2020) about the transformative nature of AI:

> This will all require bringing new activities and skills into the sector, and it will change healthcare education – shifting the focus away from memorising facts and moving to innovation, entrepreneurship, continuous learning and multidisciplinary working. The biggest leap of all will be the need to embed digital and AI skills within healthcare organisations – not only for physicians to change the nature of consultations, but for all frontline staff to integrate AI into their workflow. This is a significant change in organisational culture and capabilities, and one that will necessitate parallel action from practitioners, organisations and systems all working together.

13.1.2 From Inside Out to Outside in: Applying AI

According to this report "the biggest leap of all will be embedding digital and AI skills within healthcare organizations" (EIT Health 2020). Not only will changes permeate the day-to-day lives of doctors and frontline staff (integrating AI into workflow), but this leap requires significant development in organisational culture and in widely collaborative efforts between staff, practitioners, relevant organisations, start-ups, and healthcare systems.

To be reductive, we can simply say that collaboration and partnership in AI hold the potential to *design artificial intelligence solutions together in a structured manner that could improve care pathways, increase efficiency workflow at provider level, and result in superior clinical outcomes*. An entity alone, however, cannot make great headway in AI. For example, Europe, as a whole, ranks well in AI research and investment but in reality is limited by fragmentation (region- or country-specific approaches). A targeted EIT Health survey even found that "44 percent of the healthcare professionals had never been involved in the development or deployment of an AI solution in their organization" (EIT Health 2020).

The focus then of this chapter is to shift the perspective. How can we move away from hyped interest for a technology like AI and from the business perspective (the inside-out approach), and instead towards an outside-in process that prioritises healthcare professionals, providers, and patients?

This shift hones in on the problems of these end users in all aspects of disease management and in regard to workflow inefficiencies when moving from hospital-based to home-based care. Just as importantly, this shift creates space for MedTech and established organisations to *collaborate from the beginning on meaningful, multidisciplinary AI-based innovation*, enabling superior outcomes for patients, superior UX experience for healthcare professionals, and increased sustainability due to effective and widespread introduction.

13.2 The Power of Design and Open Innovation

The private sector continues to play a significant role here. Venture capital (VC) funding for the top 50 firms in healthcare-related AI has reached $8.5 billion (EIT Health 2020). AI has attracted a variety of relevant players, including health insurers, start-ups, pharmaceutical and medical device companies, and, as expected, big tech. The fastest growth, however, is currently seen in Asia, where consumer-focused healthcare AI has already taken off. Ping An's Good Doctor (a visionary, unicorn start-up spun off from a large insurer in China) ranks as the leading online health management platform with more than 300 million users.

The different perspectives between the private sector and healthcare (HC) professionals, however, can be problematic. Among start-up executives, less than 15 per cent considered the input of healthcare professionals as critical during the early design phase. On the other side, healthcare professionals generally viewed the private sector as holding a minimal or "non-existent" role in aggregating and analysing data, providing a secure space for data lakes, or training staff (EIT Health 2020). The start-up perspective combined with a low involvement of users (clinicians, patients, and hospital administration) in the design of AI enabled or assisted solutions makes it clear that the different languages spoken in healthcare must be harmonised into new pathways of collaboration.

Another major obstacle for successful implementation of AI solutions is the current lack of *communication regarding clinical evidence*. Healthcare practitioners are understandably reluctant to utilise tools related to patient care without understanding how an AI solution functions, how data are collected, and if there are biases in the algorithms. This is why collaboration with transparency is critical to scaling efforts.

This transparency should extend to final end users as well. Experts in the field supported the idea that user-centric design encourages better data: "If AI design delivers value to end users, those users are more likely to pay attention to the quality of data they contribute, thereby improving the AI and creating a virtuous circle" (EIT Health 2020).

Thus, partnerships between hospital researchers, MedTech start-ups, and VCs must be established with a cooperative focus on a human-centred design process that applies the principle of value-based healthcare. This focus serves as a blue sheet for the challenges being addressed (e.g. clinical needs and/or operational inefficiencies) and the methods to solve these challenges or produce superior patient/doctor experiences using AI solutions (which by design should be user-friendly and scalable within/between organisations).

13.2.1 A Closer Look: The Need for Meaningful Collaboration

Data are the foundation for AI solutions, firstly to identify and target problems (critical needs and/or operational inefficiencies), and secondly to develop and train

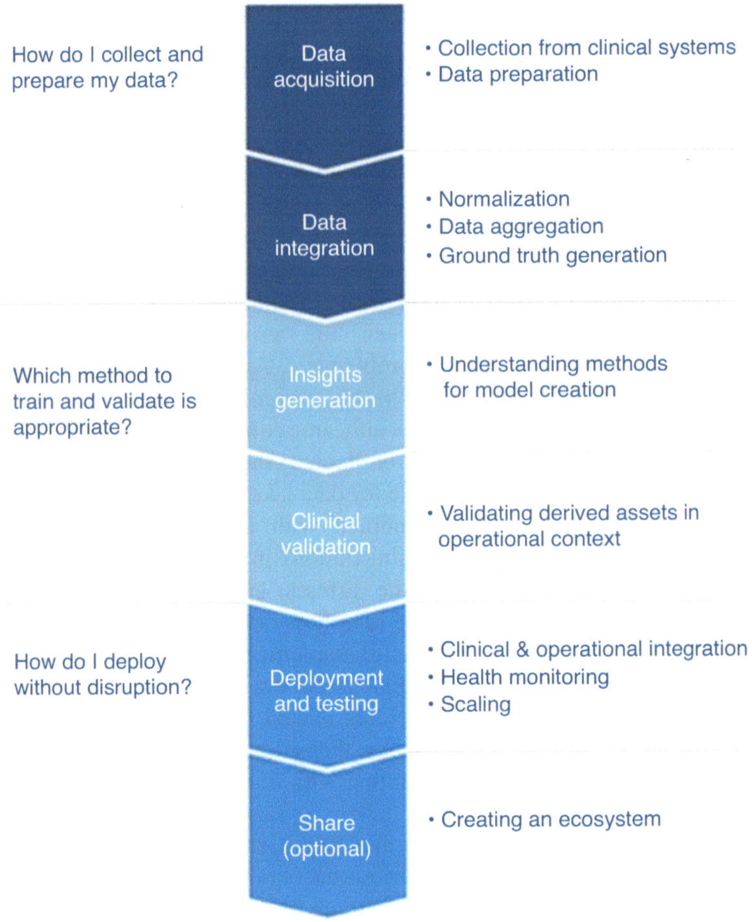

Fig. 13.2 Philip's framework for AI-enabled solutions development/deployment (Huffman 2018)

algorithms. Meaningful collaboration across start-ups and established organisations is hugely critical to collect and mobilise quality data into valuable solutions.

To understand the nuance of this collaboration need from start to finish, we offer a working example of the AI development process. Taking a very vested interest in AI, Philips leadership has finalised a framework for AI solutions from data acquisition to deployment to testing as shown in Fig. 13.2 (Huffman 2018).

As is suggested by this framework, involving healthcare providers and clinicians is at the core of the partnership. Further, this involvement must be present from the beginning; otherwise it will likely create "major barriers to addressing quality issues early on and [to] adopting solutions at scale" (EIT Health 2020).

Additionally, there are several other areas and roles (non-vertically integrated at HC provider level) necessary for effective collaboration. According to the EIT/McKinsey report (2020):

> Clinical engagement will also be required in product leadership, in order to determine the contribution of AI-based solutions within broader clinical protocols. Designers specializing in human-machine interactions on clinical decision making will help create new workflows that integrate AI. Data architects will be critical in defining how to record, store and structure clinical data so that algorithms can deliver insights, while leaders in data governance and data ethics will also play vital roles. In other data-rich areas, such as genomics, new professionals would include 'hybrid' roles, such as clinical bioinformaticians, specialists in genomic medicine, and genomic counsellors. Institutions will have to develop teams with expertise in partnering with, procuring, and implementing AI products that have been developed or pioneered by other institutions. Orchestrating the introduction of new specializations coming from data science and engineering within healthcare delivery will become a critical skill in itself.

While these AI solutions seem to create entirely new ecosystem within themselves, we must interweave existing areas for integration and further development purposes. Many of the technical skills and innovative areas mentioned in the report are ones already found in MedTech start-ups today. However, while these start-ups have the vision for an AI solution, its integration, and its future, what they lack is the data to validate their vision and the components to fine-tune concepts with patients (in their day-to-day real life) and in real-world healthcare delivery.

Regardless of the need to collaborate, there are different advantages and drawbacks for each stakeholder involved in partnerships (large organisations, healthcare groups, and the start-ups with the technical skills to create health solutions enabled by AI algorithms). In 2018, the World Economic Forum (WEF) validated and published a white paper entitled "Collaboration between start-ups and Corporates: A practical Guide for mutual understanding", which aimed to provide guidelines that would accelerate collaboration between the various stakeholders. This chapter offers insights into the field of health and the challenges for those involved in developing AI-enabled solutions. The authors firmly believe, however, if these challenges are overcome, it will lead to transformational health solutions and personalised medicines by emerging technological trends, the most relevant of which is AI.

13.2.2 The Advantages

As iterated in the WEF paper (2018) and seen in Fig. 13.3, there are different benefits for each stakeholder in start-up–corporate partnership.

Advantages from the start-up perspective:

- Revenues and independence from external capital (VC)
- Success story for the future sales
- Scalable customer base (especially in case of collaboration with a large group of hospitals)
- Riskless internationalisation
- Attractive retail channel

Fig. 13.3 Advantages of collaboration between corporates/established organisations and start-ups (WEF 2018)

- Access to proprietary asset
- Market knowledge and mentoring

Advantages from the corporate and healthcare provider/hospital group perspective:

- External Innovation and disruption
- More innovative suppliers
- Customer focus
- Entrepreneurial and more agile culture
- Staying on top of market development
- New revenue streams and business lines

While the advantages are indeed attractive, the challenges can act as failure-causing obstacles. For example, one challenge start-ups often face is gaining support of the "decision-makers, policymakers, and economic buyers" during early phases. Instead of attracting attention based on patient outcomes and return of investment (ROI), many start-ups focus on superficial features or how the technology works, which are actually less important factors for decision-makers (i.e. corporate and organisations) (WEF 2018).

13.2.3 The Challenges

Next, we examine the challenges to be overcome to create successful start-up–corporate partnership (WEF 2018) (also seen in Fig. 13.4 below).

Challenges from the start-up perspective:

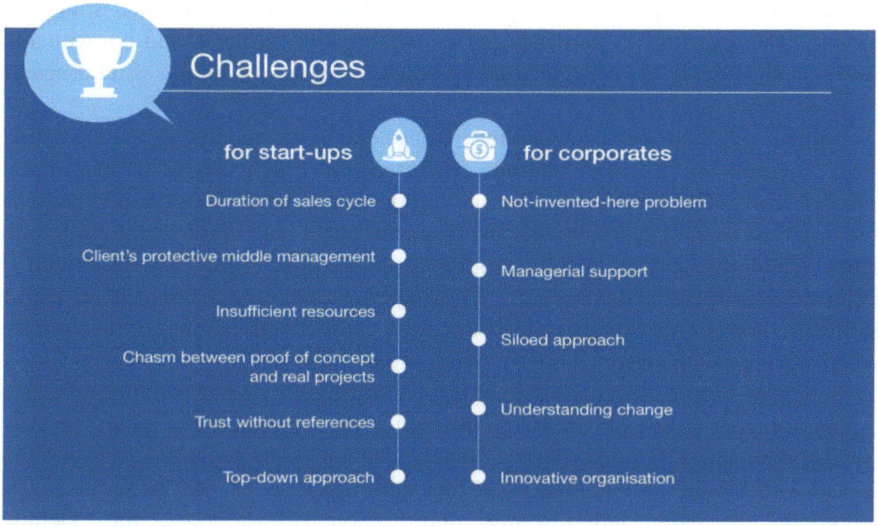

Fig. 13.4 Challenges faced in collaborative efforts between corporates/established organisations and start-ups (WEF 2018)

- Duration of sales cycle
- Client's protective middle management
- Insufficient resources
- Chasm between the proof-of-concept and real projects
- Trust without references
- Top-down approach

Challenges from the corporate and healthcare provider/hospital group perspective:

- Managerial support
- Not-invented-here problem
- Siloed approach (i.e. between hospital departments, administration, IT)
- Understanding change
- Innovative organisation

13.2.4 Beyond the Challenges: Understanding the Risks

Projects that possess the potential to radically change personalised and precision medicine through the use of AI are not without risks, and the nature of these risks is different for each stakeholder in a collaboration.

First, we can look from the view of an innovative and early-stage start-up or MedTech company. Without additional financing, or even if VC-mediated, the window of time to identify and gain new customers or new revenue sources is limited. This can lead to poor choices in collaborative partners, and consequently being narrowed into developing siloed products without scalability, wide usage, or room to grow.

Another risk, start-ups can encounter delays due to the bureaucratic and administrative complexity present in larger organisations, which can drain money and energy/efforts. These risks are especially important to be aware of and to mitigate because time to scale remains a pivotal factor in success or failure. Solutions require quick scalability, but many also require certification of safety and validation of impact based on clinical research. Considerations of time stand out as critical elements, and start-ups can risk losing their agility and failure-culture mindset when working with larger, slower-moving organisations.

Next, we observe the perspective of large companies, by which we mean complex organisations of various kinds (large hospital groups, government bodies, pharmaceutical companies, or other players in the healthcare system) that have the ambition to use AI for better personalised and precision medicine as well as for increased workflow efficiency and resource optimisation (please note: this includes optimisation of resources, not reduction in costs to treat/cure specific conditions). For these large organisations, the phrase "safety first" is indeed first and maintaining company reputation runs a close second. A corporate or large group must evaluate upfront every aspect that could lead, both in the short and long term, to potential reputational damage deriving from the development of an AI solution. Research and development of any solution in healthcare, AI or otherwise, is by its nature expensive not only for direct development costs but also for validation, certification, and scaling resources. Despite planning, and as evident from the number currently on the market, not all solutions are truly ready. If this is the case, corporates and organisations risk substantial or total investment losses. This is acknowledged as one of the great paradigms of innovation, which must always be taken into consideration.

Organisations may not be ready for this paradigm. There is always risk in innovation, in a transformational unknown. However, acceptance of this unknown (which is part of the start-up mindset) is generally at odds with the business models of many corporations. Whether employees are too siloed to understand the importance of an AI project, whether they worry too much about failure, or whether they feel threatened by new processes or a business culture that encourages rocking the status quo, it can take a large leap in mentality for both leadership and employees. This general uncertainty can be mitigated by understanding the business model first and foremost (how a solution will generate value for all stakeholders). However, lack of the right mindset and lack of readiness to adopt or adapt the solution and its business model "inherently lead to a fruitless collaboration," otherwise known as maturity misalignment (WEF 2018). For the full risk comparison, see Fig. 13.5.

Learning the benefits, challenges, and risks is only the first step. As is clear and to put it simply, it is not easy to create and structure an effective collaboration between several complex parts (small and medium-sized enterprises in MedTech, VC, HC organisation, clinicians, and patients). In conjunction with the benefits, risks, and

Fig. 13.5 Risks in collaboration for corporates and start-ups (WEF 2018)

challenges, the following themes need to be addressed by all players in the healthcare ecosystem in order to build, validate, and scale AI solutions.

- Quality of the partnership
- Cross-contamination of skills and education (a guide for mutual understanding between corporate, start-up, and HC organisations)
- Data quality, governance, security, and interoperability
- Working at scale
- Risk management and regulations
- Funding
- Intellectual properties (IP) and future monetisation models present in the term sheet agreement between all the parties.

It is quite easy to discuss these in theory, but effectively addressing them in practice is not. This leads us back to the question of *how innovations in personalised medicine and AI can happen at the intersection of start-ups and corporates.* As previously clarified, these two players cannot be the only stakeholders at the intersection. However, for an effective multidisciplinary collaboration to succeed in creating high-impact and sustainable solutions, it needs to make sense from a business point of view.

After investigating and observing various collaboration frameworks from several real-life use cases involving AI-based solutions and considering theoretical models, we have identified the most responsive solution to that pressing question: "How can we develop an AI-driven biomedical product from conception to market launch?"

13.3 A Decision Perspective Framework

The *decision perspective framework* provides clear decision-making junctions and marks where agreement is needed between parties (founders, investors, and other key stakeholders). This framework was recently published by David Higgins (Berlin Institute of Health) and Vince L. Madai, (Charitè Lab for Artificial Intelligence) in an article entitled *From Bit to Bedside: A practical Framework for Artificial Intelligence Product Development in Healthcare*, which aimed to close the translational gap that currently restricts deployment of AI/machine learning (ML)-based tools. Despite the innovative progress of the tools themselves, implementation continues to drag behind. The authors of this chapter go so far as to classify this gap as a "major public health challenge," and hope that the framework will guide digital health entrepreneurs in their efforts and resource allocation.

13.3.1 Framework for Artificial Intelligence Clinical Product Development

This decision perspective framework contains *three consecutive phases* in each of the *four domains*. We will briefly discuss both the phases and the domains.

13.3.1.1 Phases

1. *Form.* This phase encompasses forming a small group around a solution, investigating its feasibility, and formulating its journey to market. The last of which requires not only validating its clinical need but "understanding of regulatory and clinical validation paths" (Higgins and Madai 2020).
2. *Build.* This phase involves the newly formed, cross-functional team offering sufficient time/efforts and committing to an *18-month to 5-years collaboration.* With this team in place, they then work towards building a solution that is solid from a clinical and regulatory standpoint.
3. *Launch.* Solutions must pass certification and be proven effective through clinical studies. The end goal of this phase is readiness for product deployment.

13.3.1.2 Domains

Next, we examine the domains to be taken into consideration to manage expectations between the different stakeholders in an AI solution collaboration. According to Higgins and Madai, the framework should be applied to the following four domains (2020):

- Clinical domain (clinical validation)
- Regulatory domain
- Data domain
- Algorithm domain (or model development)

Clinical Validation

	Form	Build	Launch
Risks	• There is no clinical need • False assumptions about the clinical need • Inability/failure to form a team which can solve the clinical need	• Inability to show clinical benefit • PoC product is not robust to clinical conditions • A mismatch between pilot product and actual clinical need	• Clinically validated performance is considerably lower than pilot studies • Users don't understand or by-pass the solution • Time and cost of clinical validation are underestimated
Objectives	• Demonstrate clear evidence for a clinical need • Demonstrate that the clinical need is technically solvable (Proof of concept)	• Demonstrate a clinical benefit in several pilot environments • Evaluate the level of (one-off) customisation needed in delivering a solution to the clinic • Demonstrate that users are actually willing to use the product • Discover potential clinical barriers to deployment	• Clinically validate the efficacy of proposed product • Ensure that clinical validation resembles the actual clinical setting • Demonstrate clinical safety of product • Evaluate degree to which 'naive' clinical users are willing to engage with product • Develop a system for monitoring changes in clinical practice
Key Results	• Basic market research / interviews completed indicating a real clinical need • Promising initial results, derived from a robust approach, suggesting the clinical need can be solved (see also data section) • Proof of concept	• Deployment of prototype systems working in several clinical environments • Evidence that clinical need is solved across patient populations • Interest from clinical partners to continue using the software	• Extensive UI/UX testing • Third-party validation of product in a non-controlled environment (usually RCT) • Procedures for detection / tracking of updated clinical standards
Advice	• Develop market-fit / clinical-need hypotheses • Test these hypotheses early • Cycle through them on an ongoing basis • Test the bigger (cost weighted) risks first! • Demonstrate clearly proven clinical benefits (doctors are forced(!) to use the solution) or reduce costs. If you will do neither, nobody will pay for your solution	• Observe how technically naive users interact with pilot product • Do not rely on testimonials • Gather use statistics • Be open to surprising results of your validation that might change your product direction • Make a validation plan, iterate over it continuously	• Be aware that you cannot sell a product into a non pilot study environment unless it is demonstrably safe • Users will avoid change unless you provide clear, visible, proven benefits to them • Choose the trial with smallest efforts and lowest costs. It might be beneficial to work with a contract research organization

Fig. 13.6 The clinical domain centres around validation of clinical needs (Higgins and Madai 2020, originally published under the terms of the Creative Commons Attribution License; © 2020 The Authors)

Clinical Domain (Clinical Validation of an AI Product)

The clinical domain connects use cases of a solution with real-world needs in the clinical setting, and involves continual validation of these needs throughout project development. Usability and interoperability also come into play here. See details in Fig. 13.6.

Regulatory Domain
AI-enabled healthcare products will likely be subject to medical device requirements and regulations. In this domain, entrepreneurs and developers need to take the steps to fully comprehend and comply with regulatory processes in order to minimise risks (Fig. 13.7).

Data Domain
As is clear, the right data is a make-or-break condition in the development of an AI solution. This domain concerns obtaining the right data, access to that data, and ensuring data sources are sufficient for certification and clinical validation, as shown in Fig. 13.8.

Algorithm Domain, or Model Development
This domain addresses the choice of AI methods, i.e. the algorithms. Special attention should be given to amount of data and the scalability and biases of algorithms. Please note in Fig. 13.9 below that considerations for the right collaboration between different parties (partner clinics and pilot environments) are essential in this framework.

13.4 Discussion and Conclusion

In conclusion, we can say that there is no magic formula or standard rules to activate successful collaborations on developing a biomedical product based on artificial intelligence from conception to market launch. If that had been the case, artificial intelligence in the life sciences would be much more widespread. However, we still have a lot to do and many challenges to overcome. One thing, however, is clear: like in other areas of healthcare, we need methodological rigour and incremental scientific research grounded in robust evidence to which we can apply new technologies. Our success will hinge on our ability to overcome the challenges that all involved parties face in the collaborative project. We need to proactively address the risks that slow down or prevent the birth of promising ideas of transformative healthcare solutions and personalised treatments.

Start-ups and incumbents cannot be the only stakeholders at the crossroads. To build an effective multidisciplinary collaboration from a business design point of view, they must make sense not only in the beginning but also throughout their lifecycle. The best way to achieve that is by keeping the eyes close on the ever-evolving needs of doctors and patients who are ultimately the users who will co-design and adopt these new solutions. A project therefore, like experiments, has the imperative to start small but already with the idea of being scalable. Toward this end, dynamic collaboration agreements based on incremental milestones make the difference. In other words: don't overcommit to certain goals early on and forcibly stick to it. Instead, create a structure that allows for flexibility as things evolve. This means making agreements revisable and flexible and combining with a well-defined exit strategy to avoid wasting resources and losing focus, which is, of

Regulatory Affairs

	Form	Build	Launch
Risks	• Lack of awareness of future regulatory demands which will impact on development methodology	• Wrong assumptions about required certification levels and associated costs and timelines • Late discovery that key technology was not developed in a medically certifiable manner • Not using certifiable development methods now for a piece of technology which will ultimately be a part of the product	• Key assumptions about the certification process are wrong and delay certification massively • Lack of preparation for the post-launch phase lead to a "frozen" product which cannot be deployed
Objectives	• Understand regulatory demands for the Build phase • Develop initial regulatory strategy	• Demonstrably follow best practices for development of 'safe' (ie. medically certifiable) software and devices	• Fully documented, certification compliant software development process • Develop product surveillance plan • Develop product update procedures
Key Results	• Initial regulatory strategy	• Clear process pathway to certification, or initial certification • Internal processes which are appropriate for selected level of medical certification	• Medical product certification • Procedures for deployment of software updates • Procedures for surveillance and reporting
Advice	• Seek professional legal assessment of the regulatory requirements for possible future product	• Understand the certification process of your region fully • Understand and implement the correct medical software development process • Understand the specific AI-focused regulatory requirements of your country/ region • A key decision for your product is to know when to start with the implementation of medical device development tailored to medical device regulations • UI/UX designed and tested to prevent overstatement of / misleading results • Make contact to regulatory bodies early in the process • Work with a partner specialized on medical regulatory affairs particularly medical device certification	• Do not underestimate the complexity and burden of the final implementation of regulatory requirements • Consider automated detection of non-handleable cases • Work with a partner specialized on medical regulatory affairs

Fig. 13.7 Regulatory domain involves proactively avoiding pitfalls of high regulation (Higgins and Madai 2020, originally published under the terms of the Creative Commons Attribution License; © 2020 The Authors)

Data Strategy

	Form	Build	Launch
Risks	• Lack of access to data • Not enough data/signal in data to develop proof-of-concept • Assumptions about the statistical characteristics of data are wrong • Lack of understanding what kind of data is needed for later stages (beyond proof-of-concept)	• Site-to-site variability in clinical data is greater than expected (data harmonization fails) • Increased access to data does not lead to hoped for improvements in signal detection • Failure to acquire extended data access rights	• Real-world data has lower fidelity than that obtained from prototyping sites • Real-world patients are less homogenous than those in early data sets
Objectives	• Acquire access rights for initial data set • Understand the structure of the clinical data (width (D), depth (N), regularity, signal/noise, bias) • Determine appropriateness of data set to solve clinical need • Construct a data plan for data you need in later stages	• Data harmonization of different data sources • Obtain legal access to multiple new datasets	• Ensure that data collected in the clinical validations is equivalent to the real clinical setting • Determine whether the general (clinical) population have the same data characteristics as those used to build the algorithms
Key Results	• Fitting data source identified • Data is acquired / collected • Data structure is understood • Hypotheses formed, and validated, about ability to solve clinical need • Data plan developed	• Sufficiently sized and diverse data has been obtained • Harmonized database including all data sources	• Sufficient data with the pre-defined characteristics are gathered to allow clinical validation • Data is similar to previous pilot studies
Advice	• Explore free datasets • Partner with clinical institutions (study data or RWD) • Utilize an interdisciplinary approach (understand what each feature and label exactly means) • Be aware that you have little to no control over what is captured • Focus in this stage should be on whether the insight actually makes sense • It is appropriate to develop the PoC on potentially biased data, but be aware of it • Use your PoC to determine how many data points you are likely to need for a product	• Avoid, identify and mitigate bias • Look out for corner cases • Make sure you have the data for all populations that are covered by your clinical need • Hardware from different manufacturers produce different data sets. Make sure you have data from all • Remove locally specific identifiers. If not, your model might identify those instead of the pathology • At every point of the process, make sure that predictive value holds up • Make sure your team has enough people dedicated to data processing and harmonization	• Caveat: data heterogeneity can be a real problem • Ask yourself this question often: How good is your system in the real world?

Fig. 13.8 Data domain focuses on the quality and applicability of data (Higgins and Madai 2020, originally published under the terms of the Creative Commons Attribution License; © 2020 The Authors)

course, the need of the clinicians and patients. However, to the people and organisations involved, there should also be a palpable benefit that justifies everybody's investments (financial, reputational, or otherwise).

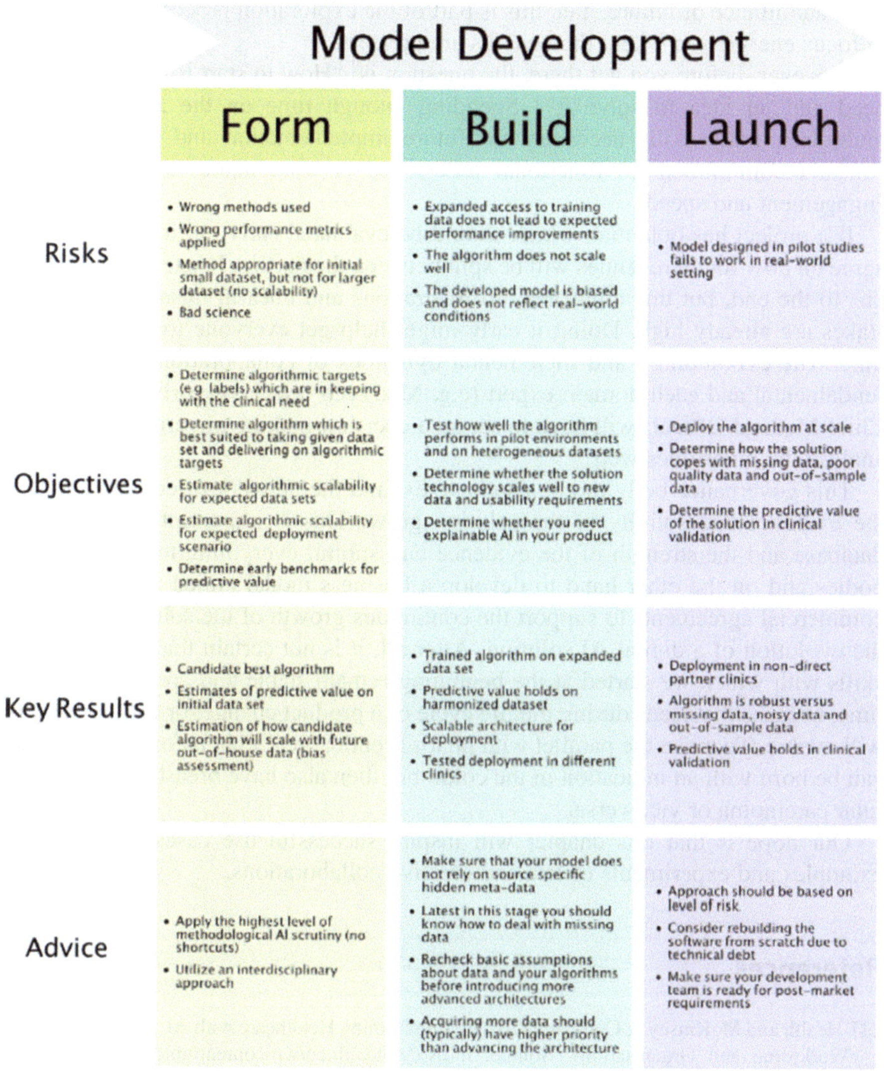

Fig. 13.9 Algorithm development requires an interdisciplinary approach and constant attention to assumptions and biases (Higgins and Madai 2020, originally published under the terms of the Creative Commons Attribution License; © 2020 The Authors)

We recommend adopting a long-term view rather than rushing to deliver a solution. Setting the right foundation means creating milestones based on clinical evidence and efficiency to speed up stop-and-go decisions. It also reduces efforts and capital needed to scale the solution further down the road. Therefore, it is perfectly fine to stop or change the development of one project and bringing the insights into the others. Too often, managers cling to existing projects because they would view it as an admittance of failure. Yet, this is part of the exploration process and allows us to focus energies on where the benefits are greatest.

However, before you get there, the question is: "How to start if I have a clinical need and an idea to solve it?" Spending enough time on the analysis of the multidisciplinary skills necessary for future implementation and guaranteeing a constant commitment to functional milestones is what makes a difference for engagement and speed.

If a project has potential, which should be evaluated early on, it is also right to agree on how future revenues will be split between the parties. It is tempting to leave this to the end, but this might lead to frustrations and unearth misalignment when stakes are already high. Doing it early might help get everyone to walk the extra mile. The evolutionary and incremental dynamics of collaborations are therefore fundamental and each domain expert (e.g. MedTech lawyers or privacy experts or Chief Medical Officer) will have to set up a decision-making governance body and analyse the milestones well.

This governance body might also have expand the network of collaborations as the evidence and solidity of the solution grow. This is important to expand the database and the strength of the evidence and submit everything to the regulatory bodies and on the other hand to develop a business model aimed at entering into commercial agreements to support the continuous growth of the solution as well as the evolution of a digital AI solution. After all, it is not certain that the assets and skills with which we started at the beginning remain stable and are sufficient over time, because the needs during the life cycle of a product change; or its intended use will evolve. To make the parallel with pharmaceutical research, an oncological drug can be born with an indication in the colon but then also have breast and hepatocellular carcinoma or vice versa.

Our hope is that this chapter will inspire successful use cases but above all examples and experiments of fast and creative collaborations.

References

EIT Health and McKinsey & Company (2020) Transforming Healthcare with AI: The Impact on the Workforce and Organisations. Source: https://eithealth.eu/wp-content/uploads/2020/03/EIT-Health-and-McKinsey_Transforming-Healthcare-with-AI.pdf Last consultation: 24/08/2022

European Parliament (2016) Artificial Intelligence: Potential Benefits and Ethical Considerations, Source: https://www.europarl.europa.eu/RegData/etudes/BRIE/2016/571380/IPOL_BRI(2016)571380_EN.pdf Last consultation: 24/08/2022

Higgins D, Madai V (2020) From bit to bedside: a practical framework for artificial intelligence product development in healthcare. Adv Intell Syst. https://onlinelibrary.wiley.com/doi/full/10.1002/aisy.202000052

Huffman J (2018) AI in healthcare: a six-step framework for getting the most out of your data. Source: https://www.philips.com/a-w/about/news/archive/blogs/innovation-matters/20180709-ai-in-healthcare-a-six-step-framework-for-getting-the-most-out-of-your-data.html. Last consultation: 24/08/2022

World Economic Forum (WEF) (2018) Collaboration between Start-ups and Corporates A Practical Guide for Mutual Understanding, Source: http://www3.weforum.org/docs/WEF_White_Paper_Collaboration_between_Start-ups_and_Corporates.pdf. Last consultation: 24/08/2022

Artificial Intelligence Augmented Medtech: Toward Personalized Patient Treatment

14

Daniel Neumann, Stefan Schlichting, Klaas Rackebrandt, and Eric Klasen

Abstract

Artificial intelligence (AI)-augmented medical device technology provides an important opportunity to optimize the detection, diagnosis, and treatment outcome for individual patients. By combining AI-augmented medical devices with closed loop feedback from continuous sensor measurements performed on the patient, AI-augmented Medtech has the potential to deliver truly personalized patient treatment. This chapter provides an overview of common AI definitions and highlights current misconceptions of artificial intelligence. It describes ongoing efforts by regulators to develop a regulatory framework for medical software and discusses a classification for AI learning schemes. Usage examples providing added patient value, in observation, visual analysis, diagnostics, or treatment, are provided. Data sources to train AI are discussed as well as data quality, a critical prerequisite for the development of safe and effective medical devices augmented by AI. The ethical aspects of AI in healthcare are presented as well as the changes and adaptations envisaged for the medical profession and physician education. The chapter concludes that the field of personalized AI-augmented Medtech will be one of the core applications for AI in healthcare and is expected to grow continuously over the next years, despite potential challenges such as product liability and data privacy regulations.

Keywords

AI MedTech · MedTech regulation · Personalized MedTech

D. Neumann · S. Schlichting · K. Rackebrandt
UNITY AG, Munich, Germany
e-mail: daniel.neumann@unity.de; stefan.schlichting@unity.de; klaas.rackebrandt@unity.de

E. Klasen (✉)
Waypoint Life Sciences Consulting, Féchy, Switzerland

A. Cesario et al. (eds.), *Personalized Medicine Meets Artificial Intelligence*,
https://doi.org/10.1007/978-3-031-32614-1_14

14.1 A Definition of Personalized Medtech

The term "personalized medical devices" has been defined by the International Medical Device Regulators Forum (2018), as "*a generic term to describe any of the types of medical devices that are intended for a particular individual, which could either be a custom-made, patient-matched, or adaptable medical device*." Manufacturing technologies such as additive and subtractive manufacturing have brought the classical custom-made devices within reach of specific patients on a much greater scale and IMDRF (2020) issued a guidance document on regulatory pathways for different types of personalized medical devices. However, in this context the term *personalized* is different from the meaning when used to define personalized medicines, as the term personalized medical devices refers to hardware devices that are matched to each individual patient based on their physical anatomy or adapted to suit an individual patient's specific anatomic-physiologic features prior to use.

This chapter will use the term AI-augmented Medtech for either medical devices that incorporate AI software, so-called Software in a Medical Device (SiMD), or for AI software used as a stand-alone medical device, so-called Software as a Medical Device (SaMD).

14.2 A Definition of Artificial Intelligence for Medtech Applications

Artificial intelligence (AI) is already ubiquitous today, even if many people are currently confronted with it unconsciously in their everyday lives—whether through chatbots, targeted advertising, smart home devices, or other means. AI in many industries is already causing a paradigm shift today. The use of new AI applications has a profound impact on a multitude of areas within business and public organizations. Slowly but surely, AI is also impacting healthcare. In many cases, AI can be utilized to provide added value if well specified and when developed on a valid data basis.

In addition to AI, the medical domain also has embarked on developing machine learning-driven approaches to speed up reproducible processes and steps in patient management and care (e.g., assisted imaging analysis in radiology).

Overall, progress to date is still limited—the main hurdles are gaps in the regulatory framework, lack of regulatory clarity and understanding, and the caveat raised by physicians that not knowing what AI means hampers its use. These hurdles tend to delay a breakthrough in adopting AI in medicine and Medtech.

In general, AI refers to the attempt to reproduce complex decision-making structures (of humans). It is often described as a system that interacts independently with its environment and uses the collected and aggregated information to process/solve complex problems. The core attribute of AI is its inherent learning ability. In the medical area, AI is being used in medical devices, medicines, complex workflows, and non-linear decision-making processes. Compared to algorithmic approaches of solving problems, AI applications are not defined by the exact series of procedures/steps that must be solved, but instead by defining the desired outcome. The algorithm then "trains" based on real-life data to identify the most suitable and efficient way to come to a desired clinical outcome. This is contrary to a procedural approach which relies on exact input-output relations and a well-defined transition. This must be ensured in a different way using AI approaches.

Despite the apparent differences of AI's learning behavior, the term AI is often being used wrongly for any algorithm that solves a complex or not easy to describe problem – independent from the actual solution approach. Kaplan and Haenlein (2020) define AI as "*the ability of a system to correctly interpret external data, learn from that data, and use those insights to achieve specific goals and tasks through flexible adaptation*."

A more general definition of AI, encompassing all product groups including medical devices and other healthcare products, and that also includes the AI techniques and sub-disciplines that are currently used to build AI systems, was provided by the High-Level Expert Group on Artificial Intelligence (2019a). The definition below is a succinct version of the AI-HLEG definition, not including the current scientific techniques and approaches as these are described in detail in section 14.6:

> Artificial Intelligence (AI) refers to human-designed systems that act on a complex goal in the physical or digital world by perceiving their environment, interpreting the structured or unstructured data they collect, reasoning about the knowledge derived from that data, and deciding the best action(s) (according to predefined parameters) to achieve the given goal. AI systems can also be designed to learn to adapt their behavior by analyzing how the environment is affected by their past actions.

As to machine learning-enabled medical devices (MLMD), the International Medical Device Regulators Forum (IMDRF) (2021) issued a draft definition for MLMD. At the time of writing this chapter, the IMDRF proposed document is still under public consultation and may be amended based on comments received. However, as the authors state in the introduction of this document, the purpose of this document is to establish internationally accepted terms and definitions across the medical device Total Product Life Cycle to promote consistency, support global harmonization efforts, and provide a foundation for the development of future guidelines related to MLMD. The IMDRF definition of MLMD, as provided in the proposed document, is "*A medical device that uses machine learning, in part or in whole, to achieve its intended medical purpose*." The document includes the standard definition of a medical device with some notes, e.g., to clarify that in some jurisdictions, products are considered medical devices but not in others, and it includes definitions for different machine learning approaches.

14.3 Misconceptions of Artificial Intelligence

Even though healthcare has adopted AI technology in medical devices, workflows, and decision-making processes, there are still misconceptions in the operational adoption of AI that may hamper or delay innovative activities and projects. We present here, without deeper discussion, popular misconceptions in the usage and implementation of AI.

High complexity in the use if AI applications is an often-expressed remark hindering its implementation. Considering human-centric innovations, such as Amazon's Alexa or Apple's Siri, which do not need any kind of training or additional software, these are applications that are easy to adopt. Today, there are several applications which serve as *invisible AI* in medical practices, as many are naturally integrated with electronic health records (EHRs). Examples are products produced by CareCloud or Athenahealth (https://www.athenahealth.com/), healthcare technology companies that market a suite of proprietary, cloud-based solutions for healthcare organizations.

High cost of investment is also a known misconception, resulting in an aversion of investment even before the development of a proper business case. Many AI-focused services (e.g., appointments, invoices/billing) have adapted and either provide subscriptions for a standard fee, or apply usage-based billing, or charge customers a standard subscription fee for a plan with a set usage limit. The risks of inaccurate billing are still a challenge in this field and the large amount of data involved is prime territory for AI applications. Companies are developing machine learning [ML] and natural language processing (NLP) to automatically recognize and extract data from medical documents for proper coding and billing for medical applications impacting the patient directly.

Replacement of human employment is a third issue raised in the adoption of AI concepts and applications. Creativity, patient empathy, compassion, understanding, honesty, competence, commitment, and humanity are profoundly human qualities required of health professionals. Such distinctively human qualities and abilities can never be replaced in the medical field. And for end-of-life situations, it is almost impossible to imagine compassionate care being delivered in a hospital by algorithms. Instead, bots assist with time-consuming and repetitive activities, freeing up skills of human resources for more useful work. Rather than replacing the human component of healthcare delivery, AI will become a vital tool to improve efficiency of care and consequently freeing up time for human caretakers to improve patient satisfaction. Improving patient satisfaction is key to building a strong patient–provider relationship, boosting patient engagement, and improving the overall health outcomes. As Davenport and Kalakota (2019) stated: "*Perhaps the only healthcare providers who will lose their jobs over time may be those who refuse to work alongside artificial intelligence.*"

14.4 Toward a Regulatory Framework for AI in MedTech

As technology continues to advance every aspect of healthcare, software incorporating AI, and specifically the subset of AI known as ML, has become an important part of an increasing number of medical devices. One of the greatest potential benefits of ML resides in its ability to create new and important insights from the vast amount of data generated during the delivery of healthcare every day.

Over the past decade, the US FDA has reviewed and authorized a growing number of devices legally marketed (via 510(k) clearance, granted De Novo request, or approved PMA) with ML across many different fields of medicine—and expects this trend to continue. Over time, the FDA has issued numerous non-binding guidance documents on various software-related issues and made several attempts to develop rules for medical device software; however, the FDA has not yet released a separate regulatory framework, nor classifications, for software devices. It should be noted that this includes hardware medical devices with embedded software, so-called SiMD as well as SaMD, as defined by the IMDRF, who formed the Software as a Medical Device Working Group (SaMD WG) to develop guidance supporting innovation and timely access to safe and effective Software as a Medical Device globally. The term "Software as a Medical Device" is defined by IMDRF as "*software intended to be used for one or more medical purposes that perform these purposes without being part of a hardware medical device.*"

Medical device guidance on software was issued by the US FDA for public consultation on November 4, 2021 (FDA 2021a). This guidance document was intended to provide information regarding the recommended documentation sponsors should include in premarket submissions for the FDA's evaluation of the safety and effectiveness of device software functions, which are functions that meet the definition of a device under section 201(h) of the Federal Food, Drug, and Cosmetic Act (FD&C Act). The recommendations in this guidance document pertain to device software functions, SiMD and SaMD. As with all guidance, it does not establish any rights for any person and is not binding on the FDA or the public. One can use an alternative approach if it satisfies the requirements of the applicable statutes and regulations, in this case of the FDA. To discuss an alternative approach, one can contact the FDA staff or Office responsible for this guidance.

Muehlematter et al. (2021) searched governmental and non-governmental databases and identified 222 devices approved in the USA and 240 devices in Europe. They highlighted that the number of approved AI/ML-based devices has increased substantially since 2015, with many being approved for use in radiology. Only very few devices were qualified as high-risk devices. As stated above, in the US, EU, or any other jurisdiction, there is currently no special regulation for medical devices with AI-based software components; however, this may soon change. Considering the potential of AI and machine learning-based software to transform healthcare and have individual patients benefit from the iterative improvement of their therapy that such learning-based software could offer, there is a critical need for a regulatory framework to develop such innovations. In response to this need, the US FDA published on April 2, 2019, a paper entitled *Proposed Regulatory Framework for Modifications to AI/ML-based SaMD – Discussion Paper and Request for Feedback*. This proposed regulatory framework will be briefly discussed in the section AI Learning Schemes further down in this chapter.

In the current absence of a regulatory framework, all products that meet the definition of a medical device in their intended purpose are required to fulfill the requirements for a certain market related to their classification. For example, in the EU, the typical requirement for medical devices is defined in the EU Medical Device Regulation, (EU) 2017/745 in Annex I, Chapter 1, Clause 1 of the MDR:

> Devices shall achieve the performance intended by their manufacturer and shall be designed and manufactured in such a way that, during normal conditions of use, they are suitable for their intended purpose. They shall be safe and effective and shall not compromise the clinical condition or the safety of patients, or the safety and health of users or, where applicable, other persons, provided that any risks which may be associated with their use constitute acceptable risks when weighed against the benefits to the patient and are compatible with a high level of protection of health and safety, taking into account the generally acknowledged state of the art.

The same regulation states in Clause 2 that risks must be reduced as far as possible without adversely affecting the benefit–risk ratio. Consequently, the MDR requires the manufacturer to demonstrate for any medical device, with or without software incorporated, that:

1. The *performance/effectiveness* of the product is achieved during normal conditions as designated by its intended purpose, and that the use of the product does not compromise the clinical condition or the *safety* of the patient, user, or any other person; hence the device should ensure *reliability*, defined as the probability that an item will perform a required function without failure under stated conditions for a specified period of time.
2. Risks must be reduced as far as possible without adversely affecting the clinical benefit–risk ratio; hence *safety* must be ensured without compromising clinical *performance*.

Software, incorporated in either a medical device or software that itself is a device, hence a SaMD, needs to demonstrate compliance with an additional requirement: repeatability**.** This is stated in Annex I, Chapter 2, Clause 17.1.

> Devices that incorporate electronic programmable systems, including software, or software that are devices in themselves, shall be designed to ensure repeatability, reliability, and performance in line with their intended use. In the event of a single fault condition, appropriate means shall be adopted to eliminate or reduce as far as possible consequent risks or impairment of performance.

Finding your way through the regulations, state-of-the-art standards, and interpretations applicable to health software, and in particular, SaMD, can be challenging. A recommended reference book is *Software as a Medical Device, Regulatory and Market Access Implications* (Cobbaert and Bos 2021). This book, published in 2021 by the Regulatory Affairs Professionals Society (RAPS), offers a guide through this complex landscape and provides the expertise of leading software experts.

In April 2021, the EU Commission published a proposal for a Regulation laying down harmonized rules on artificial intelligence (*Artificial Intelligence Act*) (2021). The proposal sets harmonized rules for the development, placement on the market, and use of AI systems in the Union following a proportionate risk-based approach. The draft regulation provides core artificial intelligence rules that apply to all industries, hence including the medical technology industry. It proposes a single future-proof definition of AI. The proposal lays down a risk methodology to define "high-risk" AI systems that pose significant risks to the health and safety or fundamental rights of persons. Those AI systems will have to comply with a set of horizontal mandatory requirements for trustworthy AI and follow conformity assessment procedures before those systems can be placed on the Union market. As regards high-risk AI systems which are safety components of products, this proposal will be integrated into the existing sectoral safety legislation. As regards high-risk AI systems related to products covered by the New Legislative Framework (NLF) legislation (e.g., machinery, *medical devices*, toys), the requirements for AI systems set out in this proposal will be checked as part of the existing conformity assessment procedures under the relevant NLF legislation.

As this draft Regulation is still a proposal, this chapter will not discuss the content, but this will be a very critical Regulation that will regulate the development and marketing in the EU of AI for medical technology and it will depend on how this proposal will be integrated into the existing sectoral safety legislation for medical devices.

14.5 AI Learning Schemes and Regulatory Compliance

The AI learning scheme applied by a medical device with AI-based software components should determine how to satisfy the regulatory requirements presented in the previous paragraph.

14.5.1 AI Learning Schemes

To outline the impact of the regulatory requirements, a classification of different AI learning schemes, as introduced by Cobbaert (2021), is laid out below.

14.5.1.1 Locked Learning Scheme

When a locked learning scheme is applied, learning data are collected in the field and used during the development or update of the product to train the AI. The product is released with a fixed model that is not updated in the field. The only possible way to change an output of the AI to a given set of inputs is by changing the working point on the operating curve, which will repeatedly yield the same results for the same setting and the same input.

14.5.1.2 Discrete Learning Scheme

A discrete learning scheme allows a medical device that is released with a starting model to be updated in the field with real-world data collected during usage at certain points in time based on a qualified set of in-field training data. For this purpose, either the health delivery organization or the manufacturer performs quality assurance measures that the model, updated based on the real-world data, satisfies the performance that is designated by its intended use while performing adequately, without risk of bias or nonergodicity (cf. 14.9.1 Data sources), on the target population of the manufacturer. The quality assurance mechanisms may also be performed automatically. It can be summarized that the update, hence the change, occurs within the intended use, and is restricted by predefined change boundaries, and is executed according to the manufacturer's algorithm change protocol (ACP), a protocol that explains how the algorithm will learn and change while remaining safe and effective, as intended by the manufacturer.

14.5.1.3 Continuous Learning Scheme

If a continuous learning scheme is applied, the product is released with a model as a starting point that is updated in the field based on training data collected during usage. Changes in the training data lead to a re-learning of model and therefore the output of the AI may verify over time with the same input. It is not guaranteed that the performance, hence effectiveness and safety, as designated by the intended use, can be achieved after learning. In their proposed regulatory framework, the FDA uses the term Good Machine Learning Practice (GMLP) to describe a set of AI/ML best practices and standards that should ensure the safety and effectiveness of such devices all along through post-market performance.

14.5.2 Regulatory Compliance

If a locked learning scheme is applied, the generation of evidence that demonstrates the fulfillment of the requirement is not different from any other medical device that is not incorporating AI-based software components as the performance in the field is predetermined.

The question is how to evaluate and monitor a software product that applies a non-locked learning scheme and where modifications to the SaMD occur during the post-market phase. Considering the potential of AI and machine learning-based software to transform healthcare and have individual patients benefit from the iterative improvement of their therapy that such learning-based software could offer, there is a critical need for a regulatory framework and guidance to develop such innovations. As mentioned above, in response to this need, the US FDA published a paper entitled "Proposed Regulatory Framework for Modifications to AI/ML-based SaMD"—*Discussion Paper and Request for Feedback* (2019). This paper describes the FDA's foundation for a potential approach to premarket review for AI and machine learning-driven software modifications, as the FDA's traditional medical device regulation was not designed for continuous learning scheme AI-based software components. The FDA's idea for a tailored regulatory framework that can handle these types of AI components relies on IMDRF's risk categorization principles for SaMD, the FDA's benefit–risk framework as well as the risk management principles described in the software modifications guidance, and the organization-based total product lifecycle approach.

As part of this proposed framework, the FDA described a "Predetermined Change Control Plan" in premarket submissions. This plan would include the types of anticipated modifications (SaMD Pre-Specifications) and the associated methodology being used to implement those changes in a controlled manner that manages risk to patients, referred to as the Algorithm Change Protocol (ACP). If such a predetermined learning scheme is applied, the initial generation of evidence is similar as for products that use a locked learning scheme. As the APC and the change boundaries are predetermined and included in the initial premarket submission, one could expect a reasonable assurance of safety and effectiveness while benefitting from the iterative improvement of performance and/or safety of the medical device, and this to the ultimate benefit of the individual patient, provided the manufacturer that performs the qualification of the changes adheres to the ACP and change boundaries. In principle, this is like calibrations that are currently performed in the field from service technicians during service and maintenance.

For a continuous learning scheme, the challenge is how to demonstrate that the performance and safety of the product remains achieved over time and that after continuous learning the effectiveness and safety of the patient remains as intended by the manufacturer. This kind of dynamic change is not a recognized approach within the current regulatory frameworks and it will be very interesting to see how the FDA will approach this under the proposed regulatory framework as without a specific regulatory framework for a medical device that incorporates a continuous learning-based software, the effectiveness and safety of such medical devices may only be achieved if qualified personnel will act on the patient based on the output of the system and monitor the effectiveness and safety of such action. In addition, the estimation and evaluation of the risk during normal use and of potential misuse may become very complex. In their proposed regulatory framework, the FDA uses the term Good Machine Learning Practice (GMLP) to describe a set of AI/ML best practices and standards and announced to be actively collaborating with other national and international stakeholders on standardization efforts in support of the development of GMLP.

AI and machine learning-based software have the potential to transform healthcare and offer individual patients the benefit of the iterative improvement of their therapy that such learning-based software could offer. Ultimately, AI and machine learning-based software could contribute to the development of truly personalized medical device treatment. However, as mentioned earlier, there is a critical need for a regulatory framework to develop such innovations and industry stakeholders are looking forward to collaborating with FDA and other regulatory bodies on the development and implementation of such a framework.

14.6 AI Technology Overview

As pointed out earlier in this chapter, artificial intelligence has two defining criteria: autonomy and adaptability. Although they help with differentiating from algorithmic and procedural technologies, they are not sufficient to cover potentials of federated learning and rising demands to data efficiency and generalization. Additionally, understanding why an AI application has delivered a specific outcome is not always understood and this is often discussed, sometimes (Goebel et al. 2018) ironically referring to the number 42.[1]

The European Commission's High-Level Expert Group on Artificial Intelligence (HLEG AI 2019a) differentiates the umbrella term *artificial intelligence* into technique groups which themselves are again umbrella terms for sub-disciplines and techniques. These technical groups and some of the sub-disciplines are:

1. Reasoning and decision making:
   (a) Knowledge representation—transform data to knowledge; how to best model knowledge
   (b) Reasoning—making inferences
   (c) Planning and scheduling
   (d) Searching through a large set of solutions
   (e) Optimization among all possible solutions
   (f) Decide what action to take
2. Learning:
   (a) Machine learning approaches:
      Supervised learning
      Unsupervised learning
      Reinforcement learning
   (b) Machine learning techniques:
      Neural networks, deep learning, decision trees, and other learning technique
3. Robotics

The classification of different AI learning schemes by Cobbaert (2021) will support the discussion on the development of an AI regulatory pathway, as well as discussions on ethics, regulatory oversight and control, and other policy areas. It also helps us in differentiating AI usage examples. As shown above, all techniques used

[1]In The Hitchhiker's *Guide to the Galaxy*, published by Douglas Adams in 1979, 42 was the answer to the ultimate question of life, the universe, and everything, calculated by a supercomputer named Deep Thought over a period of 7.5 million years.

for the development of AI systems can be bundled into two main groups, based on their intrinsic capabilities, learning and reasoning. The third technique group is robotics. Each group is discussed below.

14.6.1 Reasoning

This group of techniques includes knowledge representation and reasoning, planning, scheduling, search, and optimization. In this group, knowledge representation is the central building block for converting data into knowledge, resulting in a model. Once such model has been developed based on a large amount of training data, new sensor-derived data can be reasoned about (knowledge reasoning). Reasoning AI systems draw conclusions through symbolic rules, assessing the new, sensor-derived data against a large set of solutions, and ultimately identifying the most appropriate solution from all possible solutions to a problem. The reasoning of such an AI system is usually complex and can be based on a combination of the above-described techniques. Such a black-box type of approach can lead to a lack of confidence with physicians who need thorough evidence that they can rely on the predictions in the real world where real lives are at stake. In the end, the final question is who takes the clinical decision, the AI model or the physician? We will come back on this question in the next paragraph on Learning.

14.6.2 Learning

ML enables a system to learn how to independently solve problems that cannot be specified precisely or whose solution method cannot be described by symbolic reasoning rules. Typical examples of human perceptual abilities, like speech and language comprehension as well as computer vision, are relatively easy for most humans to learn and apply, but challenging for the AI system, especially when unstructured data must be interpreted. ML can be classified, with a reasonable degree of fuzziness, into supervised learning, unsupervised learning, and reinforcement learning.

In supervised ML, the system is trained using input/output data, but no system behavior rules are specified. If the system has access to enough training examples of good data quality, and "good quality" meaning non-biased, complete, and preferably ergodic (cf. 14.9.1 Data sources), it learns to generalize and will be able to solve the "problem" for similar content without knowing the output in advance. Besides data quality in structure and semantics, the risk of using biased data from historical datasets must be avoided. Bias is any prejudiced or partial personal or social perception of a person or group. Bias can enter the AI development chain at different points (one is training data). Algorithm bias could pose a risk for a racially and ethnically diverse intended patient population. Therefore, to ensure that medical devices augmented with AI/ML are well suited for a racially and ethnically diverse intended patient population, the FDA (2021b) emphasizes the crucial importance of applying methods for the identification of bias and improvement of machine learning algorithms.

This approach of supervised machine learning is already used in healthcare where medical devices equipped with learning AI can effectively assist the clinical expert in screening large quantities of images from scanning equipment or data from different sensors. The clinical expert ultimately makes the decision while the AI potentially improves the accuracy of the decision as it is not subject to any form of human distraction or fatigue. Moreover, the expert clinician can continue to train AI as the quantity and quality of clinically relevant data increases over time. In this chapter, the section "AI learning Schemes and Compliance with Regulatory Requirements" discussed how such AI incorporated in a medical device can be updated once it has been placed on the market.

Some ML approaches use neural network-based algorithms: small processing units, mimicking the functioning of the human brain where connected neurons interact in a weighted-input manner. During computation, these weights are adjusted to best match the training examples. The quality of such a neural network machine learning approach is defined by the accuracy of the result, expressed as the percentage of correct answers. Deep learning (DL) in this context describes a neural network approach that has multiple layers between input and output, which allows learning in single sequential steps. This increases the accuracy and reduces the need for human guidance. In the field of healthcare research, deep learning AI could be a useful tool to study the relationship between output and input parameters, specifically in case of multi-parameter correlations. This provides an opportunity to refine or redefine the trained existing relationships, with objective to improve the healthcare, whether diagnostic or treatment, of the individual patient. As already mentioned in the paragraph on Learning, the challenge of machine learning algorithms is that the connections between the inputs (i.e., data) and the outputs (e.g., predictions) can be very hard to understand. This black-box type of approach can lead to a lack of confidence with physicians who need thorough evidence that they can rely on the predictions in the real world where real lives are at stake (Schork 2019). In case the AI technique is only capable to identify correlations between input and output, and does not identify causal relationships, such an AI technique could potentially support the research and development of such causal relationships, hence being used as research tool and not directly as a clinical decision support system (CDSS).

Whether the AI is based on reasoning or learning, in the end, the final question is: who takes the clinical decision, the AI model or the physician? Van Baalen et al. (2021) approached this question and discussed the intricacies of the development and implementation of a class of AI for clinical practice, the so-called CDSS. They suggest that it is more appropriate to think of a CDSS as a "clinical reasoning support system" (CRSS) and conclude that developing CRSS that support clinicians' reasoning process requires that: a) CRSSs are developed based on relevant and well-processed data and b) the system facilitates an interaction with the clinician, and, that this can only be achieved if medical experts collaborate closely with the AI experts developing the CRSS. Moreover, they state that responsible use of a CRSS requires that the data generated by the CRSS are empirically justified through an empirical link with the individual patient. This conceptional approach of Van Balen et al. can optimize AI to achieve personalized, truly individualized medical treatment. Finally, van Balen et al. advocate proper implementation of CRSS by combining human and AI into hybrid intelligence, where clearly delineated and complementary empirical tasks are assigned to the human and to the supporting CRSS, where the CRSS assists with statistical reasoning and finding patterns in complex data, while the clinical expert focuses on interpretation, integration, contextualization, and ultimately, taking the clinical decision.

Another type of ML is reinforcement learning. The AI system receives feedback for each decision it makes, and the system is designed to maximize positively evaluated decisions. This type of AI could be useful in healthcare to provide feedback from patients to healthcare providers and institutions regarding human factors that, in conjunction with the clinical effect of specific medicinal or medical device treatment, could improve the overall well-being and clinical treatment experience of patients. Reinforcement learning could be a useful tool for those who implement value-based health care (VBHC). Since its introduction by Porter and Teisberg (2006), VBHC has received growing attention, and healthcare organizations in several countries are targeting strategies toward VBHC.

14.6.3 Robotics

Robotics is concerned with the development and training of robots that interact with humans and the world at large in predictable ways and, as stated by the European Commission's High-Level Expert Group on AI (HLEG AI 2019a), it can be understood as "AI in action in the physical world," a physical machine that must cope with the dynamics, uncertainties, and complexities of the physical world, while perception, reasoning, action, learning, and interaction capabilities with other systems, and particularly with physicians, are typically built into the control architecture of the robotic system. However, current efforts also revolve around the use of deep learning to train robots to manipulate situations and act with some degree of autonomy. Besides AI experts, it is indispensable that physicians play a key role in the design and operation of robots applied to healthcare. In addition, mechanical engineering and control theory in biomedical engineering are essential disciplines that play a critical role in the design and operation of medical robotic applications. In day-to-day life the public is aware of certain robots, such as robotic manipulators, autonomous vehicles (e.g., cars, drones, flying taxis), humanoid robots that serve as help desk, and moreover, vacuum cleaning robots. In the healthcare environment, there are four types of robots to improve the current standard of care while also helping humans to do things that they may not have been able to do in the past or do things quicker and with fewer errors. These four types of robots are surgical robots, exoskeletons, care robots, and hospital robots.

14.7 Successful AI Applications

The success of adopting AI concepts in medical device technology (and healthcare in general) depends on the combination of the following factors:

- *The benefit(s).* Clearly defined patient treatment objectives, hence clearly defined benefit/risk ratio improvement(s).[2]
- *The application.* The identified benefit(s) needs to be incorporated into the clinical context to become an application.[3]
- *The technology.* The usage of the appropriate and adequate medical technology to fully benefit from the identified added value of the AI application.[4]
- *The user(s).* The actual user/patient and the healthcare provider being involved with the application, and the user/patient benefiting from the output of the application.[5]

By appropriately aligning the benefit, the application, the technology, and the user, AI concepts can be successfully adopted in healthcare.

14.8 Examples of AI Usage in Medical Technology

The examples below are all showing an appropriate alignment of benefit, application, technology, and user.

14.8.1 Optimization of Patient's Medication Administration

Minimizing harm and optimizing the effect of patient-individual dosing by shortening the observation-treatment cycle is a lever toward individualized and improved medicine. Especially for data-intensive treatment journeys such as with type 1 and type 2 diabetes, the digitalization of logbooks and therefore the assistance in medication adjustment has proven beneficial to individual diabetes management.

---

[2]The benefit can be an improved treatment, safety, or outcome, and/or an increased efficiency of utilization of resources, and/or a measurable relief of personnel resources.

[3]One frequently observes AI applications which only provide a command-line interface (CLI) and that rely on manual integration in every clinical pathway or remain in the state of an academic result.

[4]AI as a data-driven approach does not necessarily need to be the right option for every digitalization initiative.

[5]This is where the added value is being realized. Understanding the context and perspective of both the clinical problem and the application is critical for obtaining the benefit, hence providing value to the user/patient. Not involving user/patient and healthcare provider, hence not fully understanding the clinical performance objectives and risks, is a major factor of failing AI projects in medicine.

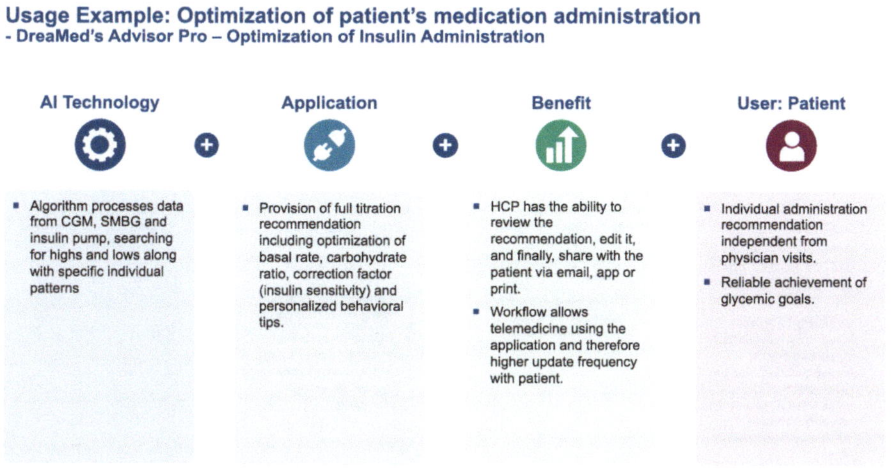

Fig. 14.1 Usage example. Optimization of patient's medication administration

As example, DreaMed's Advisor Pro[6] (Fig. 14.1), an AI-based clinical decision support system, assists healthcare providers in the management of diabetes patients who use insulin pumps or injections and monitor their glucose using a continuous glucose monitor (CGM) and/or blood glucose meters. The clinical objective of this AI technology is to optimize the glucose levels in blood, hence, to optimize the *time-in-range* (TIR) (Battelino et al. 2019). This TIR concept has been defined by the International Consensus on TIR as the time spent in the target range for blood glucose between 70 and 180 mg/dL while reducing time in hypoglycemia for patients using CGM. TIR was validated as an outcome measure for clinical trials complementing other components of glycemic control like blood glucose and HbA1c.

DreaMed Advisor pro is indicated for use by healthcare professionals when analyzing CGM monitoring or self-monitoring blood glucose (SMBG) and pump data to generate recommendations for optimizing a patient's insulin pump settings to improve the management of patients with type 1 diabetes. This application has proven to be beneficial to personalize diabetes management, although it cannot be used for automated dosing as the AI was not originally designed to support a closed loop system.

14.8.2 Improvement of Visual Analysis

AI applications in visual analysis primarily focus on an increase of detection success for specific diagnoses carried out by physicians. Currently known beneficial areas

[6]FDA Re: K191370, Trade/Device Name: DreaMed Advisor Pro, Insulin therapy adjustment device, Regulatory class: Class II, Product code: QCC.

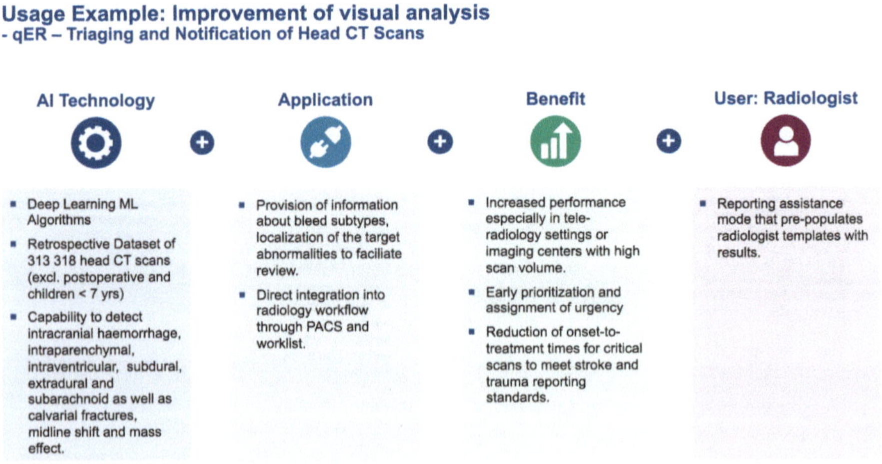

Fig. 14.2 Usage example. Improvement of visual analysis

are radiology in general, but also other image-based diagnostics techniques (e.g., in coloscopy) are generally of interest for AI applications.

The presented usage example is an existing product by Qure.ai Technologies for computer-assisted triage and notification. The product qER-Quant[7] (Fig. 14.2) is a medical image management and processing system intended for automatic labeling, visualization, and quantification of segmentable brain structures from a set of non-contrast head CT (NCCT) images. The algorithms for head CT scans are based on deep neural networks that have been trained on over 300,000 head CT scans. The product is device agnostic, meaning it may be used with any non-contrast scan device. Through a picture archiving and communication system (PACS), a computerized means of replacing the roles of conventional radiological film, images are acquired, stored, transmitted, and displayed digitally. When such a system is installed throughout the hospital, a filmless clinical environment results. PACS combined with worklists, the program interfaces directly with the radiology workflow, providing information about bleed subtypes and target head abnormalities to facilitate review. The program pre-populates radiologist templates with data from the results.

The underlying study used to validate the algorithms exhibits accuracy against a 3-radiologist majority on 500 images and a 25,000-scan validation dataset, demonstrating that it can detect important anomalies.

[7] FDA Re: K211222, Trade/Device Name: qER-Quant, Medical image management and processing systems, Regulatory Class: Class II, Product Code: QIH

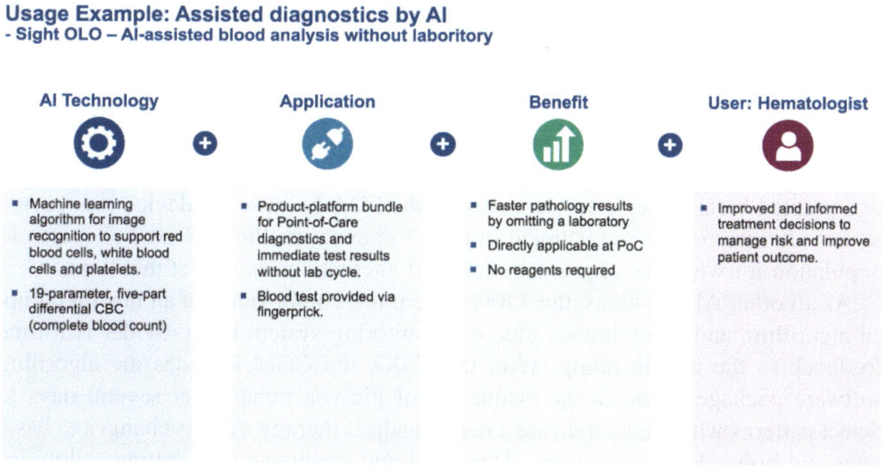

Fig. 14.3 Usage example. Assisted diagnostics by AI

14.8.3 Assisted Diagnostics by AI

A third use case which will benefit from AI is the pattern recognition of patient's health data to indicate diseases. The provided example is called Sight OLO[8] (Fig. 14.3) which is a quantitative multi-parameter automated hematology analyzer intended for in vitro diagnostic use in screening capillary or venous whole blood samples collected in K_2EDTA blood collection tubes, or fingertip samples collected using the Sight OLO test kit micro-capillary tubes. This device provides a blood analysis in combination with a central data platform and AI algorithms able to diagnose anomalies in red blood cells, white blood cells, and platelets.

The following example is an approved medical technology treatment system that can provide treatment in automated mode based on an algorithm connected to a continuous monitoring system that provides feedback to the algorithm.

14.8.4 Automated Insulin Delivery System

Systems comprised of an insulin pump connected to a continuous glucose monitor, which delivers doses of insulin determined by a software algorithm, have for a long time been called "artificial pancreas" or "closed loop system." Recently, manufacturers and the US FDA have been using "automated insulin dosing" system or "automated insulin delivery" system (both abbreviated as AID). There are several

[8] FDA Re: K190898, Trade/Device Name: Sight OLO, Regulation Number: 21 CFR 864.5220, Regulation Name: Automated Differential Cell Counter, Regulatory Class: Class II, Product Code: GKZ.

manufacturers of such AID systems. The example presented here is the MiniMed 770G System,[9] manufactured by Medtronic.

The MiniMed 670G System, identical to the MiniMed 770G System except for the lack of Bluetooth communication capability, was first approved by the FDA in 2016 for use in diabetes type-1 patients of age ≥14 years. In 2018, the indications for use were expanded to include users 7 to 13 years and up. In 2019, the 770G system was granted breakthrough device status by the FDA because the device is expected to provide more effective treatment of type 1 diabetes mellitus in the 2–6 years old population for which no approved or cleared alternatives existed at that time.

As all other AID systems, the 770G system is a combination of an insulin pump, an algorithm, and a continuous glucose monitoring system that provides real-time feedback to the insulin pump. With the 770G, the CGM includes the algorithm software package to aid in the evaluation of glucose trends over several days to detect patterns which may indicate a need to adjust therapy such as changes to basal rates and bolus dose instructions. Threshold and predictive alert settings allow for high alerts, low alerts, and alerts regarding insulin delivery suspension. The system can be used in manual or in Auto Mode. When Auto Mode is active, the device can automatically adjust basal insulin by increasing, decreasing, or turning off basal insulin delivery based on sensor glucose levels. The Auto Mode algorithm will determine when to deliver safe basal or safe basal low, depending on the patient's sensor glucose value. Verification and validation testing was conducted to confirm that the Auto Mode algorithm used in the MiniMed 770G System meets all specified requirements and that the software will operate reliably and safely under normal or abnormal use conditions. The software verification and validation were carried out in accordance with the FDA Guidance Document, *General Principles of Software Validation: Final Guidance for Industry and FDA Staff* (2002).

Automated insulin delivery (AID) systems close the loop between a glucose sensing device and an insulin delivery device to compute and deliver insulin (typically every 5 min) to achieve a desired glucose level while reducing the risk of extreme glucose variations below (hypoglycemia) or above desired range (hyperglycemia) in individuals with type-1 diabetes. There is increasing evidence that AID systems improve outcomes over conventional open-loop therapy for adults and children. The core of the AID technology is the algorithm that analyzes, predicts, and automatically adjusts basal insulin with objective to ensure TIR (see example 1 in this paragraph). The use of AI has enabled improved outcomes for many type-1 diabetes patients.

However, several limitations still need to be addressed, such as requiring user-initiated meal and correction insulin boluses, and challenges remain in improving these systems for different subpopulations (e.g., young children, athletes, pregnant women, seniors, and those with hypoglycemia unawareness).

[9]FDA—Premarket Approval Application (PMA) Number: P160017/S076Device Trade Name: MiniMed 770G System, Device Procode: OZP, Medtronic MiniMed, Inc.

14.9 Data Sources and Data Quality

The use of AI-augmented medical device technology can facilitate the development of models to predict patient treatment outcomes. As in most applications, the use of AI comes with limitations and risks if not implemented cleanly. In healthcare, implementation often hinges on the availability of valid and representative historical clinical datasets used to train the AI algorithm.

14.9.1 Data Sources

The increasing number of digital health devices generates huge amounts of data, but most of it is in silos. The data are often unstructured and non-standardized and follow proprietary protocols. This leads to difficulties in harnessing (cross-manufacturer) data.

Considering that the training of AI currently involves deriving vast amount of data generated during the delivery of healthcare to a large number of patients, both their personal parameters and their treatment input data and their clinical results data over time, a generally recognized limitation of such an approach is that combining information from a large number of individuals to identify patterns that reflect population-level relationships between data points does not necessarily result in relevant individual-level relationships. This potential lack of *ergodicity* (Fisher et al. 2018), specifically, the *lack of generalizability from group to individual statistical estimates*, could result in treatment models that are not useful for making adequate individual treatment decisions. Fisher et al. stated that as part of the validation of large datasets, one needs to demonstrate the consistency between individual and group variability before generalizing results across levels of analyses, that is, from group to individual statistical estimates. Adolf and Fried (2019) commented on the paper from Fisher et al. and stressed that ergodicity is sufficient, but not necessary, to draw inferences across levels. In our view, considering the critical importance of patient safety, it is advisable to apply tests for (non)ergodicity as part of the validation of the data used for the training of AI algorithms. This ergodicity challenge might be overcome in practice by applying a machine learning approach as this process will be based on a single patient using the medical device where the initial AI algorithm is based on a large dataset, and where the algorithm can be adapted based on the patient's own data generated during use.

A second limitation in data sources used for training of algorithms would be incomplete or unanticipated sources of bias in the datasets, as such datasets could lead to suboptimal, misguided, and in some instances even harmful recommendations for patient treatment. Moreover, once the AI system is launched to the market, a total product lifecycle approach must be adopted and real-world data should be collected and analyzed to identify opportunities for improvement of the algorithm, and to proactively respond to safety or usability concerns.

14.9.2 Data Quality

Every medical decision in diagnosis and therapy relies on good information. AI also requires a good quality database. Much of the data available today does not meet the required standards for clinical evidence and having access to big data that does not meet the required standards for clinical evidence will not help.

14.9.3 Sensitivity of Data

Health data is deeply personal, sensitive, as well as complex, and therefore subject to strict privacy and data protection regulations. Those EU member states that have established electronic health record systems have gained the trust of users by putting in place strong safeguards that limit or completely prevent the availability of the data for research or development of AI by commercial entities. It makes sense to reflect on whether public–private partnerships can be established where such sensitive data are anonymized by the public partner and provided to the private partner as aggregated data to train the AI. Ultimately, the development of AI-based systems that can be used efficiently and safely in the healthcare sector precisely requires this medical data as a basis. There are reservations (especially in some countries, e.g., Germany) among the population about making such data available. There are methods for anonymization or pseudonymization with which one can certainly do justice to both, the preservation of the privacy of patients while at the same time using this data to the benefit of future patients.

14.9.4 Digression Data Sources

One of the largest sources of data in medicine is imaging systems. The spectrum of systems ranges from X-ray machines and CTs to MRIs. All systems have different (physical) characteristics and have different processes of image acquisition. Each of these systems may also come from different manufacturers, so in principle, although there are certain standards, the first step is to establish comparability of data.

To build a valid data pool for the development of an AI application, the following steps are always necessary:

- Preprocessing to make the data comparable, align formats, etc.
- Qualification of the data by medical experts
- Validation of the selected datasets
- Anonymization or pseudonymization of the data
- Training of the AI application and clinical testing

14.10 Ethics, Acceptance, and Liability

14.10.1 Ethics

Ethics generally comprises the basic values and principles for human coexistence and human behavior in a society. Therefore, every country differs in its organization, according to its moral and ideological preferences, adapted to the political structure and its method of financing healthcare. Computer scientists have the responsibility to think about the ethical aspects of technologies they are involved in and mitigate or resolve any issues.

For the use of artificial intelligence in healthcare, ethical issues are of particular importance because, as already shown, the decisions of AI applications usually have a direct impact on the diagnosis and therapy of patients. The limited controllability and traceability of AI systems entail a limitation in terms of transparency and the decision-making authority of the users. Ethical discussions also arise from the economic incentives and the ethical values of the manufacturers of the AI applications, in the sense of a "bias": the tendency of a statistic to overestimate or underestimate a parameter (Frederking et al. 2019).

In addition to the impact on the individual patient, the introduction of AI into healthcare could be accompanied by a change in the role of healthcare professionals as it could lead to the displacement of competencies. Of course, the opposite can also occur, and core competencies could be strengthened by relieving the health system of certain repetitive tasks using AI. Considering that most healthcare systems today are running at full capacity, which during 2020 and 2021 was not necessarily driven by the coronavirus pandemic only, AI can increase efficiency of operational activities, and improve the effectiveness of patient treatments. This could support physicians and other healthcare operators to focus better on the duty of care and use more effective methods to treat patients. Against this backdrop, exclusion of AI systems from healthcare would be considered ethically questionable.

14.10.1.1 Ethics Guidelines & Recommendations

The papers issued by the Ethical Framework for a Good Artificial Intelligence Society (Floridi et al. 2018) and the High-Level Expert Group on Artificial Intelligence of the European Commission (AI-HLEG 2019b) both provide a comprehensive framework of ethical consideration for the field of AI in medicine. The following key requirements emerge from an ethical perspective for AI. These are classified in the field of medicine (technology).

- *Priority of human agency and oversight.* In the case of AI applications without approval as a medical product and without the involvement of medical specialists, the patient decides individually on the use of the results. If the AI tools are integrated into the treatment process for decision support, the healthcare professional makes the decision. Only for autonomous (closed loop systems) is the system able to transform acquired information into an action. This system class may only be used if there are demonstrably major advantages.

- *Technical robustness and safety.* The medical device must be safe in all conceivable events, hence there is no difference to standard medical devices.
- *Privacy, data quality management, and data governance.* From an ethical point of view, a decisive point is the handling of sensitive patient data. In the EU, the General Data Protection Regulation (GDPR, EU2016/679) forms the basis for this. For example, in Germany, this is also regulated through the Federal Data protection Act, as amended (BDSG). Nevertheless, the explicit consent of the individual patient is required if his or her data are used for a machine learning system.
- *Transparency.* The focus here is on the ability to explain, also toward the patient, who must understand the diagnosis to explicitly agree to the therapy based on it (informed consent—duty to explain). This means that a critical examination is necessary as to whether the AI application provides explainable results and in which type of application the patient must be informed about the use of AI. If the AI is a "black box," at least the basis for the decision must be verified and clarified.
- *Diversity, on-discrimination, and fairness.* Training data must be prevented from containing a bias, and from leading to a preference for certain patient groups.
- *Environmental and societal well-being.* These aspects must also be considered for the use of AI in healthcare. Here, the relationship between doctor and patient is the central aspect, which is discussed below.
- *Responsibility and accountability.* Accountability is of particular interest in case of error that occurs and has been discussed from a legal perspective in the field of liability. It is important that any error that occurs due to an insufficient database, mechanisms are in place to detect such a faulty database. This should include methods for the identification and elimination of bias and ergodicity.

In conclusion, the same ethical requirement applies to the use of AI-based medical devices as to all medical devices. However, there are also the aspects of data protection, traceability, and the special requirements for autonomously acting systems (Floridi et al. 2018; AI-HLEG 2019b).

14.10.2 Acceptance

The potential of AI in healthcare lies in the use of data and the application of models that enable predictions and classifications. The greatest added value could be provided by AI applications where data availability is ensured due to standardized procedures and medical technology (e.g., radiology). However, despite this value adding potential, there is still the question of acceptance of using artificial intelligence technology among healthcare workers and patients (Frederking et al. 2019). On the patient side, acceptance is already high today (despite the currently low penetration of corresponding AI applications). Sixty percent of respondents would like to see AI introduced as a mandatory second opinion if AI has a higher probability of making the right diagnosis (Frederking et al. 2019). Even 75% see huge potential in the digital twin concept for the patient. However, the level of approval always varies with the specific area of application. A diagnosis that is made exclusively by an AI is currently not popular according to these surveys (Thielscher and Antes 2019; Arnold and Wilson 2017). The debate among physicians is controversial. The advantages that can be achieved with AI are undisputed. Nevertheless, skeptics see sudden failures of the technology, hacker attacks, and the lack of human attention as points of criticism of the technology and its application in everyday clinical practice. There are no reservations about the use of AI in non-patient processes, e.g., to optimize inventories or automate billing and documentation processes.

> The medical profession will change: The question is not whether or not healthcare be improved as a result of Big Data and AI — it undoubtedly will in some cases — but whether the physician in the future will be the user of an auxiliary device or the machinist at the computer, stated Prof. Thielscher and Antes (2019).

Today, however, the available AI applications only map individual fragments of the patient journey; the autonomous takeover of entire diagnosis or therapy processes is still a long way off. In addition, the area of patient interaction (empathy in speech and action) has not yet been addressed at all. Accordingly, the role of the physician and other healthcare providers will not change soon; however, over time the use of AI applications will create free time that can be spent on other healthcare-related activities. In the somewhat distant future, the physician could primarily assume a governance and control function, linking analysis results and connecting them to the individual patient situation, as well as developing the appropriate therapy plan together with the patient.

In summary, it is expected that not only the daily work but also the role of the specialist in the care process will change in the long term if it can be ensured that the relationship between physician and patient is not damaged by the introduction of AI in healthcare. Today in most cases the AI system functions as an advisor or as support to the physician. The relationship with the patient is preserved, the quality of treatment increases, and the medical professionals are relieved. It is not expected that physicians will be replaced by AI, but it is almost certain that doctors who use AI will replace doctors who do not use AI. An important aspect in this context will be the training of physicians, who must have a much better understanding of AI. What knowledge and skills will be needed by such a workforce? Better technical and computational skills will be required—something that has already been recognized in academic medicine. AI basics need to be integrated into the training of physicians. They should be able to apply the technology, question it, interpret it, and know the respective limitations. Furthermore, a basic understanding of data security and quality should be taught. Since many practicing physicians will become confronted with AI systems, a corresponding basic knowledge must also be given a permanent place in continuing medical education and training [Bundesministerium für Bildung und Forschung (BMBD) 2019].

14.10.3 Liability

In principle, product liability (e.g., ProdHaftG in Germany, and the European Council Directive 85/374/EEC from 25 July 1985) also applies to medical devices. This mainly covers manufacturing defects, design defects, and instruction errors. A defect is defined here as a lack of justified expectation of safety, taking all circumstances into account, including (a) the presentation of the product, (b) the use to which it could reasonably be expected that the product would be put, and (c) the time when the product was put into circulation. A product shall not be considered defective for the sole reason that a better product is subsequently put into circulation.

Transposed to SaMD and AI, this means that the defined purpose (intended use) of the medical device dictates the scope of the manufacturer's liability. For example, a CDSS is to be considered in the sense of a second opinion. Thus, in case the output of the CDSS is clinically correct, and the physician decides to ignore this output, then liability clearly lies with the physician using it.

All AI applications approved today can at most be regarded as CDSSs, which means that the liability issue is relatively simple to clarify. These systems are still locked (cf. 14.5.1.1: Locked Learning Scheme), so that no intentional further development of the system can have led to the misinformation unless the algorithm has a performance issue under specific patient circumstances that leads occasionally to misinformation. In such a case, the CDSS provides misinformation, and if the physician relies on the output from the CDSS, liability lies with the manufacturer.

Furthermore, the overarching EU Medical Device Regulation (MDR) (2017) applies to the legal manufacturer of the SaMD, AI, or a machine learning device. The legal manufacturer is responsible to comply with all relevant legal obligations dictated by the MDR, e.g., performance and safety of the medical device, its design, manufacturing, associated labeling and instructions for use, as well as post-market product monitoring, and relevant post-market surveillance duties to prevent, detect, and mitigate the use of any hazardous device distributed to the market.

Additional liability regulations for artificial intelligence are not currently deemed necessary. The current law contains comprehensive liability provisions for damage caused using technical devices. It thus should sufficiently safeguard against risks that can be caused by AI systems. However, potential gaps in the current liability regulations are not clearly identified (AcaTech 2020; Kriesel 2020). The EU Parliament confirmed this position when it concluded that, although it believes that there is no need for a complete revision of the well-functioning liability regimes, the complexity, connectivity, opacity, and vulnerability, the capacity of being modified through updates, the capacity for self-learning, and the potential autonomy of AI systems, as well as the multitude of actors involved, represent nevertheless a significant challenge to the effectiveness of Union and national liability framework provisions. It considered that specific and coordinated adjustments to the liability regimes are necessary to avoid a situation in which persons who suffer harm or whose property is damaged end up without compensation. The EU Parliament also stated that it firmly believes that in order to efficiently exploit the advantages and prevent potential misuses of AI systems and to avoid regulatory fragmentation in the Union, uniform, principle-based, and future-proof legislation across the Union for all AI systems is crucial, and is of the opinion that, while sector-specific regulations for the broad range of possible applications are preferable, a horizontal and harmonized legal framework based on common principles seems necessary to ensure legal clarity, to establish equal standards across the Union, and to effectively protect our European values and citizens' rights. On October 20, 2020, the European Parliament[10] issued a *Resolution 2020/2014(INL)* with recommendations to the EU Commission on a civil liability regime for artificial intelligence. The EU Commission adopted the Resolution.

In April 2021, the European Commission published a *Proposal for a Regulation* that lays the groundwork for addressing the risks associated with the use of artificial intelligence. The proposed regulation applies to AI systems, broadly defined as systems that are "*developed with one or more of the techniques and approaches listed in Annex I of the proposed Regulation and can, for a given set of human-defined objectives, generate outputs such as content, predictions, recommendations, or decisions influencing environments they interact with.*" The proposal, while confirming that the EU institutions remain clearly focused on innovation, analyzes and proposes potential solutions to the challenges arising from AI applications. The proposal provides a detailed and balanced risk-based approach to the regulation of new technologies, with particular attention paid to the liability regime of the subject involved in the whole value chain.

As it will take some time before the proposed EU Regulation becomes effective, the technical, ethical, and legal evaluation of a Medtech product incorporating AI (SiMD), or a SaMD AI product, should in the meantime be done based on the existing Medical Device Regulation (EU) 2017/745 (2017) and the existing Product Liability Directive 85/374/EEC.

14.11 Conclusion

Despite various open questions, AI has the potential to transform healthcare. The constantly increasing digital capabilities to analyze a vast amount of real-world data generated during healthcare delivery offers tremendous opportunities to develop treatment plans to achieve better patient outcomes. Real-world performance monitoring over the lifecycle of the medical device will facilitate the implementation of iterative improvements to the algorithm, provided that any proposed regulatory framework includes a methodology that allows such modifications to be implemented timely. And, by combining AI-augmented medical devices and using

[10]P9_TA (2020)0276 Civil liability regime for artificial intelligence. European Parliament resolution of October 20, 2020, with recommendations to the Commission on a civil liability regime for artificial intelligence (2020/2014(INL)).

closed loop feedback from sensors, performing continuous measurements on the patient, to the AI component of the medical device, AI-augmented medical device technology is expected to further develop into truly personalized patient treatments, optimizing individual patient treatment outcomes.

Real-world data collection during disease treatment should enable to prevent or mitigate health-related suffering or complaints through medical or therapeutic access at any time. Therefore, wearables and health apps are important to capture such data and transfer this to the physician, who can initiate personalized individual patient-oriented medical care.

There are already various applications bridging between computer models and real-world data, building virtual representations, or "digital twins," of medical devices to predict how the individual patient will respond under various circumstances, e.g., dosing for diabetes patients.

An absolute prerequisite for the development of innovative personalized AI applications is the availability of a regulatory framework for medical technology augmented with AI or ML. Another prerequisite is the regulation of authorized data access. Compliance with data protection is a clear requirement; however, this arch must not be overstretched judicially, as the medical importance speaks for itself. Nevertheless, the field of personalized Medtech will be one of the core applications for AI and ML in healthcare and is expected to grow continuously over the next years.

References

AcaTech (2020) Machine Learning in der Medizintechnik, Source: https://www.acatech.de/wp-content/uploads/2020/04/acatech_POSITION_Medizintechnik_WEB.pdf Last consultation: 25/08/2022

Adolf JK, Fried EI (2019) Ergodicity is sufficient but not necessary for group-to-individual generalizability. PNAS 116(14):6540–6541

Arnold D, Wilson T (2017) What doctor? Why AI and robotics will define New Health, Source: https://www.pwc.com/gx/en/industries/healthcare/publications/ai-robotics-new-health.html. Last consultation: 25/08/2022

AI-HLEG (2019a) A definition of AI: main capabilities and scientific disciplines, Source: https://digital-strategy.ec.europa.eu/en/library/definitionartificial-intelligence-main-capabilities-and-scientific-disciplines

AI-HLEG (2019b) Ethics guideline for trustworthy AI - https://digital-strategy.ec.europa.eu/en/library/ethics-guidelines-trustworthy-ai

Battelino T, Danne T, Bergenstal RM et al (2019) Clinical targets for continuous glucose monitoring data interpretation: recommendations from the international consensus on time in range. Diabetes Care 42(8):1593–1603

Bundesministerium für Bildung und Forschung (BMBD) (2019) Was KI für die Medizin bedeutet, Source: https://www.bmbf.de/bmbf/shareddocs/kurzmeldungen/de/was-ki-fuer-die-medizin-bedeutet.html Last consultation: 25/08/2022

Cobbaert K (2021) Artificial intelligence, Source: https://www.raps.org/RAPS/media/Publications-Resources/Sample-Chapter/Software-as-a-Medical-Device-Sample-Chapter.pdf Last consultation: 25/08/2022

Cobbaert K, Bos G (2021) Software as a medical device – regulatory and market access implications. RAPS Publications

Council of the European Communities (1985) COUNCIL DIRECTIVE 85/374/EEC of 25 July 1985 on the approximation of the laws, regulations and administrative provisions of the Member States concerning liability for defective products. Source: https://eur-lex.europa.eu/legal-content/EN/TXT/?uri=celex%3A31985L0374 Last consultation: 25/08/2022

Davenport T, Kalakota R (2019) The potential for artificial intelligence in healthcare. Future Healthc J 6(2):94–98

European Parliament (2017) Regulation (EU) 2017/745 of the European Parliament and of the Council of 5 April 2017 on medical devices, amending Directive 2001/83/EC, Regulation (EC) No 178/2002 and Regulation (EC) No 1223/2009 and repealing Council Directives 90/385/EEC and 93/42/EEC, Source: https://eur-lex.europa.eu/legal-content/EN/TXT/PDF/?uri=CELEX:32017R0745

European Parliament (2021) Proposal for a Regulation of the European Parliament and of the Council Laying Down Harmonised Rules on Artificial Intelligence (Artificial Intelligence Act) and Amending Certain Union Legislative Acts: COM/2021/206 final, Source: https://eur-lex.europa.eu/legal-content/EN/TXT/?uri=CELEX%3A52021PC0206 Last consultation: 25/08/2022

FDA (2002) General Principles of Software Validation; Final Guidance for Industry and FDA Staff, Source: https://www.fda.gov/files/medical%20devices/published/General-Principles-of-Software-Validation%2D%2D-Final-Guidance-for-Industry-and-FDA-Staff.pdf Last consultation: 25/08/2022

FDA (2019) Proposed Regulatory Framework for Modifications to Artificial Intelligence/Machine Learning (AI/ML)-Based Software as a Medical Device (SaMD) - Discussion Paper and Request for Feedback, Source: https://www.regulations.gov/document/FDA-2019-N-1185-0001/comment?filter=artificial%20intelligence&sortBy=postedDate&sortDirection=desc Last consultation: 25/08/2022

FDA (2021a) Content of Premarket Submissions for Device Software Functions, Draft Guidance for Industry and Food and Drug Administration Staff, Source: https://www.fda.gov/media/153781/download Last consultation: 25/08/2022

FDA (2021b) Artificial Intelligence/Machine Learning (AI/ML)-Based Software as a Medical Device Action Plan, Source: https://www.fda.gov/medical-devices/software-medical-device-samd/artificial-intelligence-and-machine-learning-software-medical-device Last consultation: 25/08/2022

Fisher AJ, Medaglia JD, Jeronimus BF (2018) Lack of group-to-individual generalizability is a threat to human subjects research. PNAS 115(27):E6106–E6115

Floridi L, Cowls J, Beltrametti M et al (2018) AI4People - an ethical framework for a good AI society: opportunities, risks, principles, and recommendations. Mind Mach 28:689–707

Frederking A, Krumm S, Schaat S, Schürholz M (2019) Ein Policy Paper der wissenschaftlichen Begleitforschung des Technologieprogramms Smart Service Welt II gefördert vom Bundesministerium für Wirtschaft und Energie, Source: https://www.digitale-technologien.de/DT/Redaktion/DE/Downloads/Publikation/SSW_Policy_Paper_KI_Medizin.pdf;jsessionid=74D1CAF1E9322F053FA1BFF4200AD388?__blob=publicationFile&v=4 Last consultation: 25/08/2022

Goebel R, Chander A, Holzinger H et al (2018) Explainable AI: The New 42. In: Holzinger A, Kieseberg P, Tjoa A, Weippl E (eds) Machine learning and knowledge extraction, Lecture notes in computer science, vol 11015. Springer, Cham

IMDRF/AIMD WG (PDI)/N67 (2021) Machine Learning. enabled Medical Devices – A subset of Artificial Intelligence-enabled Medical Devices: Key Terms and Definitions, Source: https://www.imdrf.org/sites/default/files/2021-10/Machine%20Learning-enabled%20Medical%20Devices%20-%20A%20subset%20of%20Artificial%20Intelligence-enabled%20Medical%20Devices%20-%20Key%20Terms%20and%20Definitions.pdf Last consultation: 25/08/2022

IMDRF/PMD WG/N49 (2018) Final Document. Definitions for Personalized Medical Devices, Source: https://www.imdrf.org/sites/default/files/docs/imdrf/final/technical/imdrf-tech-181018-pmd-definitions-n49.pdf Last consultation: 25/08/2022

IMDRF/PMD WG/NSS (2020) Personalized Medical Devices - Regulatory Pathways, Source: https://www.imdrf.org/sites/default/files/docs/imdrf/final/technical/imdrf-tech-200318-pmd-rp-n58.pdf Last consultation: 25/08/2022

Kaplan A, Haenlein M (2020) Rulers of the world, unite! The challenges and opportunities of artificial intelligence. Bus Horiz 63(1):37–50

Kriesel T (2020) Wirtschaft: Haftung für Systeme Künstlicher Intelligenz (KI), Source: https://www.bitkom.org/Bitkom/Publikationen/Rechtsfragen-der-digitalisierten-Wirtschaft-Haftung-fuer-Kuenstliche-Intelligenz-KI Last consultation: 25/08/2022

Muehlematter UJ, Daniore P, Vokinger KN (2021) Approval of artificial intelligence and machine learning-based medical devices in the USA and Europe (2015-20): a comparative analysis. Lancet Digit Health 3:e195–e203

Porter ME, Teisberg EO (2006) Redefining health care: creating value-based competition on results. Harvard Business School Press, Cambridge

Schork NJ (2019) Artificial intelligence and personalized medicine. Cancer Treat Res 178:265–283

Thielscher C, Antes G (2019) Der Arzt behält die Deutungshoheit trotz KI. Dtsch Arztebl 116(1–2): A-18/B-18/C-18

van Baalen S, Boon M, Verhoef P (2021) From clinical decision support to clinical reasoning support systems. J Eval Clin Pract 27:520–528

Quantum Computing: Promises and Challenges

15

Stefano Patarnello

Abstract

Progress in digital healthcare and evidence-based medicine implies constantly growing demand for high-performance computing, and a strong cooperation between medical research, IT experts, and data scientists. In this scenario, the potential of new paradigms and technology solution such as quantum computing (QC) can boost progress and technical feasibility of such advanced software solutions. This chapter provides a short overview of the key principles of functioning of QC, some examples of the benefits for computing methods, and specific use cases for healthcare and life science.

Keywords

Digital healthcare · Quantum computing · Machine learning · Real-world evidence

15.1 Introduction

The role of information technology (IT) to support medical research and practice is constantly expanding, thanks to two driving factors: new algorithms and software techniques becoming available in many domains (medical imaging, genomic research, AI-based decision support system, etc.), which are exploited through easier-to-use packaged solution and the availability of technical resources more and more focused on digital healthcare segment, and increasing computing power "at reach" both in terms of user-oriented devices and workstation (i.e., more power at

S. Patarnello (✉)
Generator Research Facility, Fondazione Policlinico Universitario Agostino Gemelli IRCCS, Rome, Italy
e-mail: stefano.patarnello@gemelligenerator.it

A. Cesario et al. (eds.), *Personalized Medicine Meets Artificial Intelligence*,
https://doi.org/10.1007/978-3-031-32614-1_15

the desk of the clinical researcher), as well as high-performance computers which can help solving fundamental research questions.

This steady progress has proven already to bring material impact to daily practices and quality of care. One example is that of interventional radiotherapy. In the pioneering stage, the use of artificial intelligence (AI) techniques to analyze medical imaging was embraced as a tool to test scientific hypothesis within research projects. With the increasing accuracy of algorithms for multiple uses (contouring, feature extraction and interpretation, support to dosimetry, analysis of response to treatments, etc.) this domain has reached full maturity, in terms of technical readiness and scientific validation, such that AI-driven techniques are an integral part of electromedical apparatus and adopted in daily practices as supporting tools for medical staff.

The learning curve goes together with the exponential increase of healthcare data. These are very diverse in terms of sources and formats (electronic health records, image files from diagnostics, laboratory data, real-world data from wearable, etc.), and their growth rate is among the largest across industries, since every day new IT systems or supporting technologies are adopted from healthcare providers, and these deliver a huge amount of insightful data on a daily basis.

It is very reasonable to expect that any new inflection point in IT can trigger major steps forward in some of the algorithms that researchers are experimenting in clinical research, these being artificial intelligence algorithms to help predicting disease evolution, or number-crunching software packages for sequencing or molecular modeling.

Quantum computing (QC) is certainly a new paradigm that can significantly expand the art of the possible for what concerns the impact of new IT techniques in many domains of healthcare. There is an increasing consensus that the so-called quantum advantage (when a quantum computer will perform better than classical ones in a realistic application) can take place in the next 2 to 5 years. While the impact of this much advocated inflection point may need to be better qualified under many angles (extent of usefulness, economical affordability, skills augmentation, etc.), it certainly means that now is the right time to better understand the implications of quantum technology for many domains of application, healthcare being on the forefront for the benefit and acceleration that this may bring. In doing so, one can also understand what would be required in terms of preparedness, skills growth, conceptual experiments, and prototyping.

This chapter aims at providing some insight in this domain and provide some elements to understand the implications in some of the application areas where IT for healthcare is widely adopted, with a specific focus on data-driven techniques and AI. We will start by providing some elements to understand the basic mechanism of quantum computing; then we describe some of the building blocks (software algorithms) where quantum computing is expected to provide material advantage; we continue by showing some examples of the impact that this can bring to specific healthcare applications. We conclude the chapter by giving our perspective on some of the attention points (and potential gaps) that are worth considering, to make sure that the promise of such increased capability will transform into real advances that ultimately bring benefit to research and quality in healthcare.

15.2 Quantum Computing: Basic Concepts

The fundamental difference of quantum computers versus classical ones is the way in which data are represented, stored, and computed. The idea at the base of QC is that to solve problems in physics and chemistry which are intrinsically bound to quantum physics phenomena (including wave-particle dualism) the best solution was to build computers based on such subatomic physical entities. From these original needs, in the theory of computational models it was made clear that there are also other problems of increasing complexities in other application areas, such as other fields of simulation of complex physical systems, optimization, machine learning, cryptography, and many others to come.

The first conceptual design of quantum computers dates to the work of Richard Feynman (1982), Jurii Manin, and David Deutsch (Deutsch and Penrose 1997), who were able to demonstrate that for a simulation of physical quantum systems with complex interactions, a computer based on quantum particles (and their changes of state) would have been more effective than classical computers based on binary logic. The concept of "universal quantum computer" was introduced by Deutsch, as a quantum Turing machine able to simulate any physical system, provided the right level of isolation (very low temperature) and stability was established. Starting from those novel design points, research efforts have constantly progressed to be able to develop and manage a quantum computer, to the extent that today several implementations are available from different industrial and research institutions.

The basic idea for QC is to apply the principles of quantum mechanics to the way a system performs computation. This means that a quantum system becomes a computing device that can move from one quantum state to another (thus performing a basic operation) through an external forcing agent that triggers a transition of state in the sense of quantum mechanics. The two most important quantum phenomena at the base of the functioning of QC are the quantum superposition principle and the mechanism of entanglement.

Quantum superposition principle describes the fact that any two quantum states can be added together ("superposed") and the result will be another valid quantum state, and conversely, that every quantum state can be represented as a sum of two or more other distinct states. This leads to the introduction of the quantum bit or *qubit*, which is the basic entity in quantum information processing, and it is a quantum superposition of deterministic states. Contrary to a classical bit that can only be in the state corresponding to 0 or the state corresponding to 1, a qubit may be in a superposition of both states. This property implies a significant advantage for complex computational models, since the same operation can be performed on different instances or samples, each corresponding to a different state/combination of the two "basic" states 0/1. This means that a quantum computation can simultaneously produce different elementary computations with a resulting overall increased speed of execution.

Entanglement is a characteristic of multiple quantum subsystems which coexist and are intrinsically correlated. This implies that any measurement or basic operation on a subsystem has also an impact on other subsystems that are connected even at larger distances. Therefore, the principle of entanglement is the base for very effective "collective computation" since two subsystems which are in an entangled state perform computing operation at same very high speed.

Due to the non-binary nature of qubits, they can handle information in a continuum fashion; a computation on a qubit is a state change, which happens through a series of operations defined as quantum circuits. A state of a qubit is a multidimensional vector and an operation on a qubit is a rotation in this virtual space. The multidimensional, continuum nature of qubits implies that certain complex operations can be performed more effectively than with a classical computer. This advantage may imply to achieve fast progress in many industries and use cases where high-performance computing is required, provided that the problem to be solved can be mapped into quantum-compatible algorithms and software solutions.

The two quantum mechanisms (superposition and entanglement) are unique features of quantum computation that result in a much faster computation execution than for classical computers. On the other hand, a quantum system can be unstable (due to interference with external systems) due to the interaction with the environment and the state of a qubit can stay unchanged for a very short time (this is known as the *decoherence* issue). This problem worsens with the increasing number of qubits, which is the complexity that arises when building a complete computing system that must be based on a very large number of computing units. To solve the issue, sophisticated and costly manufacturing methods are being used, such as cryogenic cooling or error connection code techniques.

Notwithstanding these challenges, the pace of progress from many vendors is remarkable: in November 2021, IBM announced the new Eagle 127-qubit quantum computer, so far the largest implementation release in the market. Other important IT players such as Microsoft and Google are delivering their QC solutions, and some companies such as D-Wave Solutions, ColdQuanta, or Regetti are "born in the quantum era" and completely focused on this breakthrough technology.

15.3 Quantum Computing Models and Algorithms

15.3.1 Computing Models Examples: Quantum Gate Array and Quantum Annealer

The Quantum gate array (QGA) model is based on computing modules which are organized as quantum "gates" much like the design model of classical electronics, where basic operations are performed from modules organized in a standard way, and the results of such basic operations are then combined for more complex computation. In the QGA model, these computing modules (quantum circuits) are the main components based on quantum execution mechanisms. While on a classical computer the logical gates performing basic operations are designed and built using transistor-based circuits, quantum gates are implemented via electromagnetic fields with a characteristic frequency associated with each qubit and dependent on the specific manufacturing technology. QGA systems can be implemented with quantum components such as superconductors, trapped ions, or photonics. QGA model can also be simulated using traditional hardware, introducing quantum logic in the way these systems compute, and using noise models to reproduce QC instability.

The other computing model for the design and implementation of QC is based on simulated annealing principle, which is an algorithm developed in the 1980s to solve complex optimization problems and can be broadly characterized as an approach to optimize cost functions, by searching the global minimum in a "landscape of solutions" with several local minima (which is why the corresponding problem to be solved has a high degree of complexity). In the context of quantum computing, the design methodology has been defined as "quantum annealing" (Kadowaki and Nishimori 1998) and the computing model for this class of quantum computers is the so-called quantum annealer (QA). The peculiar feature of QA is that they can represent many possible solutions (energy level), and it is possible to explore many such energy configurations in a short time by taking advantage of superposition and entanglement properties of qubits. Therefore, the search for an optimal solution (i.e., the lowest energy state) is performed very efficiently.

It has been demonstrated that this general mapping into a multi-particle physical system that gets into a minimum state (as it happens in a physical annealing process) is equivalent to the cost function optimization for the class of problems known as combinatorial quadratic unconstrained binary optimization (QUBO) which can be solved only with non-polynomial algorithms. For this overview, this implies that many complex optimization problems, in many research fields and industry, where the optimal solution depends on the interaction of many elementary entities, can be mapped into such an annealing methodology—and for some of them a QA type of computer can be the most effective (i.e., fastest) solution.

As it is the case for QGA, also for QA computing model, there are solutions in the market of "quantum-inspired" annealing computers, which emulates on a traditional architecture the way in which qubits operate in the search process of a minimum cost function in a quantum computer based on annealing.

15.3.2 Algorithms Implemented on QC: Some Examples

Fundamental physics and chemistry are by definition among the fields where new simulation approaches and algorithms are being developed. In chemistry, through quantum computers it is possible to model energies related to ground and excited states at molecular level, and this will enable the investigation of many reaction pathways. Other fundamental topics include the understanding of electronic distribution between atoms and the nature of chemical bonds.

Likewise, the adoption of QC to model the behavior of subatomic particles and forces is a natural research path in fundamental physics, covering areas such as quantum field theories, quantum electrodynamics, and strong and weak interactions. As an example, quantum chromodynamics (QCD) has always been a research field where supercomputers and highly sophisticated simulation methods were used to test some of the most important theories and models. The introduction of quantum computing will certainly open new avenues and will trigger a step change in this research domain.

It is also relevant to see some examples in computer science and algorithms, which can be used in many use cases and industries. It has been already described how a QA type of computer can help addressing those problems where the goal is the minimization of an "energy" or "cost" function depending on many variables, with many similar solutions competing in a search process for the best (optimal) solution. This affinity for the annealing-based architecture is very promising for a wide range of applications in optimization and operational research, including search for shortest paths, fleet and shift management, and financial modeling.

An algorithmic area where quantum computers are expected to deliver faster performances is the search process in very large sets of data. This problem is very general, with several applications and use cases, and can be formulated in simple terms as the time required by a system to search for an element within a list of items not organized sequentially. The intuitive description is that of the steps required to find an item "hidden" in one drawer out of N identical drawers. This is also referred to as the Grover algorithm within the class of NP problems (classically solved in a non-deterministic polynomial time). With standard methods the time for the algorithm to complete the task would grow with N. It has been demonstrated that a quantum computer will solve the problem in a time proportional to $\sqrt{N}$, thus leading to a much faster execution time for very large N. The Grover algorithm can be therefore efficiently implemented in quantum gate array system, and this implies that adopting such new approach to perform the equivalent of an exhaustive search opens new possibilities.

This can have important implications in many domains of IT. If a quantum computer can perform this search so fast, then trying to break the protection of cryptography by testing all possible instances of a key would be a more affordable procedure. It means that the whole domain of cybersecurity needs to consider this type of threats, and new ideas to design highly secure protection systems are needed.

In the area of AI and machine learning (ML), the intrinsic nature of quantum computing—able to handle very efficiently the analysis of multiple states and configuration of several interacting systems (the qubits)—can be a major enabler to achieve faster speed of execution, which in turn means a much-increased ability to scale problems in size and accuracy. In the last two decades, one of the AI methods, which has shown the largest success, is that of artificial neural network (NN). This can be formulated in a very general way and with a wide range of applicability. NN systems can manage large quantities of data, analyze them to find correlations and key interactions, and then produce inferences and so-called learning capabilities.

These methods can therefore build rules with a significant degree of generalization from the datasets analyzed.

The most common NN architecture is based on a large set of communicating nodes, typically organized in subsequent layers, which exchange information through basic operations among connected nodes, leading to a relationship between input (a set of the network "external" nodes) and output, which is the result of these cascading elementary steps and the weights of the links between communicating nodes. These systems can be trained on a specific problem, meaning that weights among nodes (i.e., the sequence of cascading elementary operations that produce an output as the result of such composite computations) can be iteratively adjusted to learn a given task, on the basis of a large number of configuration examples which are specific values of input and output nodes, which are provided to the system as the knowledge base to be trained on the target task. A NN that can generalize is a computing system that, after being trained on a large set of examples, will calculate an output for a new case which is consistent with the training process and produce results which were not known a priori. Obviously, the organization of such computing nodes, and the process that leads to determine the connecting weights and composite output, mimics the conceptual scheme of a brain portion, with neurons connected through synapses, and where the connection network and intensity evolve as a result of learning/cognitive processes.

The way in which a computation is executed on a quantum computer bears many analogies with the process of building a learning neural network in the classical approach which has been just described. In fact, qubits can be organized in connected layers and their elementary computations are executed through the quantum circuits or gates. Therefore, an iterative learning process can be fulfilled by executing many instances of computation across the qubits, and the gates can be adaptively defined with parametrized weights or circuit components. Given the speed at which qubits perform elementary operation, the collective process of training a large set of qubits with many different configurations (due to the superposition principle) can be very fast, and so would be the generation of a neural network for a specific task.

With the speed of progress in testing and improving training methods of different nature, it is not unreasonable to consider that very soon a new generation of neural networks built on quantum computers will be made available, each specialized for specific class of problems (in fact, one of the most promising techniques is referred to as *parametrized quantum circuits*; Humble et al. 2018), much like in previous steps of information technology "special-purpose processors" were attached to classical computers, to perform a specific task with unparallel speed of execution.

While many of the promises in such algorithmic areas have yet to be fulfilled, these and many other use cases have raised expectations and triggered new energies with regard to QC for every industry, also in conjunction with exponential growth of data from consumers and enterprises and front-end and back-end processes. In the next section, some of the use cases from healthcare will be described.

15.4 Healthcare Use Cases

15.4.1 Genome Analysis

The rapid and reliable whole-genome analysis would be a very significant and promising area for various applications including personalized medicine. This requires solving tasks of high computational complexity to manage techniques such as de novo genome assembly, which is used for the analysis of genomic rearrangements, chromosome phasing, and genome reconstruction.

De novo assembly is a method for constructing the original DNA sequence from the unstructured set of reads without any prior knowledge of the source DNA sequence length, layout, or composition, which is essential for studying new species and structural genomic changes that cannot be detected by reading mapping. The complexity of de novo assembly depends on the genome size, abundance, length of repetitive sequences, and possible polyploidy and a task of such build on human genome can take some days of computation. This time scale is acceptable in research tasks, but it is a limitation for emergency applications (including the clinical use). De novo assembly is currently used in transcriptome and cancer analysis, as gene fusions and genome rearrangements are common causes of malignant tumors. Decreasing the costs of sequencing makes whole-genome sequencing an irreplaceable part of personalized medicine and cancer treatment. The utility of sequencing technologies requires improved workflows with de novo assemblers to uncover significant genomic rearrangements in cancer and normal tissues.

From the computation modeling standpoint, de novo assembly process can be mapped into well-developed graph techniques known as overlap layout consensus (OLC) methods. These methods map the read and matches into "connected" vertex and therefore graphs. Then genome reconstruction is achieved by finding the sequence of connections (paths) that touches all vertex only once, which produces the complete sequence. It has been demonstrated (Boev et al. 2021) that this graph configuration search can be mapped into an optimization problem where the goal is to explore the landscape of possible configuration for a collective system of binary states (so-called Ising spins, which are magnetic states with two states up or down) to minimize a given cost function (or energy function, in the analogy of a complex systems of Ising spins). Once the mapping of a graph of reads (in the context of genomic de novo assembly) is reformulated as an optimization problem, a QA type of computer can be very effective, since it is able to explore the landscape of possible solutions (i.e., the assembly graph) at very high speed, since the energy minimization of an Ising Hamiltonian can be transformed into a QUBO problem, which is the type of algorithms where quantum annealer type of computer is highly specialized.

The simulations have been performed until now on short sequences, since to reach the scale of the realistic size of sequences, the underlying quantum technology needs to reach a higher level of stability (improving the decoherence time). As these technical limitations will be overcome, the proof of concepts performed until now will evolve into real-life cases providing a material impact to the advance of assembly techniques.

15.4.2 Machine Learning for Precision Medicine

ML- and AI-based methods are more and more contributing to gain new insights and support decision-making in clinical research and practices. In general terms, these methods are based on three steps: first, the retrospective knowledge base for a specific pathology is created, by organizing all available past patients data which describe demographic data, biomarkers, disease progression, and treatment outcomes. This is a complex integration task which involves collecting, transforming, and validating large quantities of data; then, a wide range of computing techniques are exploited to analyze the different factors that link patients' health status over time, evolution of disease, and outcome of treatments for those patients/ groups of patients. This is generally achieved by adopting statistical methods to discover correlation among concurring factors and outcomes and applying ML to create evidence-based discovery, such as predictive algorithms that provide risk assessment for negative disease progression. This step is commonly known as the "training" phase where the correlation and prediction is built using a subset (yet very large) of patient cohorts (the training set). After the most effective algorithms and related parameters are achieved for the training set, the validation of the overall methods is achieved by testing the validity of the developed algorithms on a new set of patients, allowing us to evaluate in a rigorous fashion the predictive generalization power of the method, and the accuracy this method can provide when treating a new patient in a perspective approach. These methods are usually more powerful (i.e., the degree of validation much stronger), if data from different medical sites are exploited with a multi-centric setup.

With the upcoming availability of mature QC technologies, all the steps outlined for the end-to-end cycle of predictive methods in support of decisions and personalized treatments may improve in a material way.

Firstly, data collection and build-up of multimodal data models (study ontologies) will get more efficient throughput; in particular, the integration of genomic, clinical, demographics data may potentially be accelerated; real-world data collection from IoT and wearable devices will also become more feasible for building retrospective knowledge, which includes behavioral data, quality of life, and environmental critical information.

The process of building up the predictive algorithms from retrospective knowledge entails several steps demanding high computational powers: correlation analysis to understand the weight and relevance of different factors and patients' data require multimodal data collection and organization. Research progress is leading rapidly to the integration of high quantity data (such as those from medical imaging or different omics) with demographic and clinical data, and sometimes this integration requires intermediate computational steps (e.g., features in radiomics or in genomic data classification) which can be very intensive. To add to that, the goldmine of information that stems from medical reports in free text format needs also very sophisticated processing (through natural language processing, which will be discussed in the next use case). Therefore, the development of the input data model per se is a numerically intensive task, which can greatly benefit from the availability of mature quantum computing implementations.

When it comes to the definition and execution of training algorithms, these will get material improvements from the development of robust quantum computing models. Unsupervised learning algorithms are the most suitable approach when very little a priori knowledge is available about the predictive mechanism and the type/range of outcomes must be extracted from the learning process. These are the kinds of problems which can be mapped into the cost-optimization techniques comparable to the minimization of the energy landscape which depends on many variables concurring to define different similar configuration—as for the genomic use case, a QA or quantum-inspired annealer can be considered as a special-purpose computer that can greatly accelerate the optimization process and therefore the identification of prediction mechanisms.

In the case of supervised learning, where the possible outcomes (e.g., the prediction of a disease critical condition or death) are known a priori, the training of the predictive model involves the determination of several weights that connect a decision system based on neural networks, and this is also a high-performance computing task that can be hosted on a quantum-based system.

The final validation step, where the accuracy of the predictive method is tested for new patient data, combines the complexities of the data collection and prediction stages, and can be mapped onto a quantum-based approach with expected strong benefits in the overall response time. This will be extremely useful when these methods will shift from research domain to clinical practice, where the need of a real-time (or quasi real time) response system is critically important.

15.4.3 Natural Language Processing for Medical Reports

One of biggest challenges for data-driven healthcare research is the extraction, interpretation, and exploitation of the huge amount of information that is delivered daily from different staff in the clinical centers in the form of free-text medical reports. These may include consultancies from clinical specialists; nurse diaries which include very relevant information such as measures, symptoms, and vital functions; diagnostic reports from instrumental exams delivered by diagnostics specialists and technicians; and many other examples. The goal of natural language processing (NLP) is that of transforming these free-text datasets into actionable information in the form of categorical variables (presence/absence of a given biomarker or outcome, classifications of outcomes with different grading levels, etc.); in some other use cases researchers are confronted with the even more complex task of extracting key concepts that can be expressed in very different formats.[1] To

[1] For example by providing an overall assessment of the health status of the patient, by pointing to a collection of symptoms or evidence from instrumental exams, and maybe combining the same observations with patient-specific risk factors, familiarity of disease, and lifestyle specific features.

add to this, typically a data-driven study must extract and process this type of information from medical reports produced by different doctors, who may adopt different terms, or different style in the sentences they compile for the medical report.

A very representative use case is the rigorous characterization of disease progression in cancer-focused studies. To define repeatable criteria from real-world evidence (e.g., by analyzing all medical reports for a retrospective cohort of several thousand patients), the data scientist and clinical joint team have to compare and combine information from many sources, and the majority of these are located in medical reports of different kinds: CT-scan diagnostic reports analyze the size of tumor, numbers and sites of metastasis, and status of lymph nodes; oncologist reports address topics ranging from disease evolution to general patient's symptoms analysis of response to treatments, plus all evidence from diagnostics; sometime this multimodal set of information includes structured, quantitative data from laboratory exams which are easier to interpret. When looking at this diverse yet all relevant unstructured and structured information, these must be analyzed and integrated to discover and validate rules/criteria to define (and possibly quantify) disease progression in the different domains of cancer treatment. NLP techniques are the key building blocks that help the understanding and classification of this information.

With supervised NLP methods, expert knowledge can be injected into the recognition system, in the form of annotation or basic rules to identify sentences that are clear markers of critical information (the next step for a solid NLP approach is that of generalizing such criteria, and therefore identify in all medical reports other forms of expression and sentences that amount to the same outcome). With unsupervised algorithms, the more general problem of extracting concepts and classes (which translates into outcomes such as disease progression) is managed without injecting external expert knowledge, enabling the AI algorithms to find the concepts and associate them with the different report instances. Once NLP methods have helped transforming unstructured information into categorical data, downstream learning algorithms (statistical methods, neural networks, optimization techniques) can be used to develop systems predicting outcomes from the input data in an automated fashion.

Considering the amount of information, which is available in free text in medical reports, and the goal to integrate these computing tools in the daily clinical practice as supporting decision tools, the development of highly performant NLP algorithms is a trigger factor that can transform NLP-based research prototypes into validated tools and potentially decision support medical devices. Quantum computing can provide a great impulse to the development of such high-performance versions for such medical reports "intelligent search engines." In broad terms, the affinity between NLP techniques and quantum computers architecture lies in the fact that a string of words (sentence) and/or strings of sentence can be seen as interacting entities, connected at different distances and with different intensity, which is a similar logic to the organization of qubit as previously described (Di Sipio et al. 2021). One example of implementation of this mapping from NLP to quantum computers is *lambeq*, a development toolkit implemented by Cambridge Quantum

(2021), which helps transforming sentences into strings of symbols which are then mapped onto quantum circuits (or simulator). Once this transformation is achieved, it will be possible to use the high power of a quantum architecture to implement language models that combine statistical methods with symbolic/semantic methods. Such hybrid methods have proven as the most successful to deliver multi-language NLP systems. Though this approach is certainly at an early stage, it is very promising since it combines high-performance computing with multi-level modeling of the language understanding domain.

15.5 Discussion

There is little doubt that quantum computing will have a significant impact in the advances of many domains of industry and research. The momentum is also increasing in designing fit-for-purpose algorithms, and the community of researchers from many domains that is tackling unsolved issues is increasing and progressing at very high pace. Yet, there are many aspects that will require special focus and efforts that are worth mentioning as food for thought.

Firstly, the efforts to make the technology stable and functioning are still ongoing, and a lot needs to be done. The fact is that systems of qubit are still exposed to noise from external sources and therefore they may lose information and need to be corrected (the common way to characterize this early development phase is NISQ which stands for noisy intermediate scale quantum technology). The good news is that all technology providers and base research players are working to solve these issues which may require a few years at least for intermediate scale systems. And it is worth underlying that these base technology challenges do not impede progress in developing realistic use cases and put in place robust development frameworks (algorithms, libraries and complete simulation environments) that will allow researchers from all domains to build their applications while the base technology challenges are being solved.

It will also important to get a thorough understanding of where the adoption of a quantum computing model can fit the goals and bring advantage, and sometimes it is worth comparing this with more "classical" approaches where, by exploiting in a smart way what is already available, similar or better results can be achieved in terms of computing speed, time to resolution, and accuracy of results (Bernaschi et al. 2021). Simply stated, the ongoing research efforts in benchmarking classical (and very well-tuned) algorithms designed for the current technologies versus new quantum computing based ones are very relevant and will result in better software design for the larger user communities in both scenarios.

While there are research domains where a complete mapping into a quantum computing architecture is a natural choice, as it is the case for computational chemistry or material science, where quantum algorithms can be used to predict electronic structures or configurations, we should expect that some of the applications of such new technologies will happen in a hybrid fashion, where the overall algorithmic setup to solve a complete problem is managed on a traditional computing system, and part of the computation/algorithmic step is executed on a quantum computer which will act as a "special-purpose processor" very suitable for a specific task. This is not very different from what happens already today in the usage of graphical processing units (GPUs) within AI or image processing type of applications, including healthcare (Bernaschi et al. 2021).

Finally, the importance of expanding skill and knowledge of these techniques across the value chain of research and industries cannot be stressed enough. We are witnessing an increasing weight of quantum computing in basic and advanced education, which is certainly an enabling factor for the progress in the field. Likewise, industries and applied research institutions should take the bold steps of investing in prototypes and proof of concepts, even if these bring along significant efforts in time and skill upgrade. This will certainly shorten the cycle to allow for this new, promising inflection point to bring material benefit to multiple applied domains, where healthcare could be in the "lead pack" for this new frontier of computation.

References

Bernaschi M, Bisson M, Fatica M et al (2021) How we are leading a 3-XORSAT challenge: from the energy landscape to the algorithm and its efficient implementation on GPUs. EPL 133: 60005. https://doi.org/10.1209/0295-5075/133/60005/meta

Boev AS, Rakitko AS, Usmanov SR et al (2021) Genome assembly using quantum and quantum-inspired annealing. Sci Rep 11:13183. https://doi.org/10.1038/s41598-021-88321-5

Cambridge Quantum (2021) Lambeq, Source: https://cambridgequantum.com/first-qnlp-toolkit-lambeq/ Last consultation: 25/08/2022

Deutsch D, Penrose R (1997) Quantum theory, the Church-Turing principle and the universal quantum computer. Proc R Soc London A40097-117-https://doi.org/10.1098/rspa.1985.0070

Di Sipio R, Jia-Hong H, Yen-Chi CS, et al (2021) The dawn of quantum natural language processing. arXiv, Source: https://arxiv.org/pdf/2110.06510.pdf Last consultation: 25/08/2022

Feynman R (1982) Simulating physics with computers. Int J Theoret Phys 21:467–488

Humble TS, Thapliyal H, Munoz-Coreas E et al. (2018) Quantum computing circuits and devices. arXiv, Source: https://arxiv.org/pdf/1804.10648.pdf Last consultation: 25/08/2022

Kadowaki T, Nishimori H (1998) Quantum annealing in the transverse Ising model. Phys Rev E 58(5):5355

Metaverse Means Better: How the Metaverse Continuum Is Evolving Healthcare to the Next Level

16

Andrea Pagliai

Abstract

The Metaverse is bringing the next digital era in our society and economy. It is shaped by several attributes (e.g., physical and virtual places, multidimensional human experiences, multisensorial creativity, extended communities, life-centric markets, cross industries collaborations, new social and business values, products, and services) and enabled by new technologies (cloud and artificial intelligence, extended reality, blockchain, digital twins, and edge). This chapter focuses on the attributes to understand how to manage them to create the Metaverse future. The Metaverse will encourage us to reimagine healthcare, since it also provides the healthcare ecosystem with a virtual–real collaboration space in which to create robust, meaningful experiences for all value chain stakeholders. Looking at the first actions into the global markets we envision how to bring together people, spaces, and things in both the virtual and physical worlds, and how to enable users to "inhabiting" the new digitally enhanced healthcare worlds. The future of healthcare depends on how we will be able to effectively improve the healthcare delivery and patient experiences thanks to the Metaverse. We see that there is a strong demand from people/patients to leverage new digital healthcare solutions—technologically advanced and able to actively and intelligently use patient data. But no single player can build and lead the future meta-care; therefore, nowadays there is an urgent need for the healthcare ecosystem participants (both new entrants and incumbents) to speed up the process of collaboration in a pre-competitive environment to make meta-care a concrete reality for a better and healthy world.

A. Pagliai (✉)
Life Sciences Industry Lead, Digital Health Market Unit Lead Italy, Central Europe, Greece

A. Cesario et al. (eds.), *Personalized Medicine Meets Artificial Intelligence*,
https://doi.org/10.1007/978-3-031-32614-1_16

Keywords

Metaverse continuum · Meta-care · Healthcare ecosystem · Disruptive healthcare technologies · Digital healthcare twins · Digital health · Digital therapy · Digital diagnosis · Digital trials · Health and life sciences experience · Patient data

16.1 Introduction

The Metaverse is, at its root, an evolution of the Internet. It started with an Internet of Data in the 1990s and moved to the Internet of People in the 2000s. In the 2010s, we saw the Internet of Things with connected phones, devices, and machines. The 2020s have brought with them two key mutations that give the Metaverse its distinctive character:

- The *Internet of Place* brings together people, spaces, and things in both the virtual and physical worlds.
- The *Internet of Ownership* (typified by the crypto community and non-fungible tokens) enables unique, portable, durable digital products to be created, exchanged, and valued in virtual and real markets.

Then there is Web3—a way for users and developers to own what they produce on a platform and also own the platform itself. Technologies like blockchain and tokenization will enable more technologies to be incorporated into the Metaverse. The Metaverse heralds an even greater era of digital transformation shaped by several attributes: physical and virtual place, multidimensional human experiences, creativity and utility, community, identity, market structure, value, and product and service. All of them should be considered in detail and managed to create the Metaverse future. They are:

- *Creativity and utility*: the *catalyst* to combine all attributes into meaningful experiences that deliver on brands' purpose and value and cater to people's needs.
- *Multi-dimensionality*: the Metaverse must consider the many dimensions of human experience, the most prominent being *spatial health* but including many more senses including hearing, touch, and smell.
- *Virtual place*: a *synthetic world* with spaces, things, and people to explore using a computer, console, mobile, wearable technology, and more.
- *Physical place*: the *physical world* we live in (spaces, people, and things). It is enhanced with virtual layers.
- *Community*: the Metaverse is *social*. Value is driven by group consensus and amplified by *network effects*. This affects both financial and social capital.
- *Identity*: people's ability to *express themselves*, *claim identity* (or multiple identities), or have social belongings and communicate a status.

- *Market structures*: a shift to a *virtual structure is needed to enable transaction of virtual value* for both real and digital worlds in primary, secondary, and tertiary markets.
- *Virtual value*: value in the virtual economy is tied to many factors, including *uniqueness, exclusivity, status, utility, history,* and *ownership*.
- *Product*: products can be experienced in your physical world (AR), in virtual world (WR) or *paired* with real-world goods.
- *Portability and persistence*: *digital elements* that are persistent across time, either in a virtual place or the physical world, with a history of exchange or use.

16.2 Enter the Metaverse Continuum

The Metaverse should be seen as a continuum. It brings together people, spaces, and things in both the virtual and physical worlds and enables a user to move from "browsing" to "inhabiting" the Internet in an ongoing, shared experience that spans all digitally enhanced worlds, physical realities, and business models. From 2D to 3D and from cloud and artificial intelligence to extended reality, blockchain, digital twins, and edge technologies—every aspect of business will be transformed in the Metaverse:

- *Payments*: digital currency and embedded business logic to simplify payment processes, cash management, purchasing, and sales
- *Customer experiences*: immersive experiences delivered to customers in the context of their lives
- *Employee experiences*: seamless teaming and collaboration anywhere, with anyone via any channel
- *The products you make*: native digital products, digitally augmented physical products
- *How you make your products*: digital twins and immersive experiences enabling rapid design of new buildings, processes, products, and services
- *Supply chains*: new digital supply chains for digital objects and enhanced collaborative and transparent physical supply chains
- *Enterprise management*: digital twins of entire enterprises enabling immersive and collaborative information control and insights

The Metaverse will provide the healthcare ecosystem with a virt-real collaboration space in which to create robust, meaningful experiences for all value chain stakeholders. Those stakeholders can then implement innovation-driven transformation spanning the ecosystem from drug discovery and development to patient journey management. But as a continuum of rapidly emerging capabilities, use cases, technologies, and experiences, the Metaverse should be developed with responsibility at its core. It must protect data ownership, security, and personal safety and facilitate inclusion, diversity, and sustainability.

The Metaverse Continuum enables truly humanized digital ecosystems across the value chain—from product design and production to sales, after-sales service, training, and supply chain management. It will transform how healthcare providers interact with their customers, patients, and experts and will influence how work is done. End-to-end visibility (of patients, processes, products, and materials) creates opportunities for more accessible, worldwide, equitable healthcare. Blockchain, confidential computing, and distributed computing technology will help to solve underlying challenges in healthcare like protecting identity, data security, and health records management. Significant changes in health and life sciences will include:

- *A new reality in clinical trial participation.* Anyone could be part of ongoing global clinical trials at any point. Wearables, implantables, and other sensors could make large-scale decentralized trials a reality, with real-time trial data seamlessly accessible to the right people in the right format, across the globe.
- *Global collaboration between healthcare professionals* (HCPs) in a "world" where HCPs can "gather" to study and learn from one another's patient cases. Such "meta–tumor boards" would allow professionals to share images, blood tests, and patient medical records and even "hear" from the avatars of patients on their individual care journeys—to prevent and cure at speeds never possible before.
- *Collaboration between research and manufacturing teams* in meta-labs, which would shorten time to market and lower the total cost of going to market with a new therapy.

The Metaverse Continuum transcends time and space to simulate interactions and shorten learning cycles and complex practice procedures, such as in surgical training. It enables more life-like virtual therapeutics, empowering patients to manage their health and even perform some self-care.

16.3 It Is Time to Pause and Reimagine Healthcare

What role will ecosystem members play in each world, and what trusted patient experiences should be co-constructed in the next decade? It is time to reimagine the worlds in which they operate, to avoid ending up in worlds designed by, and for, someone else. To shape the healthcare future in the Metaverse continuum we need sufficient quality and quantity of talent. That means the right workforce skills, the right culture, the right ways of working, the right employee experience, and the right current and future customer service approach. The Metaverse Continuum's rise will be boosted and influenced by four transformative global technology trends (Fig. 16.1), described in Accenture's *Digital Health Technology Vision 2022*: WebMe, Programmable Word, The Unreal, and Computing the Impossible.

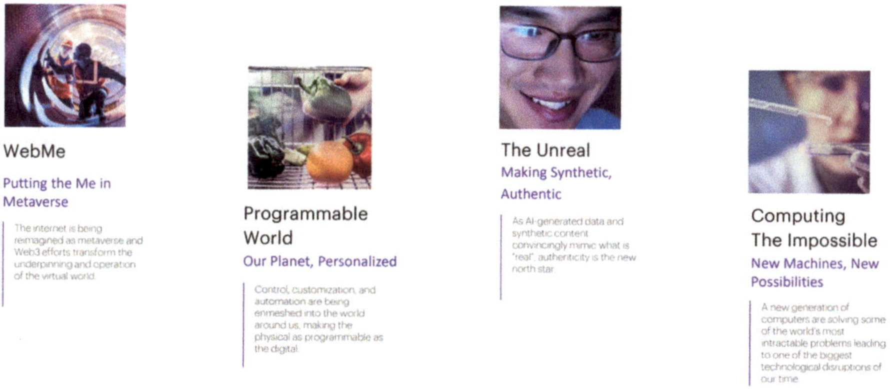

Fig. 16.1 Healthcare Technology Trends 2022 (Accenture 2022)

WebMe

For many consumers, the lines between digital life and "real life" are blurring more and more. Emerging markets consumers agree the most that their digital life is increasingly becoming their "real life".

Market Units (Agree/Strongly agree)	
Growth Markets	46%
North America	37%
Europe	31%

Top Countries (Agree/Strongly agree)	
China	72%
Thailand	65%
India	62%
Saudi Arabia	60%
Malaysia	54%
UAE	53%
Indonesia	52%
Singapore	52%
South Africa	50%

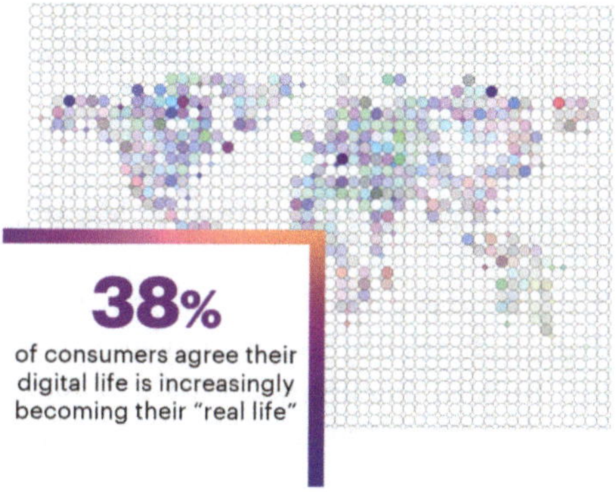

Fig. 16.2 Technology Vision 2022 Global Consumer Survey. Global $N = 24{,}000$ (Agree Net = Agree/Strongly Agree) (Accenture 2022)

16.3.1 WebMe: What It Means

WebMe allows people to live personal lives virtually—to an unprecedented extent (Fig. 16.2). Technology empowers humans by means of transparency and translation. WebMe acts as a personal interface between the physical and virtual and will address a key pain point—patient adherence. Recognizing and personalizing complex medical treatments and therapies, knowing the individual, and sharing trusted, transparent metrics among healthcare ecosystem players will increase safety and reduce relapse or comorbidities, reducing the cost of care.

16.3.2 Programmable World: What It Means

Programmable World tracks how technology is being threaded through our physical environments in increasingly sophisticated ways. Disruptive technologies like extended reality (XR) and 5G will challenge existing business models in the healthcare ecosystem by providing targeted, accurate solutions that go beyond two-dimensional technologies. Augmented reality (AR) is another potential game changer, applied across the value chain (including clinical trials, manufacturing, sales and marketing, and patient education). With the Metaverse as a foundation, XR/AR, 5G technologies, automation, and cloud computing could enable virtual, global gatherings in meta-labs to conduct experiments, share real-time insights and lab instruments, and conduct impact analysis across geographical locations and time zones.

Healthcare ecosystem companies could also extract the full potential of digital twin technology, even in the manufacturing space. The XR and digital twins for enhanced consumer and marketing engagement are already being piloted, especially in the learning and knowledge sharing area, but digital twins hold great promise in therapeutic areas requiring high-quality, multi-dimensioned clinical trial and real-world data.

16.3.3 The Unreal: What It Means

The Unreal is a trend where our environments and businesses are increasingly filled with machines that are passably human. "Unreal" qualities are becoming intrinsic to the artificial intelligence (AI) and data that enterprises aspire to integrate into mission-critical functions. With the Metaverse as the foundation real technologies will determine how competitive advantage could be enhanced by leveraging existing data strategies, algorithms, and AI. Identify where unreal content like chatbots or AI-generated images, videos, or content could help extend your brand and/or create preferred interactions with HCPs and patients.

16.3.4 Computing the Impossible: What It Means

Computing the impossible is showing us the outer limit of what is computationally possible, and how it is being disrupted as a new class of machines emerges. Quantum, biologically inspired, and high-performance computers are allowing companies to tackle grand challenges that once defined and shaped the very core of their industries. With the Metaverse as foundation these technologies will make healthcare ecosystem partnerships mandatory. We should start building relationships with next-generation computing providers now and cooperating with other ecosystem players. The industry must enable and exploit the power of AI's application to drug discovery, for example, translating complex biological problems into easier computational ones.

16.4 It Is Happening Already

Healthcare companies are pushing the boundaries of the Metaverse Continuum to deliver excellent new clinical, operational, and recreational experiences.

Health Metaverse use cases are beginning to emerge. Healthcare companies are beginning to push the boundaries of Metaverse technologies to deliver novel clinical, operational, and recreational experiences. For example (starting with proven examples and moving toward the possible):

- *Immersive training*: through virtual reality, we can create immersive learning experiences to better engage, train, and empower employees. This includes clinical social worker or customer service empathy training settings, for example.
- *Patient education*: immersive environments enable a new standard for patient education. This leads to better health literacy and care plan cooperation during disease state and treatment explanations and post-discharge recovery plan compliance.
- *Digital therapeutics*: clinical evidence continues to be built for XR therapies, delivering care for pain management, mental health, physical therapy, and more. Examples include behavioral health therapy and medication and therapy for chronic pain management.
- *Digital diagnostics*: through AR we can use movement, space, and interaction to enable different types of diagnoses, like vision assessment for glaucoma patients.
- *Augmented health*: AR can enable users to overlay relevant information while completing an activity... so everything they need is at their fingertips. This could include guidance during surgical procedures and aided medical tasks.
- *Care plan and delivery*: immersive technologies help to simulate procedures, enabling the next evolution in care planning and procedural preparation. Examples could be digital patient twins for immersive care planning and pre-operative preparation.

16.5 Immersive Training

Medical simulation is an experiential form of education, using simulated healthcare environments where "real-life" skills and experiences can be replicated. The Metaverse could enhance the way we provide patient-specific pre-op planning and education (printing your spine in 3D before surgery, for example). Current projects include:

- Case Western Reserve and Cleveland Clinic is building a 485,000-square-foot Health Education Campus to support interprofessional learning. Instead of cadavers, anatomy classes will involve HoloLens headsets to see the body's organs and systems (CWRU 2015).
- Dreamscape Learn (DL) is a partnership between Arizona State University and Dreamscape Immersive. The Metaverse technology allows students to work in a

meta-lab with a plethora of fictional animal species that adhere to biological laws (Arizona State University News 2020).
- Seoul National University Bundang Hospital is providing training to medical staff in a smart operating room in the Metaverse. The Asian Society for Cardiovascular and Thoracic Surgery (ASCVTS) has already used the smart OR for lung cancer surgery training to over 200 thoracic surgeons (Korea Biomedical Review 2021).

16.6 Patient Education

Metaverse environments can be personalized to help patients interact with situations that cause them anxiety in safe, controlled, and monitored environments. People who fear hospitals could be acclimated in advance with virtual tours of the facility and therapy path using a 360-degree video. Current projects include:

- Dr. Rob Lewis, a Californian neurosurgeon, uses advanced VR surgical tools to take patients through their own anatomy to understand tumor anatomy and the therapeutic approach he will take (Plug&Play Tech Center 2022).
- Akdeniz University pediatric nursing is developing a Metaverse-based program, the MetaHealth-Youth Project, to encourage healthy behaviors in young people. In the multi-country clinical trial, participants enter Metaverse rooms to learn about nutrition, exercise, and stress management to reduce noncommunicable disease (NCD) risk factors (Clinical Trials Arena 2022).

16.7 Digital Therapeutics and Diagnosis

HoloLens 2 devices can be based at care homes, enabling remote round-the-clock interaction and support. They can also be installed in patients' homes, so that consultants wearing HoloLens 2 headsets can share their opinions with other consultants using Remote Assist and provide patients with highly specialized, properly informed treatment. Medical notes and X-rays can be placed alongside the call in the wearer's field of view to allow in-person and remote consultants to triage collaboratively. This is where digital twins also come into play—a digital twin is a virtual model/simulation, of any object, process, or system, generated using real-world data, for the purpose of learning more about its real-world counterpart. In the case of the Metaverse, the digital twin could be of the patient themselves. Digital twins will eventually become personalized "test dummies" to forecast everything from postoperative recovery rates to patient reactions to specific medicines. This will be aided by increasingly sophisticated gene mapping and interpretation. Current projects:

- Precision Os is an FDA-approved VR tool used to "fly" through the patient's body and examine it before reconstructive surgery (PrecisionOS 2022).

- Torbay and South Devon NHS Foundation Trust are piloting Microsoft HoloLens 2 and Dynamics 365 Remote Assist in the Breast Care Unit where specialist nurses send real-time video feeds to consultants to get immediate advice on a patient's needs. Consultants can add digital markers and annotations to the videos to guide nurses (Health Tech Newspaper 2021).

16.8 Augmented Health

The Metaverse opens a new frontier for telemedicine. Moving beyond home care, it takes remote care to extremes, hosting patients and doctors in virt-real spaces available 24/7 and just a click away. Patients are no longer limited to local clinicians but can choose doctors based on public biographies that detail their expertise. Doctors could also be made available to areas with a shortage of medical professionals, or to patients located in remote regions. Current projects:

- Microsoft's HoloLens technology has already been explored in non-operative as well as surgical cases to provide medical care remotely (Healthcare Outlook 2022).
- DeHealth announced the creation of a decentralized Metaverse where doctors can interact with each other and with patients in 3D and earn virtual assets (DeHealth 2021).
- HealthLand.io, formerly known as Healthify, allows fitness/sports trainers and health experts to open their own gyms. People can join in the HealthLand metaverse and receive low-cost emotional health consultations and fitness training from the comfort of their homes.
- AccuVein is a vein imaging system company trying to eliminate bruising after injections. Cosmetic surgeons can see the enormous network of veins and blood vessels beneath the skin before treatment thanks to AccuVein's game-changing technology (Healhtcare Outlook 2022).

16.9 Care Plan and Delivery

The Metaverse gives patients access to their personal health records and care plans. It could also help patients create goals based on analysis of their personal data and provide motivation by illustrating potential therapy results using digital twins. Current projects:

- CVS Health has become the first pharmacy to enter the Metaverse. It has filed with the US Patents office to trademark its logo and establish an online store and create downloadable virtual goods ranging from prescription drugs to beauty and personal care products (CNBC 2022).
- Akuvera uses thermal cameras based on algorithms to detect and predict whether a patient will adhere to bedrest requirements (Plug&Play Tech Center 2022).

- Evolve Rehab is using a Microsoft Kinect camera to trace person's movement and gamify physical rehabilitation (Plug&Play Tech Center 2022).
- IoTeX is a project that, with its existing remote monitoring capabilities, has the potential to assist patients through the Health Blocks project, which rewards users for changing their daily habits to live a healthy lifestyle (Healthcare Outlook 2022).

16.10 Introducing Meta-Care [1]

The future of healthcare depends on a sound understanding of context with respect to effective healthcare delivery and patient experiences. The motivation for the Metaverse Continuum is clear. Digital transformation plans succeed or fail on patient readiness to adopt digital solutions and improve their interactions with the healthcare ecosystem and live healthier lives with minimum effort.

The 2021 *Accenture Health and Life Sciences Experience Survey* reveals how the healthcare experience has changed during the COVID-19 pandemic and could help to define what it will look (Accenture 2021a) like going forward. Key results of the study show:

- Satisfaction with healthcare experiences is generally low and not homogeneous among surveyed countries, but as patients move along the patient journey, they increasingly leverage emerging digital health technologies.
- The quality of access to healthcare elicited strong dissatisfaction, which has worsened since the pandemic. Given these statistics regarding access to key healthcare services, digital health is a prime candidate to provide information about who to talk to, how to understand therapies, and where to get expert guidance more easily.
- Digital technologies like virtual consultations and appointments show significant potential to boost digital care, based on patient interest. Among the Italian respondents, Accenture research says that 15% would use and pay for services or digital technologies for disease prevention, but an impressive 34% would pay if it were at a low or discounted cost—38% would only use these services or technologies if they were free, while 10% would not use these services or technologies at all. Younger generations are more willing to pay full or discounted prices for this—a trend that holds across all countries surveyed, as well. Willingness to pay full or discounted prices for these services/technologies is slightly higher in Italy (49%) than in France (37%), Germany (33%), and Spain

[1] Meta-care is the new patient experience of healthcare—shaped using opportunities created by the Metaverse Continuum. It is the patient-focused result of humanized digital healthcare ecosystems—from product design and production to sales, after-sales service, and supply chain management. A transformed interaction with patients by healthcare providers, meta-care implies accessible, worldwide, equitable healthcare that improves care outcomes while maintaining trust through effective data governance.

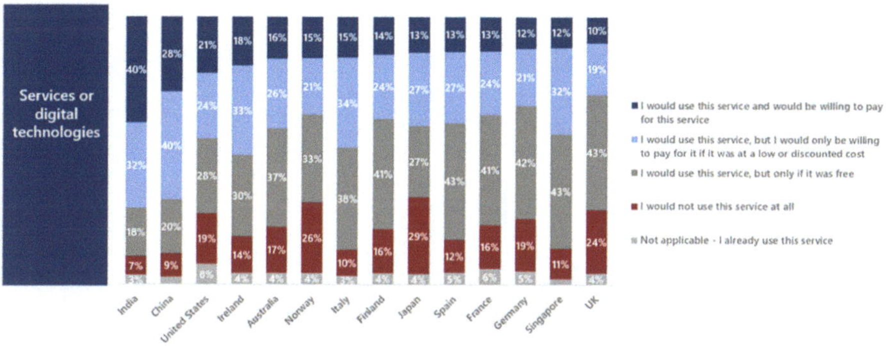

Fig. 16.3 Willingness to pay for services or digital technologies for disease prevention, per country. Q: What would the cost need to be for you to consider services or digital technologies for disease prevention (e.g., healthcare services to improve my health and wearable device to track my fitness goals)? (Accenture 2021a)

(40%). This same trend applies to ongoing disease management and regular health check-ups (Fig. 16.3).

While digital health technology is often viewed as impersonal (counter-intuitively) it can increase the frequency and quality of care and create a more connected experience for patients if used appropriately. In fact, Italian respondents are more likely to say they experience anxiety during in-person appointments than virtual ones (Fig. 16.4).

16.11 How Does One Neutralize Potential Healthcare-Related Anxiety?

For a positive experience with medical providers, respondents expect them to explain their condition and treatment clearly, listen, and provide emotional support. This is in line with other European countries surveyed, where people also expect efficient digital visits—something which could be enhanced by meta-care.

Meta-care should take into consideration the perceived benefits of a well-shaped hybrid digital–physical experience and mitigate the anxiety created by the current physical care pathways. It should also leverage the fact that Italian people are more comfortable using apps powered by AI to determine if they need a diagnosis (51%, in Italy vs. 38% in France, 27% in Germany, and 47% in Spain), and using digital technology and AI to get diagnoses/treatments (52% in Italy, vs. 35% in France, 33% in Germany, and 47% in Spain).

If you ask people whether they would consider using digital therapeutics, almost one-third say they would. That applies to disease prevention (when healthy), for symptomatic treatment or for ongoing disease monitoring. Trust is key for meta-care to grow and succeed, now and in the future. Respondent data indicates that

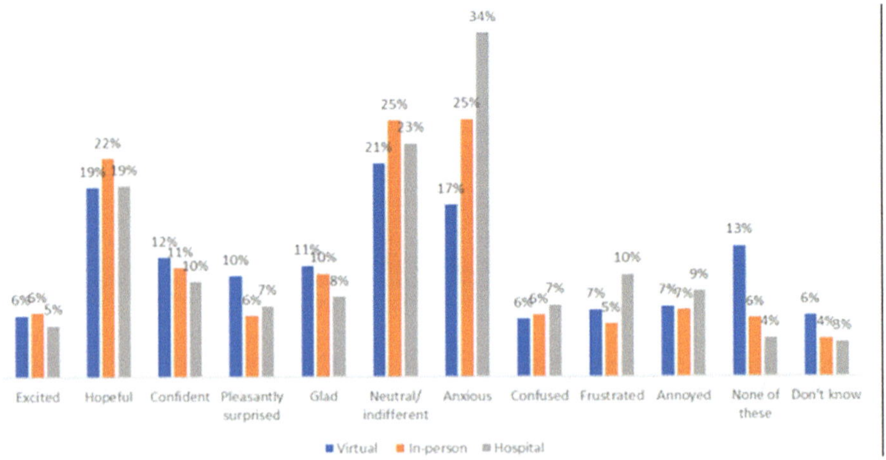

Fig. 16.4 How Italian respondents feel when accessing several types of care virtually and in-person, in-person or at a hospital. Q: Think back to the LAST time you had a virtual healthcare appointment (e.g., a medical appointment over the phone, via an app), an appointment with a medical provider in-person (e.g., with a doctor, specialist, nurse), and an appointment in-person at a hospital (i.e., either a planned appointment or an unplanned appointment). Which, if any, of the following emotions describe how each interaction made you feel? *Select all that apply* (Accenture 2021b)

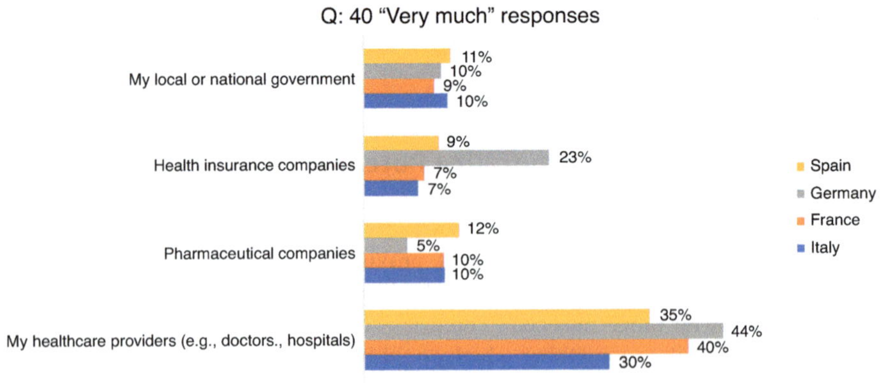

Fig. 16.5 Healthcare providers are still the most trusted source of healthcare information by a significant margin. Q: Overall, how much do you trust information provided by the following sources? *"Very much" responses* (Accenture 2021a)

traditional healthcare players (providers, pharmacists) enjoy more patient trust than non-traditional players (technology companies, government, pharmaceutical companies, health insurance) (Fig. 16.5).

This reality is vital to creating effective meta-care. Clear, transparent communication is vital to trust. Some providers have struggled to engage properly with patients, yet being able to do so is a key success factor for meta-care. Patients are changing their habits. They have changing needs and are finding new ways to manage their lives and their healthcare experiences. Whether it is buying medicines, managing their health regime, or gathering health information, people's expectations are changing faster than clinicians, pharmaceutical companies, and healthcare organizations can adapt—creating a mismatch. The gap will only grow if the healthcare ecosystem does not respond quickly, and meta-care is a swift way to close it.

16.12 Meta-Care Is Key, But It Is Not Enough

In the last few years, patients have expressed a strong demand for more digital health management tools and services across all care types and at every touchpoint. From prevention to actual therapeutic care—more responsive, agile, and accessible tools must be used to ensure higher adherence, pre-empting, limiting, and more effectively managing illnesses and comorbidities.

The human touch remains vital, though, even in meta-care. So, while it will undoubtedly form a crucial part of healthcare's future, meta-care must include human contact for patients. Meta-care does not completely replace physical care, but rather augments it. It creates better accessibility for greater and more frequent doctor–patient interaction, a smoother immersive experience for medical personnel, and better outcomes. However, meta-care also presents a variety of potential challenges—from providing equitable access to technology to keeping patient data secure and ensuring patient safety as they explore care in new realms on their own terms and in their own time. As we enter this new era, the right governance and regulations must ensure that enthusiasm for meta-care's great potential does not come at the expense of caution and care for the human at the center of the experience. Organizations need to lead with people-centric experiences and help consumers unify their digital ones (Fig. 16.6).

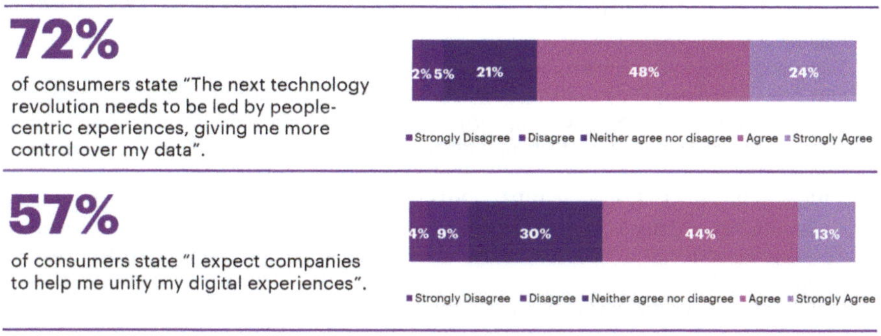

Fig. 16.6 Technology Vision 2022 Global Consumer Survey. Global $N = 24{,}000$ (Agree Net = Agree/Strongly Agree) (Accenture 2022)

It will be critical to ensure that access to meta-care is genuinely equitable. Tech companies' efforts to generate profit must be regulated, and less tech-savvy groups like older or underprivileged people may find themselves further disenfranchised without the right safeguards. Technically speaking, interoperability is a critical component of meta-care and the whole healthcare industry should be pushing for data and communication standards. Costs should also be evaluated since meta-care's full potential depends on high-tech hardware such as glasses, gloves, sensors, and other wearables capable of reading patients' vital signs. Reimbursement methods and payer support should be reshuffled and linked to the outcomes to avoid making technology available only to those who can afford it.

16.13 It Takes an Ecosystem: You Can't Do It Alone

No single player can lead meta-care. There is an urgent need for healthcare ecosystem participants (both new entrants and incumbents) to speed up the process of collaboration in a pre-competitive environment and to define new rules of engagement that allow the ecosystem to jointly build new, trusted capabilities and services into meta-care. Clear and precise terms of engagement will open the gates of investment and should reflect all stakeholders' desired outcomes so that value is demonstrated upfront and for all—and realized through disciplined implementation. The call for change is louder everywhere—and so is the willingness to adopt new meta-care solutions. The scene is set, and it is time for the actors to take the stage.

References

Accenture (2021a) 2021 Accenture Health and Life Sciences Experience Survey – Global report. Source: https://www.accenture.com/_acnmedia/PDF-161/Accenture-Digital-Adoption-In-Healthcare.pdf Last consultation: 02/02/2023

Accenture (2021b) The Italian Health and Life Sciences Experience, Source: https://www.accenture.com/it-it/insights/health/digital-adoption-healthcare-reaction-or-revolution Last consultation: 02/02/2023

Accenture (2022) Accenture Digital Health Technology Vision 2022 - Healthcare Provider Survey Findings, Source: https://www.accenture.com/_acnmedia/PDF-178/Accenture-Digital-Health-Technology-Vision-2022.pdf#zoom=40 Last consultation: 02/02/2023

Arizona State University News (2020) 'Dreamscape Learn' to transform educational experience through exploration, Source: https://live-asu-news.ws.asu.edu/2020-09-18-discoveries-dreamscape-immersive-asu-partnership-virtual-reality-learners-worldwide Last consultation: 02/02/2023

Case Western Reserve University (CWRU) (2015) CWRU and Cleveland Clinic break ground on Health Education Campus; to open summer 2019, Source: https://thedaily.case.edu/cwru-and-cleveland-clinic-break-ground-on-health-education-campus-to-open-summer-2019/ Last consultation: 02/02/2023

Clinical Trials Arena (2022) The metaverse: the next frontier in clinical trials?, Source: https://www.clinicaltrialsarena.com/analysis/metaverse-clinical-trials/ Last consultation: 02/02/2023

CNBC (2022) CVS files to trademark its pharmacy and health clinics in the metaverse, Source: https://www.cnbc.com/2022/03/04/cvs-files-to-trademark-its-pharmacy-and-health-clinics-in-the-metaverse-.html Last consultation: 02/02/2023

DeHealth (2021) The world's first healthcare metaverse from DeHealth, Source: https://www.dehealth.world/post/the-worlds-first-healthcare-metaverse-from-dehealth Last consultation: 02/02/2023

Health Tech Newspaper (2021) National pilot explores Microsoft HoloLens 2, https://htn.co.uk/2021/09/13/national-pilot-explores-microsoft-hololens-2/ Last consultation: 02/02/2023

Healthcare Outlook (2022) Top Used Cases of Metaverse in Healthcare to look out in 2022, Source: https://www.healthcareoutlook.net/top-used-cases-of-metaverse-in-healthcare-to-look-out-in-2022/ Last consultation: 02/02/2023

Korea Biomedical Review (2021) Metaverse in operating room is changing medicine rapidly, Source: http://www.koreabiomed.com/news/articleView.html?idxno=11477 Last consultation: 02/02/2023

Plug&Play Tech Center (20/05/2022) Healthcare in the Metaverse - YouTube, Source: https://www.youtube.com/watch?v=0ccMFH0H6RE Last consultation: 02/02/2023

PrecisionOS (24/06/2022) Preoperative Planning | Virtual Surgical Planning, Source: https://www.precisionostech.com/preoperative-planning/ Last consultation: 02/02/2023

Patients' Reactions to Anthropomorphic Technologies in Healthcare. The Predictor Roles of Perceived Anthropomorphism and Human-Like Interaction: A Preliminarily Study

17

Andrea Sestino and Alfredo D'Angelo

Abstract

The fascinating concept of anthropomorphic technologies refers to those digital technologies appearing as human-like in its design, and in terms of attribution of human-like characteristics to these non-human objects (e.g., chatbots, robots, virtual avatars, and so on). In the healthcare sector, new anthropomorphic technologies may revolutionize the service delivery process, contributing to a reduction in physical distance and enriching the doctor–patient relationship. Furthermore, considering the evolutionary trend toward digital business models in healthcare, these new digital technologies could lead the healthcare system to a futuristic level. Through an exploratory research design, using a sample of 382 participants, in this preliminary study we investigated the combined effect of perceived human-like interaction level and anthropomorphism in influencing individuals' reactions (intention to use) toward these new medical digital technologies. Discussions for healthcare managers and policymakers, together with food for thought for healthcare management and technology innovation in social services, are offered.

Keywords

Anthropomorphic technologies · Healthcare · Innovation · Human-like interaction level · Anthropomorphism

A. Sestino (✉)
Department of Business Studies, University of Rome Tre, Rome, Italy

LUISS Guido Carli University, Rome, Italy
e-mail: andrea.sestino@uniroma3.it

A. D'Angelo
Department of Economics and Management, Catholic University of Sacred Heart, Rome, Italy

A. Cesario et al. (eds.), *Personalized Medicine Meets Artificial Intelligence*,
https://doi.org/10.1007/978-3-031-32614-1_17

17.1 Introduction

The impact of new technologies in healthcare in recent years has been disruptive (Rahimi et al. 2018). Digital health technologies use information platforms, connectivity, software, and sensors to pursue a wide range of objectives (Lupton 2014; Edirisinghe 2022) from achieving and maintaining a general level of well-being to developing medical and diagnostic devices in order to better monitor and operate actively on the health of patients (see Värri 2020 for a review). Thus, today it is possible to use precision digital tools during surgical operations (e.g., as in Adamo et al. 2020), to monitor patients remotely (e.g., Sestino et al. 2023), or to manage the entire patient lifecycle (e.g., Triantafyllidis and Tsanas 2019). For instance, thanks to the ability to read medical records (Dinh-Le et al. 2019), artificial intelligence-based algorithms are able to formulate treatment plans and prevent serious diseases (Bhinder et al. 2021; Cesario et al. 2021a), e.g., as for cancer (Cesario et al. 2021b), develop new drugs, and even analyze samples of cancerous and non-cancerous tissue (Mak and Pichika 2019). Blockchain technologies, for example, may certify patient medical records and their hospitalization history (Hasselgren et al. 2020). Internet-of-Things devices may be exploited to monitor patients and intervene when requested (Sestino et al. 2020). Among the recent technological innovations, the Metaverse, despite in its infancy, may also deploy its maximum effects in the healthcare sector. The Metaverse consists of a virtual and parallel reality, similar to the concept of a virtual world that combines elements of many technologies including AI, immersive reality, advanced connectivity, and Web3. With this combinatorial technology, individuals may enter the digital world through virtual identity (avatars), virtually do their daily activities, socialize, work, shop, do sport, and meet friends (Sparkes 2021). Thus, the Metaverse can be considered an important value-creation technology for both consumers and industries such as healthcare, automotive, education, and luxury (McKinsey 2023; Sestino et al. 2022).

By its nature, the Metaverse could revolutionize the entire process of providing healthcare services with reference to telemedicine (Wang et al. 2022). Indeed, telehealth is a video conversation via a computer or telephone where a doctor uses a telephone or computer to assist a patient who is off site (VandenBos and Williams 2000). In addition, telehealth employs electronic information and telecommunications technologies to support and promote long-distance clinical healthcare, professional education, patient information, public health, and health systems administration (Mort et al. 2003). These technologies include video conferencing, media streaming, and wireless communications (Fong et al. 2020).

Thinking about the possibility of the Metaverse to "turn" the real world into a virtual reality by catapulting the participants into a sort of parallel reality, the whole livable healthcare experiences may be magnified (Damar 2021; Dwivedi et al. 2022). Furthermore, the Metaverse could enable precision medicine and encourage greater research and development capacity through the possibility of creating avatars, which are digital representations of human beings in the virtual world. Thus, from the perspective of the Metaverse's final users (i.e., the patients), the convergence of the aforementioned new technologies could allow medical professionals to provide a range of highly integrated, deliberate, and individualized care without being constrained by the siloed nature of current healthcare models in a "parallel" daily life environment (Sun et al. 2022). By taking the doctor–patient relationship to a future level, the Metaverse may promote quickness in both doctor–patient and doctor–doctor communication, enabling hitherto unheard-of levels of complexity in prevention, diagnosis, and treatment. Summarizing, the Metaverse is emerging as an immersive combinatorial technology with great potential to offer value-centric patient care across the healthcare spectrum (Dwivedi et al. 2023).

One of the enabling technologies of the Metaverse refers to the avatar of participants (e.g., both doctors and patients) in a virtual environment in the Metaverse. The avatar can closely resemble its user in real life or be different depending on how this is built in the design phase (Nowak and Fox 2018). Thanks to the total abstraction of the body and its transposition into virtual reality, it will be incredibly possible to even enter a virtual human body and understand its mechanisms, understand its functioning in the smallest details, be able to study its characteristics and pathologies in all evolutionary phases, and determine the most effective drug therapies. The digital transposition of patients and doctors implies the creation of their virtual "duplications" capable of interacting in a totally digital-based environment. Nevertheless, this new digital technology simulating interactions in a parallel virtual reality where avatars are digital transposition of patients and doctors in the real world cannot underestimate the importance of human-like interactions for its value-based implementation.

Given these premises, through an exploratory research design, this chapter sheds light on the role of two important variables in the human-like interaction level, and anthropomorphism, which prove to be important antecedents of the intention to use these new digital anthropomorphic technologies. An experiment has been conducted among a sample of randomly recruited 382 participants. Results highlight that the level of individuals' perceived human-like interaction (high vs. low) on their reactions (intention to use) toward anthropomorphic technologies is mediated by their perceived anthropomorphism. More specifically, results confirm that higher level of perceived human-like interaction level leads to a higher intention to use anthropomorphic medical digital technologies because of increased level of individual's perception.

This chapter is organized as follows. In the second section, we provide some foundations about the concept of anthropomorphic technologies explaining the reasoning behind our preliminarily study. In the third section, we describe the methodological approach implemented together with details on the experiment. In the fourth section, we present the results of our experiment. Finally, in the fifth section, we conclude offering some insights for healthcare marketers, managers, and policymakers in approaching medical virtual agents' design.

17.2 Overview of the Study

17.2.1 Anthropomorphic Technologies

The concept of anthropomorphism has long been studied in the literature. Such a concept refers to the interpretation of what is not human or personal in terms of human or personal characteristics, through a sort of humanization. To clarify, the concept of anthropomorphism relates to individual's propensity to see inanimate objects as having human-like characteristics (Guthrie 1993). It entails the process of inferring from things and non-human entities that have exterior traits, motives, actions, and underlying states that are typical of humans (Epley et al. 2008; Guido and Peluso 2015). Individual's perceptions of anthropomorphism may have a direct impact on whether they intend to use an intelligent agent that resembles a person, depending on the type of agent (e.g., robot gender, individual likeness, and so on) or the sort of service (Blut et al. 2021). This person's aim is mostly motivated by the resemblance they attach to digital tools like chatbots or avatars (Sheehan et al. 2020). Indeed, anthropomorphism, previously defined as the attribution of human-like characteristics, behaviors, or mental states to non-human entities such as objects, brands, animals, and, more recently, technological devices may include a wide range of characteristics, from physical appearance to the various mental states that characterize human beings, such as engaging in reasoning, making moral judgments, and forming intentions (Golossenko et al. 2020; Kim and McGill 2011), even when perceived as acting as humans (Guido et al. 2019). The metaphorical manifestation of anthropomorphism known as personification of a non-human creature is likewise a kind of this (Wang 2017). More importantly, anthropomorphism may make users feel more connected to technology due to a great sense of connectedness (Kang and Kim 2020): Thus, a greater sense of closeness leads to more favorable individuals' reactions to anthropomorphic technologies.

By considering the technology-related stream of research, an anthropomorphic technology (AT) is technology that is human-like in design and motivates anthropomorphism, reached through attribution of human-like characteristics to non-human objects (see Li and Suh 2022 for an extensive literature review on this domain). As for anthropomorphism, Kang and Kim (2020) have found that anthropomorphism increases the sense of connectedness between user and the technology. According to their study, the increased sense of connectedness in turn evokes more positive user responses toward the technology (Kang and Kim 2022). By considering the virtual agents as part of anthropomorphic technologies, such as those usable in the healthcare industries (e.g., medical chatbots or doctors' avatars interacting in the Metaverse), individuals' perceived anthropomorphism may thus have a fundamental explanatory effect in undermining their intention to use (Han 2021). For instance, in the Metaverse the presence of anthroponomic forms of virtual agents such as avatars may boost the communication between doctors and patients as well as among doctors themselves, enabling hitherto unheard-of levels of complexity in prevention, diagnosis, and treatment (Dubosc et al. 2021).

17.2.2 Human-Like Interaction Level

In the context of anthropomorphic technologies, a further fundamental characteristic that cannot be overlooked is that relating to the interaction modality between technologies (e.g., computers, virtual agents, and so on) and individuals (Dix 2016; Karray et al. 2008). Previous research shows that when a technology that mimics a person may engage and display empathy with human users, it may be more widely accepted by final users (Pelau et al. 2021).

Anthropomorphic traits are significant in the relationship between individuals and the intention to use new technologies, according to existing studies (Pelau et al. 2021; Strait et al. 2014; Wan and Aggarwal 2015). For instance, according to the computer as social actors theory (Nass and Moon 2000) humans mindlessly apply the same social heuristics used for human interactions to computers because they call to mind similar social attributes as humans. Thus, individuals may exhibit stronger intention to utilize anthropomorphic technologies if they believe that such interactions are comparable to those that may be with human peers (Heerink 2010; Kim and Sundar 2012). Indeed, such perception toward new digital technologies increases their acceptance in different situations as it gives them greater power and a sense of equality with humans.

When considering the healthcare sector and the new technology for service delivery, not all the available virtual agents exhibit the same level of human-like interaction. For example, chatbots could be perceived as more limited and less interactive as they are based on a finite knowledge base, and their ability to react is exhausted when they are no longer able to answer the set of questions known to them (Chung and Park 2019; Gentner et al. 2020). Otherwise, the avatars in the Metaverse could be perceived as more interactive since, being guided—like puppets—by the man behind them, they are able to react exactly like these, almost as if they were their "extension" in the virtual world. Despite efforts by medical chatbots to become more anthropomorphic in the healthcare industry (Bhattacharya et al. 2022; Bulla et al. 2020), individuals may be more accepting of avatars since they are thought to be more like the individuals they represent in the real world (such as doctors or clinicians) (Sun et al. 2022). Additionally, because humans are in charge of the avatars rather than computers, they may embody all the positive qualities, behave like individual, and avoid making technological mistakes. Additionally, due to the limited knowledge base on which completely automated virtual agents like chatbots are constructed, they are susceptible to mistakes, which may have a negative impact on individual's intentions to use such technologies (Sheehan et al. 2020).

Based on above literature, in this study we suggest that the human-like interaction level of the technology may positively influence individuals' intentions to use through the effect of their perceived anthropomorphism.

Thus, we formally hypothesize that:

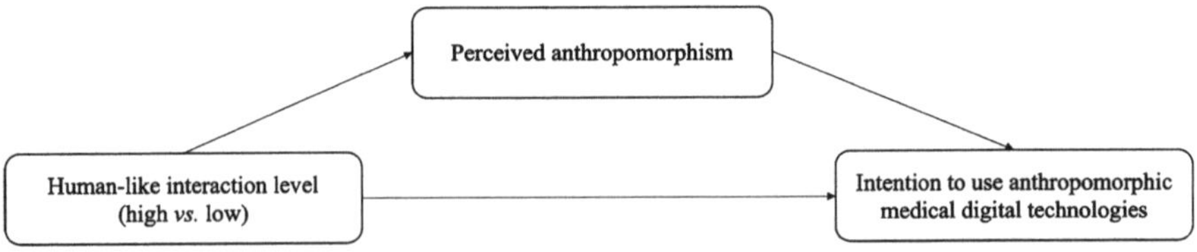

Fig. 17.1 Human-like interaction level and perceived anthropomorphism effects on individuals' intention to use anthropomorphic medical digital technologies

Hypothesis Individuals' perceived human-like interaction level positively influences their intention to use anthropomorphic medical digital technologies, through the effect of their perceived anthropomorphism.

Figure 17.1 below presents the proposed conceptual framework for which we seek empirical validation.

17.3 Methodology

To test the aforementioned hypothesis, through an explorative research design we conducted an experimental study where we manipulated the perceived human-like interaction level (high vs. low) to sage the effect of individuals' perceived anthropomorphism on their intention to use anthropomorphic medical digital technologies.

More specifically, the experimental design was based on a two-cell experiment and conducted by manipulating the level of human-like interactions in a digital-based healthcare service delivery. Building on the previous literature, explaining the different level of perceived human-like interaction levels, we manipulate human-like interaction level by exposing half sample to a scenario describing the use of chatbot for healthcare service delivery (to define the low condition of perceived human-like interaction level). The other half of the sample has been exposed to a scenario describing a Metaverse-based healthcare service delivery where users were able to interact via their avatars (to manipulate the high condition level).

Coherently with our explorative research design, a survey-based experiment was implemented. The survey was organized into three sections. In the first one, participants were welcomed and invited—on the basis of our two-cell experiment—to see the two alternative scenarios. Participants were then randomly assigned to two different settings where we manipulated the level of human-like interaction of the online digital-based healthcare service delivery (low level of human-like interaction vs. high level of human-like interaction). In the second section, we gathered information for participants' intention to use anthropomorphic medical digital technologies by using a two-items scale drawn by Fishbein and Ajzen (1977) (e.g., "I intend to use this digital healthcare service in the future"), and their perceived anthropomorphism by using the four-items scale drawn by McLean and Osei-Frimpong (2019) (e.g., "The interaction experience with the digital doctor is close to that with a human being"). Both individuals' intention to use anthropomorphic medical digital technologies and perceived anthropomorphism, together with their reported emotional receptivity, were assessed on a 7-point Likert scale.

The survey was built via the software Qualtrics and distributed through the online platform Mechanical Turk (Aguinis et al. 2021). We recruited a sample of 382 participants, aged between 22 and 71 ($M_{age} = 33.13$; $SD_{age} = 10.01$), consisting of 219 males (67%) and 108 females (33%). The large part of the sample declared to have a B.Sc. or M.Sc. (271 participants; 83%) or to hold a degree lower or equal to a high school diploma (40 participants or 12%) or a Ph.D. (17 or 5%). As for their geographical location, the sample consisted of only European participants.

17.4 Results

To test our hypothesis, we ran the simple mediation model (Model 4) developed by Hayes' (2017) using the PROCESS macro for SPSS (Table 17.2). The mediation model included the advertised travel communication focus (-1 = high level of perceived human-like interaction and 1 = low level of perceived human-like interaction defined as the independent variable *(X)*, individuals' intention to use anthropomorphic medical digital technologies as the dependent variable (*Y*), and individuals' perceived anthropomorphism as a mediator (*Me*) in explaining the effect of the independent variable on the dependent variable). Results report a significant and negative effect ($b = -0.511$, $t = -8.156$, $p = 0.000$) of the type of effects of human-like interaction level on individuals' perceived anthropomorphism, confirming how such technology characteristic (human-like interaction) is fundamental in making perceptible digital tools as anthropomorphic (Tables 17.1 and 17.2).

However, individuals' perceived anthropomorphism seems to exert a positive and significant effect on their intention to use anthropomorphic medical digital technologies ($b = 0.810$, $t = 19.111$, $p = 0.000$). More importantly, results reveal a significant and direct effect of the perceived anthropomorphism on individuals' intention to use anthropomorphic medical digital technologies through the effect of the different levels of human-like interaction ($b = 0.035$; $t = 0.620$; $p = 0.002$).

Thus, our result show that, due to higher level of human-like interaction level, individuals may perceive the digital tools as more anthropomorphic, finally exhibiting greater intention to use anthropomorphic medical digital technologies.

Table 17.1 Effects on individuals' perceived anthropomorphism

Dependent variable: Perceived anthropomorphism (*Me*)	*b*	*SE*	*t*	*p*
Constant	5.802	0.063	92.570	0.000
Human-like interaction level (*X*)	−0.511	0.063	−8.156	0.000
$R^2 = 0.149$, $MSE = 1.038$, $F\ (1, 380) = 66.519$ $p = 0.000$				

Note. $N = 382$

Table 17.2 Result of the mediation analysis

Dependent variable: Intention to use anthropomorphic medical digital technologies (*Y*)	*b*	*SE*	*t*	*p*
Constant	1.184	0.251	4.713	0.000
Human-like interaction level (*X*)	0.035	0.056	0.620	0.002
Perceived anthropomorphism (*Me*)	0.810	0.042	19.111	0.000
$R^2 = 0.537$, $MSE = 0.709$, $F\,(2, 379) = 220.187$ $p = 0.000$				

The combined effect of human-like interaction level and perceived anthropomorphism on individuals' intention to use anthropomorphic medical digital technologies
Note. $N = 382$

Conversely, when the human-like interaction level is perceived as low, the effects on individuals' perceived anthropomorphism decrease, together with their intention to use. The results generally confirm that a higher level of human-like interaction leads to an equally higher perception of the anthropomorphism of the digital medical tools. This effect positively influences the individuals' intention to use anthropomorphic medical digital technologies.

17.5 Discussion and Conclusion

In a complex set of new healthcare technologies, virtual agents today have the ability to use a virtual character created through computer generation, animation, and artificial intelligence as customer service agents. These virtual agents can be "less autonomous" as in the case of chatbots, which are able to answer questions posed on the basis of a fixed knowledge base (however capable of continuous learning), or "more autonomous" as in the event of the avatars of individuals in the Metaverse as guided by them, almost like an extended self. Such autonomy has been interpreted by research as a characteristic of these anthropomorphic technologies, also known as human-like interaction, able to measure how much a virtual agent is able to interact—interactively—with humans.

Drawing from previous studies, in this chapter we demonstrate that, by considering the digital healthcare services for teleconsulting delivery, individuals' perceived human-like interaction level (high vs. low) of these virtual agents may positively influence their intention to use anthropomorphic medical digital technologies through their perceived anthropomorphism. This means that when patients perceive the virtual agents replacing "flesh and blood" doctors as interactive as a human but also anthropomorphic, the combined effect can boost their intention to use such digital technologies.

These preliminary insights are interesting in the current technological and social landscape in which new digital technologies, such as the Metaverse, are rapidly emerging. The Metaverse could catapult individuals into a "twin" reality parallel to the real one and favor, for example, new healthcare services delivery by breaking down geographical distances enhancing the doctor–patient relationship. With regard to teleconsultation services, both doctors and patients through their avatars could access the Metaverse, meet, discuss, and act as if they were in the doctor's office. Furthermore, the doctor can "digitally show" any materials related to the patient (e.g., X-rays and body analyses), directly on the patient's avatar, in line with the augmented reality paradigm.

Our preliminary results can contribute to the research in several ways. From a theoretical perspective, they add knowledge to theories, such as that of computers as social actors' theory (Nass and Moon 2000), or to the line of research related to human–robot interaction (HRI), as a field of study dedicated to understanding, designing, and evaluating robotic systems for use by or with humans. By definition, interactions require communication between robots and humans. Indeed, we explain how the combined effect of high levels of human-like interaction and perceived anthropomorphism may lead to higher patient intention to use anthropomorphic medical digital technologies in the healthcare service delivery.

Moreover, from a managerial perspective, our findings provide important insights for healthcare marketers, managers, and policymakers. For instance, our findings suggest that in the design of new digital health services, such as teleconsultation, higher level value-centric patient care services than telephone calls or video calls embrace more immersive technologies to provide wider support to the final patient. From a healthcare policymaker perspective, our findings suggest the imperative to introduce such new technologies into the healthcare systems as soon as possible to enrich the value proposition of new digital-based services. Indeed, the fact that new technologies are pushing healthcare toward a digitization of treatments and doctor–patient relationships is now well established, especially following the health emergency of COVID-19 pandemic, telemedicine systems and, in general, connected care systems have been developed, tracing the direction toward the decentralization of care.

New opportunities, such as those considered in our experiment, could open a new frontier in this sense. Indeed, the Metaverse combinatorial technology is not limited to bringing treatment to the patient's home, but takes this concept of decentralization to extremes, hosting patients and doctors in new physical/virtual spaces thanks to their avatars.

More importantly, the delivery of healthcare services in Metaverse could provide new opportunities and unexplored solutions, giving such patients the possibility to access medical care provided by health facilities physically located in other regions, in other states, and on other continents, breaking down, on one hand, the geographical distances and, on the other hand, offering the possibility to patients with reduced mobility to reduce the inconveniences due to physical movement.

Despite the promising results of our research, this study has some limitations. First, in light of the exploratory approach used, starting from the theories offered (e.g., Nass and Moon 2000), our study considered only the human-like interaction level (manipulated through a scenario) and perceived anthropomorphism as antecedents of individuals' intention to use. Further studies should necessarily enrich this model, for example by leveraging further theoretical models, such as those referable to the acceptance of technology (Davis 1989). Moreover, future studies should consider further patient-related variables, such as empathy, emotional receptivity, or the effect of patients' tendency to human contact, as other complex elements to keep unchanged in totally digital-based environments. Although adequate for a preliminary experimental study, further studies could test the effects of the proposed model with other variables and on a larger population, or by leveraging real use cases.

References

Adamo JE, Bienvenu RV II, Dolz F, Liebman M, Nilsen W, Steele SJ (2020) Translation of digital health technologies to advance precision medicine: informing regulatory science. Digital Biomark 4(1):1–12

Aguinis H, Villamor I, Ramani RS (2021) MTurk research: review and recommendations. J Manag 47(4):823–837

Bhattacharya P, Obaidat MS, Savaliya D, Sanghavi S, Tanwar S, Sadaun B (2022) Metaverse assisted telesurgery in healthcare 5.0: an interplay of blockchain and explainable AI. In: 2022 international conference on computer, information and telecommunication systems (CITS). IEEE, pp 1–5

Bhinder B, Gilvary C, Madhukar NS, Elemento O (2021) Artificial intelligence in cancer research and precision medicine. Cancer Discov 11(4):900–915

Blut M, Wang C, Wünderlich NV, Brock C (2021) Understanding anthropomorphism in service provision: a meta-analysis of physical robots, chatbots, and other AI. J Acad Mark Sci 49(4): 632–658

Bulla C, Parushetti C, Teli A, Aski S, Koppad S (2020) A review of AI based medical assistant chatbot. Res Appl Web Dev Des 3:1–14. ISO 690

Cesario A, D'oria M, Bove F, Privitera G, Boškoski I, Pedicino D et al (2021a) Personalized clinical phenotyping through systems medicine and artificial intelligence. J Pers Med 11(4):265

Cesario A, D'oria M, Calvani R, Picca A, Pietragalla A, Lorusso D et al (2021b) The role of artificial intelligence in managing multimorbidity and cancer. J Pers Med 11(4):314

Chung K, Park RC (2019) Chatbot-based healthcare service with a knowledge base for cloud computing. Clust Comput 22:1925–1937

Damar M (2021) Metaverse shape of your life for future: a bibliometric snapshot. J Metav 1(1):1–8

Davis FD (1989) Perceived usefulness, perceived ease of use, and user acceptance of information technology. MIS Q 13(3):319–340

Dinh-Le C, Chuang R, Chokshi S, Mann D (2019) Wearable health technology and electronic health record integration: scoping review and future directions. JMIR Mhealth Uhealth 7(9): e12861

Dix A (2016) Human-like computing and human–computer interaction. In: Proceedings of the 30th International BCS Human Computer Interaction Conference 30 (pp 1–6)

Dubosc C, Gorisse G, Christmann O, Fleury S, Poinsot K, Richir S (2021) Impact of avatar facial anthropomorphism on body ownership, attractiveness and social presence in collaborative tasks in immersive virtual environments. Comput Graph 101:82–92

Dwivedi YK, Hughes L, Baabdullah AM, Ribeiro-Navarrete S, Giannakis M, Al-Debei MM (2022) Metaverse beyond the hype: multidisciplinary perspectives on emerging challenges, opportunities, and agenda for research, practice and policy. Int J Inf Manag 66:102542

Dwivedi YK, Hughes L, Wang Y, Alalwan AA, Ahn SJ, Balakrishnan J et al (2023) Metaverse marketing: how the metaverse will shape the future of consumer research and practice. Psychol Mark 40(4):750–776

Edirisinghe R (2022) Toward concept realization of digital health technologies. In: Digital disruption in healthcare. Springer International Publishing, Cham, pp 91–101

Epley N, Waytz A, Akalis S, Cacioppo JT (2008) When we need a human: motivational determinants of anthropomorphism. Soc Cogn 26(2):143–155

Fishbein M, Ajzen I (1977) Belief, attitude, intention, and behavior: an introduction to theory and research. Philos Rhetor 10(2):177–188

Fong B, Fong ACM, Li CK (2020) Telemedicine technologies: information technologies in medicine and digital health. John Wiley & Sons

Gentner T, Neitzel T, Schulze J, Buettner R (2020) A systematic literature review of medical chatbot research from a behavior change perspective. In 2020 IEEE 44th Annual Computers, Software, and Applications Conference (COMPSAC). IEEE, pp 735–740

Golossenko A, Pillai KG, Aroean L (2020) Seeing brands as humans: development and validation of a brand anthropomorphism scale. Int J Res Mark 37(4):737–755

Guido G, Peluso AM (2015) Brand anthropomorphism: conceptualization, measurement, and impact on brand personality and loyalty. J Brand Manag 22(1):1–19

Guido G, Pichierri M, Pino G, Nataraajan R (2019) Effects of face images and face pareidolia on consumers' responses to print advertising: an empirical investigation. J Advert Res 59(2): 219–231

Guthrie SE (1993) Faces in the clouds: a new theory of religion. Oxford University Press, New York

Han MC (2021) The impact of anthropomorphism on consumers' purchase decision in chatbot commerce. J Internet Commer 20(1):46–65

Hasselgren A, Kralevska K, Gligoroski D, Pedersen SA, Faxvaag A (2020) Blockchain in healthcare and health sciences—a scoping review. Int J Med Inform 134:104040

Heerink M (2010). Assessing acceptance of assistive social robots by aging adults (Doctoral dissertation, Universiteit van Amsterdam [Host])

Kang H, Kim KJ (2020) Feeling connected to smart objects? A moderated mediation model of locus of agency, anthropomorphism, and sense of connectedness. Int J Human-Comput Stud 133:45–55

Kang H, Kim KJ (2022) Does humanization or machinization make the IoT persuasive? The effects of source orientation and social presence. Comput Hum Behav 129:107152

Karray F, Alemzadeh M, Abou Saleh J, Arab MN (2008) Human-computer interaction: overview on state of the art. Int J Smart Sens Intell Syst 1(1):137–159

Kim S, McGill AL (2011) Gaming with Mr. slot or gaming the slot machine? Power, anthropomorphism, and risk perception. J Consum Res 38(1):94–107

Kim Y, Sundar SS (2012) Anthropomorphism of computers: is it mindful or mindless? Comput Hum Behav 28(1):241–250

Li M, Suh A (2022) Anthropomorphism in AI-enabled technology: a literature review. Electron Mark 32:1–31

Lupton D (2014) Critical perspectives on digital health technologies. Sociol Compass 8(12): 1344–1359

Mak KK, Pichika MR (2019) Artificial intelligence in drug development: present status and future prospects. Drug Discov Today 24(3):773–780

McKinsey Quarterly (2023) A CEO's guide to the metaverse. Available at: https://www.mckinsey.com/capabilities/growth-marketing-and-sales/our-insights/a-ceos-guide-to-the-metaverse

McLean G, Osei-Frimpong K (2019) Hey Alexa... examine the variables influencing the use of artificial intelligent in-home voice assistants. Comput Hum Behav 99:28–37

Mort M, May CR, Williams T (2003) Remote doctors and absent patients: acting at a distance in telemedicine? Sci Technol Hum Values 28(2):274–295

Nass C, Moon Y (2000) Machines and mindlessness: social responses to computers. J Soc Issues 56(1):81–103

Nowak KL, Fox J (2018) Avatars and computer-mediated communication: a review of the definitions, uses, and effects of digital representations. Rev Commun Res 6:30–53

Pelau C, Dabija DC, Ene I (2021) What makes an AI device human-like? The role of interaction quality, empathy and perceived psychological anthropomorphic characteristics in the acceptance of artificial intelligence in the service industry. Comput Hum Behav 122:106855

Rahimi B, Nadri H, Afshar HL, Timpka T (2018) A systematic review of the technology acceptance model in health informatics. Appl Clin Inform 9(03):604–634
Sestino A, Prete MI, Piper L, Guido G (2020) Internet of things and big data as enablers for business digitalization strategies. Technovation 98:102173
Sestino A, Guido G, Peluso AM (2022) Non-fungible tokens (NFTs): examining the impact on consumers and marketing strategies. Springer Nature
Sheehan B, Jin HS, Gottlieb U (2020) Customer service chatbots: anthropomorphism and adoption. J Bus Res 115:14–24
Sparkes M (2021) What is a metaverse. New Sci 251:18
Strait M, Canning C, Scheutz M (2014) Let me tell you! Investigating the effects of robot communication strategies in advice-giving situations based on robot appearance, interaction modality and distance. In: Proceedings of the 2014 ACM/IEEE international conference on human-robot interaction, pp 479–486
Sun M, Xie L, Liu Y, Li K, Jiang B, Lu Y, Yang D (2022) The metaverse in current digital medicine. Clin eHealth 5(1):52–57
Triantafyllidis AK, Tsanas A (2019) Applications of machine learning in real-life digital health interventions: review of the literature. J Med Internet Res 21(4):e12286
VandenBos GR, Williams S (2000) The internet versus the telephone: what is telehealth anyway? Prof Psychol Res Pract 31(5):490
Värri A (2020). What is digital health? Review of definitions. Integrated Citizen Centered Digital Health and Social Care: Citizens as Data Producers and Service Co-Creators 275:67
Wan J, Aggarwal P (2015) Befriending Mr. clean: the role of anthropomorphism in consumer-brand relationships. In: Strong brands, strong relationships. Routledge, pp 119–134
Wang G, Badal A, Jia X, Maltz JS, Mueller K, Myers KJ et al (2022) Development of metaverse for intelligent healthcare. Nat Mach Intell 4:1–8
Wang W (2017) Smartphones as social actors? Social dispositional factors in assessing anthropomorphism. Comput Hum Behav 68:334–344

Some Ethical and Educational Perspectives on Using Artificial Intelligence in Personalized Medicine and Healthcare 18

Andrea Manto and Marika D'Oria

Abstract

Starting from the recent issues elucidated by the European Parliament, the following contribution focuses on how algorithms are perceived by society as characterized by a "personhood" that may convey prejudice and how, conversely, the metaverse can influence human behavior by leading to interesting changes in identity. A deeper reflection explores the common root of "health" and "salvation" (*salus-ūtis*) that lies as the basis of the concept of "salutogenesis" (a sense of coherence that human beings have about their identity, in every aspect of care). This reflection helps to understand what we seek for when we ask for some posthuman solutions that may harm human dignity. Ethical and educational suggestions are provided.

Keywords

Ethics · Education · Identity · Algorithm · Metaverse · Prejudice · Artificial intelligence · Personalized medicine

18.1 Introduction

In May 2022, the European Parliament published a review encompassing the current implementations of artificial intelligence (AI) in healthcare. The document highlighted some issues, including:

A. Manto (✉)
Pontificia Università Lateranense, Rome, Italy

M. D'Oria
Fondazione Policlinico Universitario A. Gemelli IRCCS, Rome, Italy

A. Cesario et al. (eds.), *Personalized Medicine Meets Artificial Intelligence*,
https://doi.org/10.1007/978-3-031-32614-1_18

- Misuse of medical AI tools and patient harm due to AI errors
- Gaps in AI accountability and lack of transparency
- Privacy and security issues
- Obstacles to implementation in real-world healthcare, risk of bias, and perpetuation of inequities

International stakeholders and policy makers at different levels of governance should consider assess the impact of using AI in healthcare, for individual and public health. Besides, the European Parliament proposed some solutions to face these issues (European Parliament 2022):

- The reduction of the European divide in medical AI
- The creation of mechanisms for enhancing medical AI traceability, trust, and transparency
- The promotion of multi-stakeholders engagement and co-creation collaborations

All these solutions should be integrated within education programs aiming at enhancing literacy and skills in professionals as well as in citizens. However, since personalized medicine (PM) is tailored to "the person" (Cesario et al. 2021), it is crucial to consider ethical implications of using AI starting by taking into account the possible scenarios from an individual and collective perspective.

18.2 "Are Algorithms Racist?," "Is LaMDA Sentient?," and Other Wonder-Full Questions

Several examples from the literature (Zou and Schiebinger 2018; Singh 2021; Layland et al. 2022) on algorithms developing discriminatory outcomes, and the recent (retreated) news from a former Google engineer (Lemoine 2022) on the "sentient" algorithm LaMDA (Wired 2022), are keeping the global attention focused on the possible development of a "personhood" by intelligent algorithms.

In the era of personalized precision medicine, we should take care of the words that are used in explaining some outcomes of AI, especially when compared to humans, because of the possible illusion of overlapping humans with machines. In the abovementioned examples, when the algorithm is biased by design, meaning that the rules through which it learns represent an unconscious prejudice (e.g., cultural biases) of the programmer, it learns to analyze data and report information with an unfair approach, by "privileging" some variables instead of others.

The most obvious forms of discrimination related to cultural bias are (NASEM 2017):

- Racism: which distributes more resources and power over one race/ethnicity than others
- Sexism: which privileges one gender over another (in many cases, men have more privileges than women do)

- Heterosexism: which prefers heterosexual people to those with other orientations
- Classism: which gives unfair disadvantage to those who have more wealth and higher social status than others do
- Xenophobia: which gives more rights and power to non-migrants than immigrants
- Ableism: which prefers people without disabilities to those who have them

In this scenario, wonder plays a crucial role in exaggerating or, conversely, minimizing the potential of these technologies on all creatures and the environment. Believing an algorithm could be "racist" recalls a fascinating imaginary in which humans attribute human properties and behaviors to things and machines by addressing them with behavioral tags.

Consequently, the algorithm becomes "racist," "sentient," "superintelligent," etc. Indeed, some automatisms and robots have been conceived to emulate, and actually replace, humans in some fields and activities. The collective imaginary that fantasizes on the fear that the "creature" (AI) will go beyond (and, eventually, kill) its "creator" (humans): the myths of Pygmalion, Oedipus, and Cronos represented this fearful fantasy very well.

However, fantasies and fears on AI should be analyzed toward ethical and educational lenses, since humans and technologies are intrinsically different. Using a language that matches humans with machines can lead to confusion (e.g., by using metaphors and analogies that compare our memory to a computer, the AI to a superbrain, naming algorithms with the same name of neural networks, etc.).

On the other hand, the fear of being replaced and forgotten recalls the necessity of human beings to reconnect with their intimate meaning, to discover their real nature (not reduced to automatic, yet sophisticated, actions and functions) and how they are necessary and different in this world.

18.3 Behavior, Identity, and the Metaverse

If mass and social media are persuasive tools that influence people's attitudes and behaviors, the metaverse is instead a transformative technology, capable of modifying what people think reality is because it "works like our minds" (Wiederhold 2022). Specifically, the technologies behind a metaverse trigger several cognitive mechanisms, such as proprioception (the experience of being in a place and in a body), brain-to-brain synchrony, and experiencing/inducing emotions (Riva and Wiederhold 2022). By using AI, augmented reality (AR), virtual reality (VR), and connectivity (such as 5G networks), the metaverse could create more immersive, experiential, and interactive online environments, aiming at becoming "the most advanced form of human–computer interaction allowing individuals to act, communicate and become present in digital and digitally enhanced physical environments" (Riva et al. 2021).

The word "metaverse" refers to a virtual reality existing beyond reality, and it is composed by the word of "meta" (going beyond, over, transcendence) and "universe" (world, cosmos, space) (Kim 2020). After this concept appeared in a 1990s science fiction novel (*Snow crash*) for the first time, extensive efforts have been made to make the metaverse a reality (Wiederhold 2022). Today, the metaverse can be defined as "*a 3D-based virtual reality in which daily activities and economic life are conducted through avatars representing the real themselves*" (Go et al. 2021; Kye et al. 2021).

This is not a simple "combination" of the physical and virtual worlds, but an "interaction" of both, in which daily life and economic activities are "conducted in a unified way" (Kye et al. 2021). In addition, the metaverse means "*a world in which virtual and reality interact and co-evolve, and social, economic, and cultural activities are carried out in it to create value*" (Lee 2021). In other words, the avatar in the metaverse is identified with one's real self by engaging in social, economic, and cultural activities in the metaverse world.

In 2006, the Acceleration Studies Foundation (ASF) presented a metaverse roadmap that explained the existing types of the metaverse, by drawing a distinction between "augmentation" vs "simulation" technologies, and "intimate" vs "external" worlds (Smart et al. 2007; Kye et al. 2021). Specifically:

- Augmentation technology adds a new function to an existing real system, by superimposing information on the physical environment we perceive.
- Simulation technology provides a unique environment for interaction by modeling reality.

These technologies can create different metaverses if the virtual information is implemented in physical or virtual reality. Moreover, metaverses can also be distinguished from their focus on the inner or the external/outer world (Smart et al. 2007; Kye et al. 2021):

- The inner world focuses on the identity and behavior of an individual that has direct agency in the environment, by using or embodying an avatar or digital twin.
- The external/outer world focuses on aspects of external reality, by displaying information about the user environment and how to manage it.

The metaverse could impact healthcare due to the convergence of three major technological trends (Table 18.1) that could come together to create entirely new means for delivering care, potentially lowering costs, and vastly improving patient outcomes (Wiederhold 2022).

Clearly, these possibilities define new scenarios with positive and negative outcomes (Riva and Wiederhold 2022). Some of its possibilities regard the change in body perception, and therefore behavioral assets and contextual interactions: for example, VR can be used to visually substitute a person's body by a life-sized virtual one. Such embodiment results in a perceptual illusion of body ownership over the virtual body: research has shown that the form of the VB can influence implicit attitudes, such as in particular, embodying White people in a Black virtual body is

Table 18.1 Short description of the main technological trends, according to Wiederhold (2022)

Tech trends	Short description
Telepresence	The immersion (the sense of "being there") provided by VR enhances the experience of telepresence, allowing both patients and their providers to interact naturally, just as they would in person. Decades of research has shown that for many patients, lasting change happens when they are able to confront the situations that cause them distress and learn how to cope with them constructively with the help of a therapist. The metaverse can serve as a transitional stage between current in-person VR therapy sessions and real-world experiences. As the metaverse materializes, healthcare providers can begin to engage in therapy with patients in the metaverse and only after that have them practice on their own, eventually translating coping skill sets to the real world.
Digital twin	A digital twin is "a virtual model, or simulation, of any object, process, or system, generated using real-world data, for the purpose of learning more about its real-world counterpart." in the case of exposure therapy within the metaverse, the digital twin could be a version of a patient's classroom or office, or even a visual reproduction of patients themselves. In the metaverse, there is the opportunity for healthcare providers to truly accompany patients into specific individualized environments, thus enhancing the efficacy of treatment.
Blockchain	Blockchain is defined as "distributed and encrypted databases that allow data to be stored and transferred securely in a way that no one except the data owner can tamper with." In healthcare, blockchain use would be in the management and security of individual health data. Unlike reams of paper or transfer via unsecured email or online portals, blockchain could allow patients to own their medical records on a secure personal file. Purportedly blockchain is unhackable, yet it is simple to give consent to any clinician anywhere in the world to review the records with the click of a button. The metaverse, with its capabilities for immersion, customization, and security, has an important role to play in the future of healthcare, allowing for the proactive prediction, prevention, and treatment of health concerns, leading to better patient outcomes.

associated with an immediate decrease in their implicit racial bias against Black people (Banakou et al. 2016). Another research shows that participants embodying the Einstein body performed better on a cognitive task than the Normal body, considering prior cognitive ability (IQ), with the improvement greatest for those with low self-esteem. Einstein embodiment also reduced implicit bias against older people. Hence virtual body ownership may additionally be used to enhance executive functioning (Banakou et al. 2018). It would be interesting to understand whether these technologies can empower or disempower patients in different clinical scenarios (from addictions to chronic conditions, from mental to physical illness, etc.).

On the other hand, it is likely that situational and embodiment variables may lead to severe consequences similar to the Stanford prison experiment performed in 1971. In the study, some volunteers were randomly selected to be "guards" of the Palo Alto prison and were given uniforms specifically to de-individuate them while instructed to prevent "prisoners" volunteers from escaping. Over the following 5 days, psychological abuse of the "prisoners" by the "guards" became increasingly brutal (Stanford News 1997).

Further limitations of the metaverse may weaken social connections, the possibility of privacy infringement, the misuse of data, or maladaptation to the real world (Kye et al. 2021). Educating ourselves as to its promise, and the challenges it may present, is a necessity (Riva and Wiederhold 2022): policy makers and stakeholders (including healthcare professionals) should carefully analyze how people understand the metaverse, while designing environments to solve problems cooperatively and creatively (Kye et al. 2021). Likewise, research from neuropsychology, linguistics, and social sciences should broaden understanding on the effects that virtual simulation and embodiment (still recognized at the roots of cognition, emotion, and attachment) have on individual and collective lifelong learning in terms of attitudes, behaviors, and identity.

18.4 Health, Salvation, and Dignity: Are we Ready for Deep Humanism?

If PM aims at providing "*the right treatment to the right person, at the right time*" (Cesario et al. 2021), we should consider what a person really expects to receive when a personalized treatment is proposed. Even though it is impossible to give a compelling answer to this question, the personalization of a treatment leads to the implicit promise that a patient is "unique" and, therefore, special. Besides this implicit promise, we should also consider that every patient desires to regain the former equilibrium of her or his health and, in some cases, to seek for salvation. The terms "health" and "salvation" have the same origin from the Latin root *salus-ūtis*: it is interesting to recall the concept of "salutogenesis" by Antonovsky, understood as everything that "creates health" by enabling people to make conscious health choices toward their internal and external resources, to proactively increase their resilience even in severe adversity (disability, trauma, chronic illness, etc.). In this sense, "health" and "salvation," which in the Hebrew יְשׁוּעָה (yeshûʿâh), Greek *sotería*, and Latin *salus*, often overlap, and the same etymological derivation "salus" brings us back to the common meaning of wholeness, fullness.

Fundamental to *salutogenesis* is a "sense of coherence" considered as a "comprehensive and deep feeling that whatever happens in life can become understandable and be managed." Medical and technological ethics should assure the sense of coherence (or decoherence) in a patient's identity in all aspects of care (prevention, personalization, participation, prediction, etc.). To this aim, educating healthcare professionals to use AI in their practice therefore requires a salutogenic understanding of the patient that recognizes suffering as a natural condition of existence:

The terrain of human suffering is much broader, much more varied, and multidimensional. Human beings suffer in different ways that are not always covered by medicine, even in its most advanced specializations. Suffering is something even broader than illness, more complex, and at the same time even more deeply rooted in humanity itself. A certain idea of this problem comes to us from the distinction between physical suffering and moral suffering. This distinction takes as its foundation the twofold dimension of the human being and points to the bodily and spiritual element as the immediate or direct subject of suffering (John Paul II 1984 No. 5).

The first challenge that ethics and education should help address concerns the protection of the dignity of the person. While we celebrate the universal value of human dignity, we must consider how its legal recognition has been declined in different statuses (e.g., minorities, people with disability, people living in poverty, children, women) over time.

We can agree with the Italian psychiatrist Eugenio Borgna (2015) in saying that what drives the failure to recognize the dignity of the other is prejudice (individual, collective), which primarily results in discrimination. Prejudice wounds the dignity of each person even more when unable to defend themselves from the gaze of others that denies them recognition of their dignity as persons. At the root of prejudice are indifference and contempt for people who are not "like us," who are not "us." As Pope Francis said (Acistampa 2020):

> it is the distorted view of the person, a gaze that ignores his dignity and his relational character. Sometimes we look at others as objects, to be used and discarded.

How does the question of human dignity arise in connection with human life and death, which are increasingly subject to technological domination? If dignity belongs to humans regardless of their living conditions, it affects humans from the moment of conception and beyond their natural death. If, on the other hand, dignity is also manifested in his or her being capable of openness and acceptance of the other, then it also includes the possibility of self-expression whereby, by giving him or her the opportunity to do so, we respect his or her dignity. If, however, we identify the person with functions he or she performs we would exclude his or her potential capabilities, with the risk of cosigning him or her by reducing him or her to an instrument.

Considering that we are not our data but our data belong to us, it is crucial to understand that humans have the power and the responsibility of what they delegate to machines (Valentini and Cesario 2021). As suggested by Valentini and Cesario in their contribution, we can consider *Deep Humanism* (patient profiling obtained by integrating clinical data with Internet-of-Medical Things and artistic stimuli) "as an opportunity to understand emotional preferences of the patient," in order to create personalized models specifically for a patient of a subpopulation of patients.

However, the question of when human life begins, when it ends, and what is it is not new, but today it has become decisive as we increasingly advance toward posthuman models of existence that are eroding the very concept of humanity and life itself. Of course, one cannot condemn new scientific and technological discoveries if they help to defeat genetic diseases or to live better with the support of prostheses or artificial organs: but the question about human dignity remains (De Rosa 2009), as well as our inquiry about our readiness for Deep Humanism.

References

Acistampa (2020) Papa Francesco, la visione distorta della persona è una malattia sociale. La catechesi del Papa dedicata alla riflessione sulla pandemia, Source: https://www.acistampa.com/story/papa-francesco-la-visione-distorta-della-persona-e-una-malattia-sociale-14802 Last consultation: 11/01/2021

Banakou D, Hanumanthu PD, Slater M (2016) Virtual embodiment of white people in a black virtual body leads to a sustained reduction in their implicit racial bias. Front Hum Neurosci 10: 601. https://doi.org/10.3389/fnhum.2016.00601

Banakou D, Kishore S, Slater M (2018) Virtually being Einstein results in an improvement in cognitive task performance and a decrease in age bias. Front Psychol 9:917. https://doi.org/10.3389/fpsyg.2018.00917

Borgna E (2015) La dignità ferita. Feltrinelli, Milano

Cesario A, D'Oria M, Bove F et al (2021) Personalized clinical phenotyping through systems medicine and artificial intelligence. J Pers Med 11(4):265. https://doi.org/10.3390/jpm11040265

De Rosa E (2009) La dignità umana: dottrina sociale della Chiesa, Source: http://www.cssbachelet.org/wp-content/uploads/2009/12/La-dignit%C3%A0-umana-Dottrina-Sociale-della-Chiesa-S.-E.-Mons.-Michele-De-Rosa.pdf Last consultation: 11/01/2021

European Parliament (2022) Artificial intelligence in healthcare. Applications, risks, and ethical and societal impacts, Source: https://www.europarl.europa.eu/RegData/etudes/STUD/2022/729512/EPRS_STU(2022)729512_EN.pdf Last consultation: 11/07/2022

Go SY, Jeong HG, Kim JI, Sin YT (2021) Concept and developmental direction of metaverse. Korea Inf Process Soc Rev 28:7–16

John Paul II (1984) Salvifici Doloris, n. 5. Source: https://www.vatican.va/content/john-paul-ii/it/apost_letters/1984/documents/hf_jp-ii_apl_11021984_salvifici-doloris.html Last consultation: 25/12/2021

Kim S (2020) Metaverse: digital world, world of emerging items. PlanB Design, Hwaseong

Kye B, Han N, Kim E et al (2021) Educational applications of metaverse: possibilities and limitations. J Educ Eval Health Prof 18:32. https://doi.org/10.3352/jeehp.2021.18.32

Layland EK, Maggs JL, Kipke MD, Bray BC (2022) Intersecting racism and homonegativism among sexual minority men of color: latent class analysis of multidimensional stigma with subgroup differences in health and sociostructural burdens. Soc Sci Med 293:114602. https://doi.org/10.1016/j.socscimed.2021.114602

Lee S (2021) Log in Metaverse: revolution of human×space×time (IS-115). Seongnam: Software Policy & Research Institute, Source: https://spri.kr/posts/view/23165?code=issue_reports Last consultation: 11/07/2022

Lemoine B (2022) Is LaMDA Sentient? – an Interview, Source: https://cajundiscordian.medium.com/is-lamda-sentient-an-interview-ea64d916d917 Last consultation: 11/07/2022

National Academies of Sciences, Engineering, and Medicine (NASEM) (2017) Communities in action: pathways to health equity. National Academies Press, Washington DC

Riva G, Wiederhold BK (2022) What the metaverse is (really) and why we need to know about it. Cyberpsychol Behav Soc Netw 25(6):1–5. https://doi.org/10.1089/cyber.2022.0124

Riva G, Di Lernia D, Sanjo E, et al (2021) Virtual reality therapy in the Metaverse: merging VR for the outside with VR for the inside. Ann Rev Cyberther Telemed 19

Singh S (2021) Racial biases in healthcare: examining the contributions of point of care tools and unintended practitioner bias to patient treatment and diagnosis. Health 7:13634593211061215 https://doi.org/10.1177/13634593211061215

Smart J, Cascio J, Paffendorf J (2007) Metaverse roadmap: pathway to the 3D web [Internet]. Acceleration Studies Foundation, Ann Arbor. Source: https://metaverseroadmap.org/MetaverseRoadmapOverview.pdf Last consultation: 11/07/2022

Stanford News (1997) The Stanford Prison Experiment: Still powerful after all these years, Source: https://news.stanford.edu/pr/97/970108prisonexp.html Last consultation: 11/07/2022

Valentini V, Cesario A (2021) Oltre la persona: cos'è lo 'Human Digital Twin' nella medicina personalizzata. In: Anelli F, Cesario A, D'Oria M, Giuliodori C, Scambia G (eds) Persona e Medicina: sinergie sistemiche per la Medicina Personalizzata. Milano, FrancoAngeli, pp 70–76

Wiederhold BK (2022) Metaverse games: game changer for healthcare? Cyberpsychol Behav Soc Netw 25(5):267–269. https://doi.org/10.1089/cyber.2022.29246.editorial

WIRED (2022) LaMDA and the Sentient AI Trap, Source: https://www.wired.com/story/lamda-sentient-ai-bias-google-blake-lemoine/ Last consultation: 11/07/2022

Zou J, Schiebinger L (2018) AI can be sexist and racist – it's time to make it fair. Nature 559(7714): 324–326. https://doi.org/10.1038/d41586-018-05707-8

The "Human Factor" Beyond Humans: Perspectives for an AI-Guided Personalized Medicine

19

Marika D'Oria, Alfredo Cesario, Luciano Giacò, Andrea Manto, Charles Auffray, and Giovanni Scambia

Abstract

AI-guided machines can improve safety within areas where "human variables" can lead to "human errors," such as medicine. Surgical interventions require high precision, and tele-robotic surgery may help augment human capabilities in order to achieve it. The chapter focuses on understanding the possible degrees of coordination between humans and (intelligent) machines to support medical decision-making. Several robotic functions are indicated and subdivided by analytical, procedural, prosocial, and second-order functions. However, performance should not be confused with competence. Despite machines can be trained to perform actions and calculi, structural needs drive human decision-making. Decision-support systems based on personalized data can prevent adverse events or ineffective therapies, and because these systems should support in life and death decisions, it is crucial to deepen the understanding of priority computing. Although AI-guided machines and algorithms can support (but not substitute) human decisions in medicine, some practical and ethical implications should be seriously taken into account.

M. D'Oria (✉) · A. Cesario · G. Scambia
Fondazione Policlinico Universitario A. Gemelli IRCCS, Rome, Italy
e-mail: alfredo.cesario@policlinicogemelli.it; giovanni.scambia@policlinicogemelli.it

L. Giacò
Bionformatics Research Core Facility, Gemelli Science and Technology Park (G-STeP), Fondazione Policlinico Universitario A. Gemelli IRCCS, Rome, Italy

A. Manto
Pontificia Università Lateranense, Rome, Italy

C. Auffray
European Institute for Systems Biology and Medicine (EISBM), Vourles, France
e-mail: cauffray@eisbm.org

A. Cesario et al. (eds.), *Personalized Medicine Meets Artificial Intelligence*,
https://doi.org/10.1007/978-3-031-32614-1_19

Keywords

Personalized medicine · Human factor · Artificial intelligence · Medical humanities · Decision-making · Social robotics

19.1 Big Data, Machine Learning, and Complex Adaptive Systems

Several research projects contributed to stratify human complexity by collecting biomedical data and leveraging -omic sciences, thus contributing to an innovated personalized/precision medicine framework. Historical examples are the analysis of the physiome (the Physiome Project in 1989), the genome (the Human Genome Project in 1990), the musculoskeletal asset (the Living Human Project in 2002), computational physiology (the Virtual Physiological Human project in 2005), and the immunological system (the ImmunoGrid project in 2006). Current projects equally contribute to this aim by contributing artificial intelligence/machine learning solutions, such as the "GATK" on DNA sequencing for mutations identification by the Broad Institute (2021) (Peltenburg et al. 2016), the Google algorithm "DeepVariants" on genetic variants (Poplin et al. 2018), or the "AlphaFold" program on proteins folding developed by EMBL-EBI and Google's DeepMinds (2021).

All these projects, combined with evidence-based clinical and epidemiology research, are precious sources that help medicine integrating patient-specific data with population-based representations, as well as making accurate predictions, differential diagnoses, and decisions, or propose personalized treatments. Prediction can be a serious challenge when phenomena are variable and chaotic (such as the weather); despite this, the modern meteorological models are quite reliable. Nobel Prize winner in Physics Professor Giorgio Parisi conducts cutting-edge research on disorder, fluctuations, and frustration as necessary elements in complex systems while bridging biology, neurosciences, and machine learning (Hwang et al. 2020; Baity-Jesi et al. 2021; Charbonneau et al. 2021). This demonstrates that using artificial intelligence (AI) to understand complex adaptive systems is possible, albeit challenging. Because human beings are complex adaptive systems, this premise can help us imagine what could happen when medicine uses AI to create patient-specific predictive models and simulations, such as human digital twins (HDTs), and investigate whether these models are capable to process cognitive mimesis.[1]

[1]Cognitive mimesis in HDTs is the ability "*to mimic how people process information in order to design intelligent technologies*" (see Saariluoma and Karvonen 2020).

19.2 Machines, Artificial Intelligence, and Human Augmentation

In order to understand the possible degrees of coordination between humans and "intelligent" machines, we should think about the "human factors" that influence our decision-making. Human factors can be defined as "*human and individual characteristics, which influence behaviour at work in a way which can affect health and safety*" (Health and Safety Executive 2021). Studies on "human factors" have been introduced in the fields of aviation, nuclear energy, and medicine.

By analyzing human decision-making processes, several factors contribute to situation analysis. First, psychological and physiological functions and needs (e.g., digestion, differentiation, growth, metabolism, respiration, excretion, reproduction, safety, and need for survival) (National Cancer Institute 2021). Other aspects regard mammalian structural coupling with the environment (including water supply, oxygen, nutrients, heat, shelter, and pressure). In some conditions, human cannot survive while machines can; therefore, human decision-making is intertwined with Maslow's primary needs (including self-actualization and esteem).

Human factors are subdivided into (Proctor and Van Zandt 2017):

- *Perceptual*: recognition, interpretation, and evaluation of visual, auditory, olfactory, gustative, and tactile information, and proprioception (the ability of feeling where and how our own body is located within the space)
- *Actional*: response selection and aptness with the context, movement learning, action, and control
- *Cognitive*: attention and mental workload, information comprehension and retention, problem-solving and decision-making, learning.
- *Environmental*: anthropometrics, spatial ergonomics, and safety.

We should add to this list (pro)social factors (e.g., the ability to connect with people, and to empathize). Despite machines can be trained to say they love someone, sociality is crucial for humans since we are "social animals" with structural needs of belonging (e.g., social acceptance, the desire to be recognized) and relational care (e.g., the desire to be loved, mother–child symbiosis), which shape our decisions.

As stated before, our focus is to understand the possible degrees of coordination between humans and (intelligent) machines. Indeed, some human functions are empowered/substituted with powerful calculi performed by machines, conceived as systems for human aid or augmentation (Sang-won et al. 2018). A non-exhaustive list of functions is shown in Table 19.1, even though other classifications can be equally useful.

An interesting usage of mixed technologies can be seen in the world of automation. In August 2021, the announcement of the Tesla Bot © (humanoid robot) by CEO Elon Musk opened an exciting yet surprising scenario on AI-autopilot robots (Tesla 2021). Musk said the realization of a Tesla Bot is possible since we can use old technology (robots with wheels) combined with the emerging one. Just to have a

Table 19.1 List of mechanic/robotic functions (both AI-driven and not). For each function, there could be several definitions. The following are indicative

Analytical functions Functions conducted or assessed according to a set of principles of validity, in order to perform specified tasks	
Analysis	Examine data in an organized way to retrieve relevant information (e.g., analogies, hidden patterns, differences)
Assessment	Evaluate the amount, value, quality, or importance of data
Prediction	Estimate what could happen in the future (e.g., deterministic/probabilistic modeling)
Memory	Storage of information (big data), with the possibility to retrieve it when necessary
Search	Retrieve information that the operator wants to find, starting from digital (text, button)/vocal input
Execution	Execute tasks by input from coding, digital, or vocal inputs (e.g., Alexa ©, SIRI ©)
Translation	Change the words of one language into the words of another language that have the same meaning
Planning	Organize a strategy to solve a problem according to a previous analysis of the situation and/or possible alternatives (Russell and Norvig 2003; Luger and Stubblefield 2004)
Playing	Act and react in *real time*, according to a set of shared rules. It is an analytical function because machines do not enjoy a game (there is no playfulness involved)
Problem-solving	Find strategic (and unexpected) solutions to linear and complex problems (e.g., chess and Go games)
Decision-making	Take decisions according to previous training (machine/deep learning)
Simulation	Virtual reconstruction/reproduction of real-life scenarios
Procedural function Functions related to a set of actions involving physical maneuvers.	
Movement (automated activity)	Change position or enact physical movements by following instructions (e.g., robotic arms for precise intervention)
Movement (self-directed behavior)	Change position toward a real-time analysis for self-driving (e.g., the Boston Dynamics' robots able to parkour)
Emulation	Mechanically mimic something or someone by imitating it according to some recognizable features (e.g., facial expressions of human emotions)
Prosocial functions Functions related to the promotion of sociality, acceptance, and engagement	
Communication (output generation)	Give a visual or auditory feedback (e.g., speaking) regarding social interactions (Miklósi and Gácsi 2012). It also includes textual messages (e.g., chatbots) that trigger human emotions according to the user's preferences (and generating the so-called artificial empathy)
Recognition (input decoding)	Know something because it has been experienced before. Recognition can occur through images (texts, patterns, faces) (Dhamecha et al. 2014) or sounds (music, vocals, also through natural language processing) (see Spicer and Sanborn 2019)

(continued)

Table 19.1 (continued)

Analytical functions Functions conducted or assessed according to a set of principles of validity, in order to perform specified tasks	
Second-order functions Functions related to abstraction. All the following categories are recognized by the literature as "umbrella terms," so they may vary according to the source or reference	
Creativity	Develop original ideas, things, or procedures (artworks, paintings, music, jokes, texts) toward conceptual blending
Mind	Adapt one's own capabilities according to internal and external factors. In black-box models, human operators do not know what happens to the machine while processing the input, but also the output can be known
Intelligence	Acquire and apply knowledge and skills
Understanding	Grasp the meaning of something by shaping the new information to experienced data. It is different from recognition (input = output) because understanding involves framing the input also on situational/ environmental assumptions
Thinking	Formulate "ideas" or "opinions" on something
Learning	Gain knowledge through training toward specific procedures (e.g., trials and errors, reinforcement, machine learning, and deep learning)
Autonomy	Make choices rather than being influenced by external decisions

glimpse into Tesla Bot © technology, it is composed of Full Self-Driving (FSD) hardware, multi-cam video, neural networks (planning), auto-labeling, simulation tools, autonomy algorithms, deep learning training system (Dojo training), evaluation infrastructure, and the mechanical robotic engine.

19.3 Some Differentiating Challenges in Human-Machine Interaction

AI-guided machines can improve safety within areas where "human variables" can lead to "human errors" (Bogner 1994; Leape 1994). For example, machines can help when there is a lack of attention, perception, or memory, when humans need to manage stress, tiredness, and fatigue, as well as when they are dominated by boredom or worries. Machines can also help in extreme environmental conditions and to improve situational awareness. Surgical interventions require high precision, and tele-robotic surgery may help achieving it (Valentini et al. 2022). However, performance should not be confused with competence (Firestone 2020).

Another important reflection comes within communication between humans and machines (Guzman and Lewis 2019), and the field that studies the interaction among them is social robotics. Humanoids and ML algorithms (e.g., chatbots) may help people in "feeling listened" while supporting them to overcome daily struggles (Fiske et al. 2019). To this aim, these technologies are largely used in the field of medical assistance, pediatrics (ASD, diabetes), rehabilitation, and surgery. However,

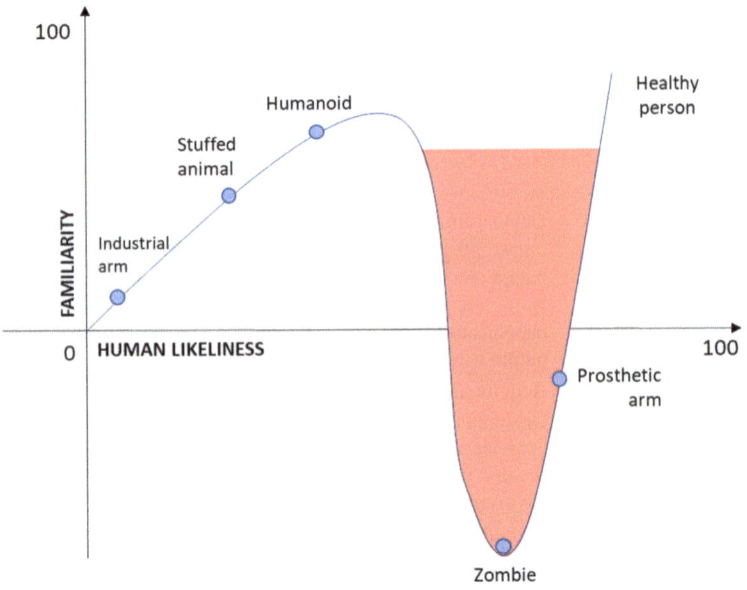

Fig. 19.1 The "Uncanny Valley" phenomenon (represented in red) shows human discomfort when seeing or interacting with robots/machines that seem like ill representations of the human body. Image adapted from Cheetam et al. (2014)

some affective and engagement issues can arise in their interaction, impacting on trust and compliance (Hertz and Wiese 2019). The "Uncanny Valley" is well known throughout the literature, as an unexpected reaction by humans who express uncomfortable reactions to android robots specifically designed to have pleasant social interaction with them (Cheetam et al. 2014; Wiese and Weis 2020). The phenomenon can be explained with Fig. 19.1.

The Uncanny Valley phenomenon indicates humans really perceive robots as social partners and expect them to have intentional stances such as thought, beliefs, and desires (Dennett 1989). For instance, research on human–machine interaction should broaden its boundaries embracing the fields of psychology, pedagogy, sociology, and anthropology.

The ability of "feeling" something is related to the human structural and functional anatomy of receptors. It is impossible for machines to "feel" like humans unless they will be programmed with organic elements equally as a human being. This means that an AI "feels" no stress, no effort, and no pain. AI does not experience emotional consciousness. Human beings struggle with their limitations, and they could desire to thrive them. Indeed, they act desiring to overcome their limitations. Human beings can physically feel desire, by perceiving a sense of lack, and emotions. Human beings act to fill the gap left by an absence. So, human actions are driven by emotions and regulated by the brain. Robots act according to a program. We can physically experience frailty, loneliness, fear of death, and rejection (as well as love, sympathy, etc.) while machines do not. Because illness or loss is associated with pain (one of the driving forces for human actions) and complex situations such as chronic pain or rare diseases involve several dimensions of human affection, emotions, and sense-making, it is difficult to match humans and machines in those situations where real empathy is required. Most human experiences can be shared through language (Di Ciaccia 1996); therefore, they can be "linguistically emulated" through modern algorithms. However, some experiences cannot be shared (ecstasies, madness, hallucinations, mystical visions) because they are too overwhelming and break the rules of language (Lacan et al. 1977).

Machines can/should learn by trials and errors. We can do errors and, despite the fact we know what the "correct" solution is, persevere in doing mistakes (e.g., addictions). We also face with the coexistence of contradictory knowledge and cognitive biases (through which we make decisions, too). Interesting examples can be retrieved from political communication. Another distinctive part of human decision-making is the instinct: we have anticipatory feelings that do not belong to machines. Last but not least, the centrality of the soul through which we are guided is a form of vocation or inner guidance. Sometimes it is linked with consciousness, even though consciousness cannot be the sole attribution to be considered as human, especially in cases where people have disorders of consciousness. Therefore, the connection with the environment as living creatures is strictly intertwined and interconnected with others. Beyond that, quantum physics shows interesting experiments on entangled particles, demonstrating how they immediately influence each other despite distance.

19.4 Modern and Future Directions

In a linear logic framework, classical machines are designed to be predictable for humans. Their planned behavior can be easily managed and supervised by people. However, it is not possible to predict what AI-guided algorithms will learn and deliver, for example, when they are asked to find hidden patterns.

The learning process could be misled if limited to an "if-then" logic. To understand this issue, it is worth mentioning a classical "if-then" error toward Bateson's "syllogism in grass" (Bateson 1991).

Syllogism in BARBARA [correct learning]	Syllogism in GRASS [wrong learning]
Men die	Grass dies
Socrates is a man	Men die
Socrates will die	Men are grass

Just to give a deterministic example, if an algorithm analyzes several databases on the demographics of patients with the same disease, it can learn that "marriage" is a risk factor for that disease only because this information regularly occurs in most patients' information. Other patterns could be found that not always represent correct evidence, and this is why corrective functions have been implemented in statistics and mathematics. With the epistemological change from biomedicine (disease-centered medicine, deterministic approach) to a progressively complex and probabilistic approach with systems medicine (finally, personalized precision medicine), the evolution of artificial intelligence not only has accelerated but also has helped medicine in seeing things with more layers of complexity in systems dynamics.

Personalized medicine simulations seem to be promising in tailoring medical treatments based on patient's specific characteristics (e.g., genetic profile, demographics, and microbiota). Although AI-guided machines and algorithms can support (but not substitute) human decisions in medicine, some practical and ethical implications should be seriously considered. According to Musk, AI will represent a serious potential threat, "*more dangerous than nukes*" (CNBC 2018). In the example of self-driving machines in a complex scenario, AI-guided cars may be put in a situation of deciding whether running over a person on the street or crashing the car with the driver inside.

Consequently, to completely delegate decision-making to AI-guided machines seems to be potentially harmful and thus machines always need human supervision, especially because their lack of common sense. Decision-support systems based on personalized data can prevent the patient experience adverse events or ineffective therapies, and because these systems should support in life and death decisions, it is crucial to deepen the understanding of priority computing (Sadiq et al. 2008). In a collaborative manner, future research and philosophical analysis should reflect on what happens when humans find a third, creative, alternative in a paradoxical situation (even absurd or unpredictable—but life-saving—like an insight, a discovery, or serendipity), probably related to human insights.

References

Baity-Jesi M, Calore E, Cruz A et al (2021) Temperature chaos is present in off-equilibrium spin-glass dynamics. Commun Phys 4(74)

Bateson G (1991) A sacred unity. Further steps to an ecology of mind. Cornelia & Michael Bessie Book, New York

Bogner MS (1994) Human error in medicine. Lawrence Erlbaum Associates, Hillsdale

Broad Institute (2021) HaplotypeCaller in a nutshell, Source: https://gatk.broadinstitute.org/hc/en-us/articles/360035531412-HaplotypeCaller-in-a-nutshell Last consultation: 29 October 2021

Charbonneau P, Corwin EI, Dennis RC et al (2021) Finite-size effects in the microscopic critical properties of jammed configurations: a comprehensive study of the effects of different types of disorder. Phys Rev E 104:014102

Cheetam M, Suter P, Jancke L (2014) Perceptual discrimination difficulty and familiarity in the Uncanny Valley: more like a 'Happy Valley'. Front Psychol 5:1219

CNBC (2018) Elon Musk: 'Mark my words - A.I. is far more dangerous than nukes, Source: https://www.cnbc.com/2018/03/13/elon-musk-at-sxsw-a-i-is-more-dangerous-than-nuclear-weapons.html Last consultation: 29 October 2021

DeepMind, EMBL-EBI (2021) *AlphaFold*, Source: https://alphafold.ebi.ac.uk Last consultation: 29 October 2021

Dennett DC (1989) The intentional stance. MIT Press, Cambridge

Dhamecha TI, Singh R, Vatsa M, Kumar A (2014) Recognizing disguised faces: human and machine evaluation. PLoS One 9:e99212. https://doi.org/10.1371/journal.pone.0099212
Di Ciaccia A (1996) Lacan e il linguaggio. L'inconscio strutturato come un linguaggio, Source: http://www.psychiatryonline.it/node/7425 Last consultation: 29 October 2021
Firestone C (2020) Performance vs. competence in human-machine comparisons. PNAS 117(43): 26562–26571. https://doi.org/10.1073/pnas.1905334117
Fiske A, Henningsen P, Buyx A (2019) Your robot therapist will see you now: ethical implications of embodied artificial intelligence in psychiatry, psychology and psychotherapy. J Med Internet Res 21(5):e13216. https://doi.org/10.2196/13216
Guzman AL, Lewis SC (2019) Artificial intelligence and communication: a human-machine communication research agenda. New Media Soc 22(1):70. https://doi.org/10.1177/1461444819858691
Health and Safety Executive (HSE) (2021) Introduction to human factors, Source: https://www.hse.gov.uk/humanfactors/introduction.htm Last consultation: 29 October 2021
Hertz N, Wiese E (2019) Good advice is beyond all price, but what if it comes from a machine? J Exp Psychol Appl 25(3):386–395. https://doi.org/10.1037/xap0000205
Hwang S, Lanza E, Parisi G et al (2020) On the number of limit cycles in diluted neural networks. J Stat Phys 181:2304–2321
Lacan J, Sheridan A, Bowie M (1977) The function and field of speech and language in psychoanalysis. Routledge, London
Leape L (1994) Error in medicine. JAMA 272(23):1851. https://doi.org/10.1001/jama.1994.03520230061039
Luger GF, Stubblefield WA (2004) Artificial Intelligence. Structures and strategies for complex problem solving. Benjamin/Cummings Publishing
Miklósi A, Gácsi M (2012) On the utilization of social animals as a model for social robotics. Front Psychol. https://doi.org/10.3389/fpsyg.2012.00075
National Cancer Institute (NCI) (2021) Body functions & life process, Source: https://training.seer.cancer.gov/anatomy/body/functions.html Last consultation: 29 October 2021
Peltenburg J, Ren S, Al-Ars Z (2016) Maximizing systolic array efficiency to accelerate the PairHMM Forward Algorithm. In: 2016 IEEE International Conference on Bioinformatics and Biomedicine (BIBM), 758–762
Poplin R, Chang P, Alexander D et al (2018) A universal SNP and small-indel variant caller using deep neural networks. Nat Biotechnol 36:893–897
Proctor RW, Van Zandt T (2017) Human factors in simple and complex systems. CRC Press, Boca Raton
Russell S, Norvig P (2003) Artificial intelligence: a modern approach. Prentice Hall, Upper Saddle River
Saariluoma P, Karvonen A (2020) Human digital twins and cognitive mimetic. In: IHIET 2020: Human Interaction, Emerging Technologies and Future Applications III, 97–102
Sadiq K, Mazzo MD, Zasada SJ et al (2008) Patient-specific simulation as a basis for clinical decision-making. Philos Trans R Soc 366:3199. https://doi.org/10.1098/rsta.2008.0100
Sang-won L, Agrawal H, Maes P (2018) Robotic symbionts: interweaving human and machine actions. IEEE Pervasive Comput 17(2):34. https://doi.org/10.1109/MPRV.2018.022511241
Spicer J, Sanborn AN (2019) What does the mind learn? A comparison of human and machine learning representations. Curr Opin Neurobiol 55:97–102
Tesla (2021) Intelligenza Artificiale e Autopilot, Source: https://www.tesla.com/it_IT/AI Last consultation: 29 October 2021
Valentini V, D'Oria M, Cesario A (2022) Artificial intelligence and AI-guided robotics for personalized precision medicine. In: Riva G, Marchetti A (eds) Humane robotics. Milano, Vita & Pensiero
Wiese E, Weis PP (2020) It matters to me if you are human - examining categorical perception in human and nonhuman agents. Int J Human-Comput Stud 133:1–12. https://doi.org/10.1016/j.ijhcs.2019.08.002

Conclusion

The chapters presented in this book show how the discipline of Artificial Intelligence (AI) has several potential applications for achieving Personalized Medicine, thus highlighting how some transversal challenges should be considered and addressed.

Since the biomedical scenario grows rapidly and beyond imagination, every progress we make could become "obsolete" in the very near future. In our effort to show how this discipline is going to be integrated within biomedical practices, we identified some consolidated experiences while underlying that the regulatory, methodological, and ethical frameworks should remain a key transversal asset for current and future evolutions in this field.

If innovation and technological progress run fast in this field, we do not expect the same fastness within the mentioned frameworks. On the one hand, there is a need for strong, consolidated evidence. On the other hand, there is the necessity for understanding the true potential and pitfalls of emerging and future technologies. Biomedical research and care aim at possibly responding to the needs of all patients with novel sustainable treatments, global stakeholders and policy makers are making significant improvements to drive together the change from different perspectives. Moreover, constraints from the market can shape how progress will develop, for example by investing more in some technologies and research lines instead of others. Thus, the outlook of future biomedicine is progressively in definition.

Above all these considerations, there is a need for further literacy and education for research and care practitioners in order to go "beyond the hype" and the fears that a still uncertain scenario carries in the collective expectations. To this aim, we hope that this book will help students, researchers, and professionals involved in the biomedical and data science fields to further question and investigate how scientific, economic, and progress will shape human care.

A. Cesario et al. (eds.), *Personalized Medicine Meets Artificial Intelligence*,
https://doi.org/10.1007/978-3-031-32614-1

MIX
Papier aus verantwortungsvollen Quellen
Paper from responsible sources
FSC® C105338

If you have any concerns about our products,
you can contact us on
ProductSafety@springernature.com

In case Publisher is established outside the EU,
the EU authorized representative is:
Springer Nature Customer Service Center GmbH
Europaplatz 3, 69115 Heidelberg, Germany

Printed by Libri Plureos GmbH
in Hamburg, Germany